Adhesion and Friction in Biological Systems

BIOLOGICALLY-INSPIRED SYSTEMS

Series Volume: 3

http://springer.com/series/8430

Stanislav N. Gorb
Editor

Adhesion and Friction in Biological Systems

Previously published in:

Naturwissenschaften Volume 88, Issue 12, 2001 and Volume 85, Issue 1, 2005

Applied Physics A Volume 100, Issue 1, 2010

Journal of Comparative Physiology A Volume 186, Issue 9, 2000; Volume 187, Issue 12, 2001; Volume 190, Issue 7, 2004; Volume 192 Issue 11, 2006; Volume 195, Issue 3, 2008; Volume 196, Issue 9, 2010

Tribology Letters Volume 8, Issue 1, 2000

Arthropod-Plant Interactions Volume 1, Issue 2, 2007; Volume 1, Issue 4, 2007; Volume 2, Issue 4, 2008; Volume 3, Issue 1, 2008; Volume 4, Issue 2, 2010

Applied Scanning Probe Methods XIII, 978-3-540-85049-6, 2009

IUTAM Symposium on Scaling in Solid Mechanics, 978-1-4020-9033-2, 2007

Springer

Editor
Prof. Dr. Stanislav N. Gorb
Functional Morphology and Biomechanics
Zoological Institute
Christian Albrecht University of Kiel
Am Botanischen Garten 1–9
D-24098 Kiel
Germany

ISBN: 978-94-007-1444-1

Springer Dordrecht Heidelberg London New York

Library of Congress Control Number: 2011926987

Cover illustration: The green dock beetle *Gastrophysa viridula* on its host plant. In the background: scanning electron microscopy image of adhesive hairs of the beetle leg in contact with the rough surface.

Printed on acid-free paper

Springer is part of Springer Science+Business Media (www.springer.com)

Series Preface

Structure and function of biological systems as inspiration for technical developments
Throughout evolution, organisms have evolved an immense variety of materials, structures, and systems. This book series deals with topics related to structure-function relationships in diverse biological systems and shows how knowledge from biology can be used for technical developments (bio-inspiration, biomimetics).

Preface

Biotribology: Diversity matters

Stanislav N. Gorb

Department of Functional Morphology and Biomechanics
Zoological Institute of the University of Kiel
Am Botanischen Garten 1–9, D-24098 Kiel, Germany
sgorb@zoologie.uni-kiel.de

The majority of edited books on biotribology deal with a diversity of approaches for microscopical, mechanical, and tribological characterization of surfaces, but most of them are restricted to very few model systems dealing with different materials of the human body or implants. However, we have to keep in mind that *Homo sapiens* is only one of 13–14 million of estimated species of organisms, and interesting adhesion- and friction-related contact problems can be found everywhere in Nature. Gecko foot, snake skin, insect joints are just a few striking examples from a tremendous diversity of biological surfaces and systems that remain rather poorly studied, when compared to medically relevant biotribosystems.

In 2001, with the publication of "*Biological Micro- and Nanotribology: Nature's Solutions*" by Scherge and Gorb we tried to set a new trend in studying tribological phenomena in diverse biological systems. The book described a few biomechanical systems with frictional surfaces or fluid secretions involved in attaching parts of the body to each other, or the entire organism to a substrate. Such systems have fascinated microscopists for centuries, but only recently the use of novel experimental and microscopical methods enabled the functioning of at least some of these systems to be really understood.

Since then, a large amount of scientific publications devoted to biotribology and bioadhesion of various biological systems have appeared in the literature. In the meantime, research on frictional and adhesive properties of very diverse biological surfaces and interfaces became quite a broad field in research on the frontier between tribology and biology (see Figure, Gebeshuber et al., 2009). During the past decade in my group, which was initially called *Biological Microtribology Group* at the Max Planck Institute of Developmental Biology, Tübingen, Germany (1999–2002), later *Evolutionary Biomaterials Group* at the Max Planck Institute for Metals Research, Stuttgart, Germany (2002–2008), and finally *Functional Morphology and Biomechanics* at the University of Kiel, Germany (2008–), we concentrated on studies of biological adhesive systems and their biomimetic prototypes, as well as on biological materials with specialized frictional properties. We followed an approach in which various microscopy methods, such as light microscopy, white-light interferometry, SEM, Cryo-SEM, TEM, and AFM were combined with force measurement techniques at the macro-, micro- and nanoscale. These studies resulted in numerous publications which, depending on the major experimental approach, were published in biological, materials science, engineering, or physical journals accessed mostly by members of particular scientific communities. Our idea was to collect these publications into one volume to provide potential readers with an overview of systems, phenomena, and approaches on a variety of non-human friction and adhesive systems in biology.

This book is a collection of experimental studies from my group originally published in various Springer journals. These studies that were carried out in collaboration with coworkers within the group and with numerous colleagues from other institutions worldwide, demonstrate structure-function relationships in biological systems having particular surface specialization to increase/decrease friction and adhesion. Studies on snake skin, adhesive pads of insects and spiders, wing-interlocking devices, and sticky mouthparts of insects, as well as anti-adhesive and adhesive surfaces of plants are included in the present volume containing four main subsections: (1) adhesion, (2) friction, (3) attachment devices, (4) attachment-related behavior.

In the first section, four different adhesion-related systems are discussed. Two of them deal with the enhancement of adhesion forces: moist mouthparts of beetles that are specialized for the collection of pollen grain (Karolyi et al.) and orthopteran attachment pads demonstrating different adhesive properties, depending on their mechanical properties (Perez Goodwyn et al). The third system, represented in this section by epicuticular waxes of plants (Gorb et al.), is specialized in reducing adhesion of insect attachment devices. Epicuticular wax crystals might be fractured by insect adhesive organs and contaminate the latter, depending on crystal geometry (Borodich et al.). The exact function of the fourth system, which is the covering wing surface in the female Colorado potato beetle, remains unresolved. However, this is an important study characterizing the female surface to which males adhere while mating. It is known that the covering wing surfaces of the female may play a role in sexual selection due to their different adhesive properties (Voigt et al.).

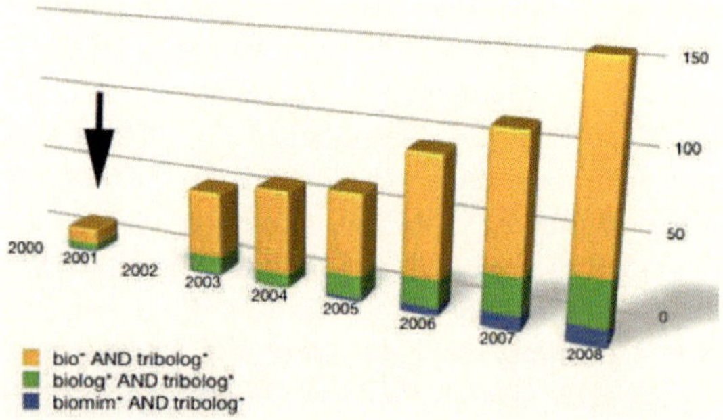

Figure. Number of publications on biotribology in the period from 2000 to 2008 (from Gebeshuber et al., 2009). Arrow indicates the publication year of the book *Biological Micro- and Nanotribology: Nature's Solutions* (Scherge and Gorb, 2001).

The second section, starting with some general statements on the microtribology of biological materials (Scherge and Gorb), is devoted to friction in various organisms such as spiders, insects, and snakes. The first two chapters deal with friction-related phenomena in arthropods, such as friction anisotropy in spider attachment pads (Niederegger and Gorb), and friction reduction mechanism in sliding joints between the fore-wing and hind-wing in bugs (Perez Goodwyn and Gorb). In the two last contributions of the section, the surface structure, frictional properties (Berthé et al.), hardness, and effective elasticity modulus (Klein et al.) of the snake skin are described.

The third section contains four papers on insect attachment devices starting with a broad overview on locomotory attachment pads of hexapods (Gorb and Beutel). The next chapter shows the relationship between the ultrastructural architecture and mechanical properties of attachment pads in the grasshopper (Gorb et al.). In chapter three, Niederegger et al. have applied a combination of methods to visualize contact behavior of adhesive setae in the fly attachment pad. Finally, an unusual micro-Velcro mechanism, called mid-coxal prong, is described for Brachycera (Diptera) in the fourth chapter of the section (Frantsevich and Gorb).

How do insects stick and how do plants prevent this? Four chapters in the final section deal with insect locomotory behavior that is related to the attachment ability of insects to different surfaces and structures of their natural terrain, which is mainly represented by plant surfaces. The general overview of surface related insect-plant relationships is provided in the first chapter (Gorb and Gorb). Insect walking techniques on thin stems are described and catalogued by Gladun and Gorb. Mirid bug locomotion on sticky plant terrain is studied in detail in the two final chapters (Voigt et al.; Voigt and Gorb).

Additionally, numerous experimental methods for characterizing tribological properties of biological surfaces at macro-, micro-, and nanoscale levels are demonstrated in the book. We are confident that this book will be an example of interdisciplinary science in action, combining approaches from biology, physics, engineering, tribology, and materials science. This work may potentially be of interest to both engineers and physicists working with biological systems, as well as to biologists studying friction and adhesion. Since biological surfaces have a great potential for technological ideas that could be used in developing new materials and systems, inspirations from biology reported here may also be used for biomimetics purposes.

I am most grateful to L.I. Frantsevich (Schmalhausen Institute of Zoology, Kiev, Ukraine), who, as my first supervisor, recognized the strong potential of studies on biological adhesion and friction and co-authored one of the chapters in this book. U. Schwarz (Max Planck Institute of Developmental Biology, Tübingen, Germany) strongly supported this branch of research during my stay at his institute. I have greatly benefited from collaboration with M. Scherge (MikroTribologie Centrum µTC, Fraunhofer-Institut für Werkstoffmechanik IWM, Freiburg, Germany), who, with his strong experience in microtribology, provided much advice on the measurements of small forces, development of equipment for micro- and nanotribology research (see his co-authored chapters in the book), and inspired me to co-author the book on *Biological Micro- and Nanotribology* (Scherge and Gorb, 2001). Collaboration with R.G. Beutel (University of Jena, Germany) resulted in an integration of diverse structural information on attachment pads of insects into a solid evolutionary context. I also wish to thank H. Schwarz (co-author of one of the chapters), J. Berger, I. Zimmermann, V. Kastner, B. Sailer, G. Scheer (Max Planck Institute of Developmental Biology, Tübingen), who helped with various microscopy techniques that contributed enormously to understanding the results obtained in mechanical tests.

As editor, I would particularly like to acknowledge the contributions of all co-authors, who have given so much to the success of the original publications: F. Karolyi, H.W. Krenn (University of Vienna, Austria); J.K. Deuschle, P. Perez Goodwyn, A. Peressadko, V. Kastner (Max Planck Institute for Metals Research, Stuttgart, Germany); Y. Jiao, H. Schwarz (Max Planck Institute of Developmental Biology, Tübingen, Germany); E.V. Gorb, D. Voigt, M.-C. G. Klein, H. Peisker (University of Kiel, Germany); S.D. Eigenbrode (University of Idaho, USA); F. Borodich (University of Cardiff, UK); M. Scherge (MikroTribologie Centrum μTC, Fraunhofer-Institut für Werkstoffmechanik IWM, Freiburg, Germany); S. Niederegger (Institut für Rechtsmedizin, Universitätsklinikum Jena, Germany); R.A. Berthé, G. Westhoff, H. Bleckmann (Zoological Institute, University of Bonn, Germany); R.G. Beutel (Friedrich Schiller University of Jena, Germany); L.I. Frantsevich, D. Gladun (Schmalhausen Institute of Zoology, Kiev, Ukraine). Z. Bernhart and E. Machado (Springer, Dordrecht, the Netherlands) strongly supported the idea of the book and helped during all stages of the production.

The editor, on behalf of the authors, is grateful for the financial support provided for numerous research projects during the last decade by the Federal Ministry of Education and Research of Germany (BMBF: Grants BioFuture 0311851; Biona 01RB0812A; Inspirat 01RI0633D; BIOKON Phase II, 01RS0411), the German Science Foundation (DFG: Grants BE 1789/1-2; GO 995/9-1 in SPP 1420 priority program "Biomimetic Materials Research: Functionality by Hierarchical Structuring of Materials"; GO 995/4-1 in the European Science Foundation EUROCORES Programme FANAS), the EC Sixt Framework Programme (ERAS-CT-2003-980409), the German National Academic Foundation (doctoral scholarship to D. Voigt E2002D0730), the State Fund of Fundamental Research, Ukraine (05.07/00194 of SFFR), and the Landesstiftung Baden-Württemberg, Germany (Grant "Plants as Source for Biomimetics of Anti-Adhesive Surfaces").

References

Gebeshuber, I.C., Majlis, B.Y., and Stachelberger, H. (2009) Tribology in biology: Biomimetic studies across dimensions and across fields. Int. J. Mech. Mat. Eng., 4(3): 321–327.

Scherge, M., and Gorb, S.N. (2001) *Biological Micro- and Nanotribology: Nature's Solutions*. Berlin et al.: Springer.

Contents

1. Adhesion

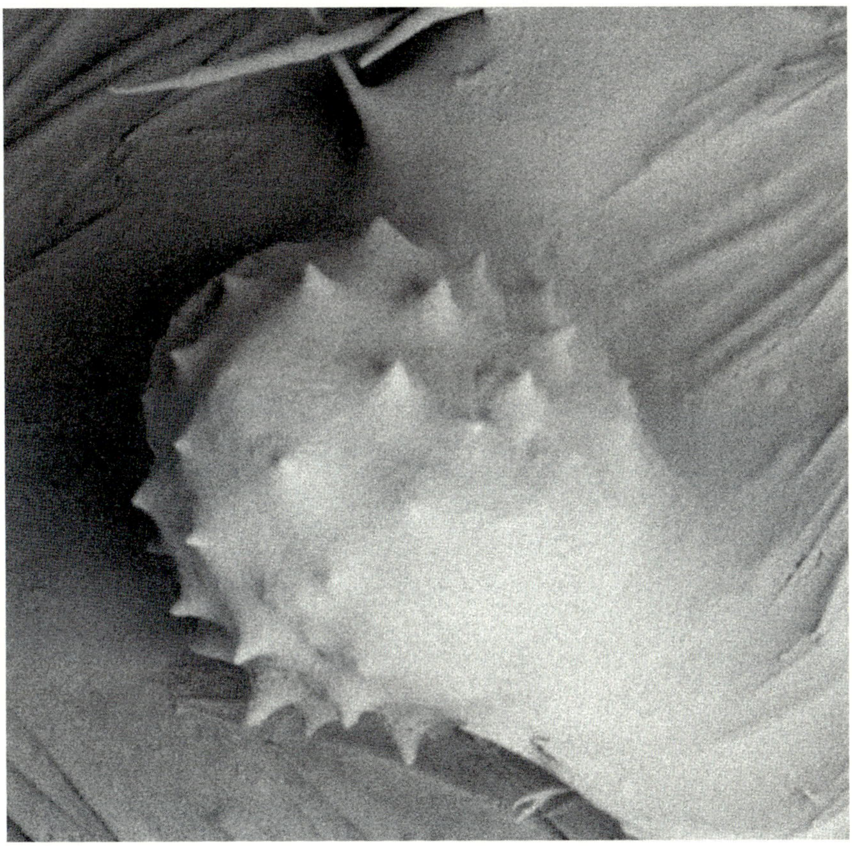

Arthropod-Plant Interactions
DOI 10.1007/s11829-008-9052-5

ORIGINAL PAPER

Pollen grains adhere to the moist mouthparts in the flower visiting beetle *Cetonia aurata* (Scarabaeidae, Coleoptera)

Florian Karolyi · Stanislav N. Gorb · Harald W. Krenn

Received: 4 June 2008 / Accepted: 28 November 2008

Abstract Flower visiting beetles possess numerous structural adaptations of their mouthparts to adhere and ingest pollen grains. Using a Cryo-SEM approach the examination of the mouthparts in rapidly frozen *Cetonia aurata* (Scarabaeidae) indicates a previously unknown technique of pollen uptake in Coleoptera. Cryo-SEM micrographs of the mouthparts reveal a fluid covering the bristles on the buccal surface. In this way the bristles of the galeae form a wet brush which represents the primary organ of pollen uptake. The fluid improves adhesion of pollen to bristles lacking any specialized adhering surface or highly sculptured cuticle as present in other pollen feeding Coleoptera. The well developed mola region of the otherwise non-biting mandibles of *C. aurata* indicates that these beetles open pollen grains mechanically before ingestion. Examination of gut content demonstrated that crushed and intact pollen occur in all regions. The Cryo-SEM method represents a new approach to study functional morphology including the interaction of microstructures and fluids on cuticular surfaces of insects.

Keywords Mouthparts · Functional morphology · Pollen feeding · *Cetonia aurata* · Coleoptera · Flower visiting

Handling editor: Gimme Walter

F. Karolyi (✉) · H. W. Krenn
Department of Evolutionary Biology, University of Vienna, Althanstrasse 14, 1090 Vienna, Austria
e-mail: flokarolyi@hotmail.com

H. W. Krenn
e-mail: harald.krenn@univie.ac.at

S. N. Gorb
Evolutionary Biomaterials Group, Department of Thin Films of Biological Systems, Max-Planck-Institut für Metallforschung, Heisenbergstr. 3, 70569 Stuttgart, Germany
e-mail: s.gorb@mf.mpg.de

Introduction

Many flower visiting insects consume pollen and therefore possess characteristic mouthpart adaptations for taking up pollen grains from flowers. Insect mouthparts are highly integrated structural units, forming interfaces with the environment (Betz et al. 2003). They often possess dense bristles for harvesting small pollen grains, as well as structural modifications for the manipulation of ingested pollen. Pollen feeding has been recorded for a number of species from various families of beetles. Their prognathous mouthparts are adequate for feeding on nectar, pollen and flower petals (Krenn et al. 2005).

Pollen grains are encased by a hard and highly resistant outer hull, the exine (Johnson and Nicolson 2000) and single grains range from 10 to 200 μm in diameter (Stanley and Linskens 1974). The challenges confronting pollen feeders are the removal of the often dry pollen and gaining access to the encapsulated nutrition. Pollen is potentially a valuable food source for flower visiting insects since it contains a wide variety of lipids, carbohydrates and different protein concentrations (Johnson and Nicolson 2000).

Characteristic features of pollen feeding Coleoptera include an elongated maxillae with variously shaped combs and bristles that function as the primary pollen harvesting devices (Fuchs 1974). The mandibles, too, are particularly adapted for pollen manipulation and consist largely of the prostheca, a soft and usually bristly lobe, and the postmola which together knead the pollen (Schremmer 1961; Fuchs 1974; Nel and Scholtz 1990). The pollen grains are conveyed into the actual mouth opening by coordinated movements of the mouthparts (Fuchs 1974).

Although few investigations have been devoted to the mouthparts of flower visiting beetles, the functional anatomy of the mouthparts of the flower visiting rose beetle *Cetonia aurata* (Scarabaeidae) was previously studied in detail (Bürgis 1986a, b, 1987; Schremmer 1961). These studies concluded that the incisive part of the mandibles has lost its biting function, the mola part is responsible for crushing pollen and that the maxillae extend outward from their resting position under the clypeus and labrum with a semi-circular movement. Furthermore, the distal parts of the maxillae, i.e. the galeae, open laterally and extend forward. They reach into flowers and with their dense and wavy bristles sweep pollen grains into the cibarium (preoral cavity) during the retreating motion of the maxillae. Simultaneously, the laciniae with their short bristles convey any previously acquired pollen between the mandibles (Bürgis 1986a, b, 1987) (Fig. 1).

Despite the above mentioned studies on this frequently encountered European rose beetle, the full feeding mechanisms are not entirely understood. One open question concerns the possible interaction between pollen grains and the variously shaped bristles of the cuticle. The aim of this study was to examine the mouthparts of *C. aurata* under near natural conditions using a Cryo-scanning electron microscope. By examining rapidly frozen specimens which have not come in contact with liquids during preparation, this method allows the detection of pollen grains on mouthpart structures in their natural position. Any liquids naturally occurring on mouthpart surfaces will remain intact. The present examination represents the first application of this method for the visualisation of insect mouthparts. It demonstrated the presence of a liquid on the buccal surfaces of the mouthparts which compels us to new functional interpretations of mouthpart morphology and ultimately a deeper understanding of the pollen feeding mechanisms in this species. In addition, examination of the gut content should also provide insights in the role of different cuticular surfaces during feeding.

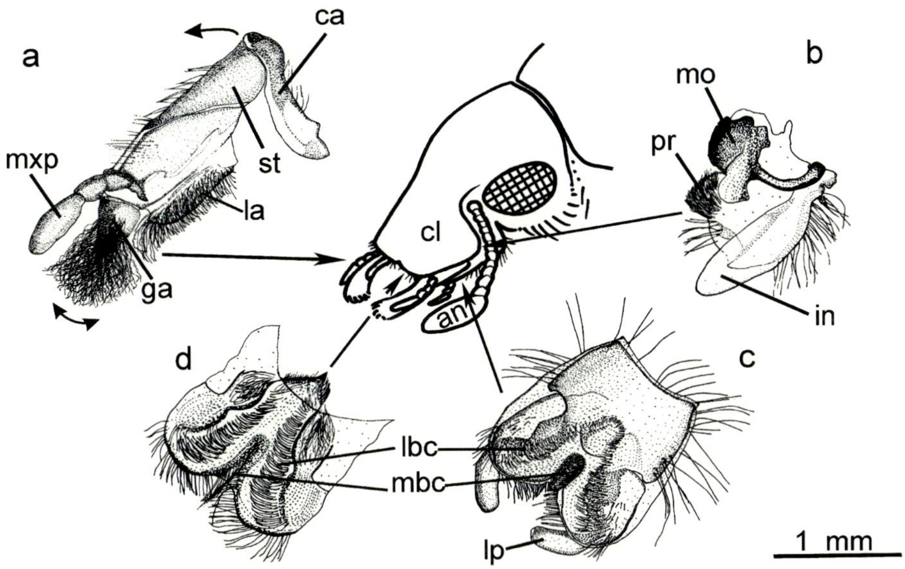

Fig. 1 Head and mouthparts of *Cetonia aurata* (Scarabaeidae). *Straight arrows* indicate position of mouthparts on the head; curved arrows indicate movements. **a** Maxilla swivels out laterally and extends forward; during its retraction the galea turns inward sweeping pollen to the lacinia. **b** Mandible with non-biting incisivus, prostheca and mola with pollen crushing surface. **c** Buccal side of labium with bristles crests. **d** Buccal side of labrum with bristle crests. The smooth dorsal side lies directly under the heavily sclerotised clypeus. an, antenna; ca, cardo; cl, clypeus; in, incisivus; ga, galea; la, lacinia; lbc, lateral bristle crest; lp, labial palpus; mbc, median bristle crest; mo, mola; mxp, maxillary palp; pr, prostheca; st, stipes. Bar scale applies to **a–d**

Material and methods

Cryo-scanning electron microscope (Cryo-SEM)

Specimens of the beetle *Cetonia aurata* (Coleoptera: Scarabaeidae, Cetoniinae), were collected from natural food plants in surroundings of Stuttgart-Vaihingen (southern Germany) and anaesthetized with CO_2 prior to preparation.

Heads were severed from the body of anaesthetized insects with the use of sharp razor blades, glued with Tissue-Tek O.C.T. fluid to brass stubs, and frozen in melting liquid nitrogen to prevent a gaseous phase building on the surface during the shock-freezing process. After a short sublimation procedure (about 3 min at the temperature difference −120 and −90°C), samples were sputter-coated in the frozen condition (temperature −140°C, thickness 6 nm) and observed in the SEM Hitachi S 4800 at accelerating voltage of 1–3 kV and temperature of −120°C. The method allowed us to visualise fluids located on the cuticle surface (Gorb 2006) and to study the surface of beetle mouthparts with high resolution and with a minimum of artefacts due to treatments by solvents and drying procedure.

Scanning electron microscope

The micrographs were taken using a conventional SEM (Philips XL 20) at the Institution of Cell Imaging and Ultrastructure Research (University of Vienna, Austria).

Beetles were collected from natural food plants in surroundings of Vienna. For the SEM, the specimens were rapidly frozen on flowers of *Crataegus monogyna* (Rosaceae) using a Cryo Freeze Aerosol (Agar Scientific Limited), stored in a can containing a nitrogen atmosphere, put in 100% ethanol, and finally stored at −54°C. To dissect mouthparts under a stereo microscope, the severed heads were then embedded in a molten wax-rosin mixture using a soldering tip (Star Tec ST 081).

All specimens were dehydrated in 100% ethanol and Hexamethyldisilazan (15 min) was used as an intermedium. After air drying over night, the objects were mounted on object holders using double-sided carbon-containing adhesive tape and sputter-coated with gold (300 s in an Agar sputtercoater B7340).

Gut examination

Beetles were collected from natural food plants in surroundings of Vienna and fixed in ethanol (70%). They were dissected under a stereo microscope, and their guts were removed. Parts of fore-, and hindgut were separately mounted in Glycerine on glass slides. Digital images were taken using a light optical microscope equipped with a digital photo camera Nikon CoolPix 950.

Results

Mouthparts and feeding mechanism

In *C. aurata*, the various structures of the mouthparts function together as a pollen harvesting and processing device. The cibarial roof is formed by the clypeus and the heart shaped labrum. This rather soft structure bears two lateral bristle combs, joining proximally to a v-shaped ridge and enfolding a median crest of bristles that perches on a median keel (Fig. 1d). The lateral bristles overlie the median part of the mandibles, while the bristly epipharynx lies between the mola surfaces. In Cryo-SEM micrographs all buccal bristles of the labrum are covered with a liquid. Pollen grains were found embedded in the liquid layer situated between moist bristles of the median crest. While the thinner, inner lateral bristles of the epipharynx point in the caudal direction, the outer lateral bristle rows align towards the median keel and the median bristles reach into the preoral cavity (Fig. 2a, b).

The distal parts of the maxillae, i.e., the galeae, represent the primary organs for pollen uptake (Figs. 1a, 2d–g). The Cryo-SEM study revealed that the long bristles on the tip of the galeae are completely saturated with fluid, forming a structure that can be functionally compared with a wet brush tip taking up pollen (Fig. 2e–g). Pollen grains were also found on the lacinia bristle crests. These bristles are likewise covered with a liquid to which the grains adhere, but they do not form a brush-like structure as described for the galeal bristles (not illustrated).

Conventional SEM micrographs revealed wavy bristles at the tip of the galea and a fine serration at the tip of otherwise smooth lacinia bristles (not illustrated). These structures presumably assist the pollen transport through the cibarium. It is not expected that they support the adherence of the liquid layer to the bristle's surface otherwise the entire shaft would be serrated.

The mandibles are composed of the thin blade-like incisivus which lies alongside the basal part of the maxilla, the soft and bristly prostheca and the well developed mola region (Fig. 1b). Cryo-SEM micrographs revealed that the bristles on the inner edge of the

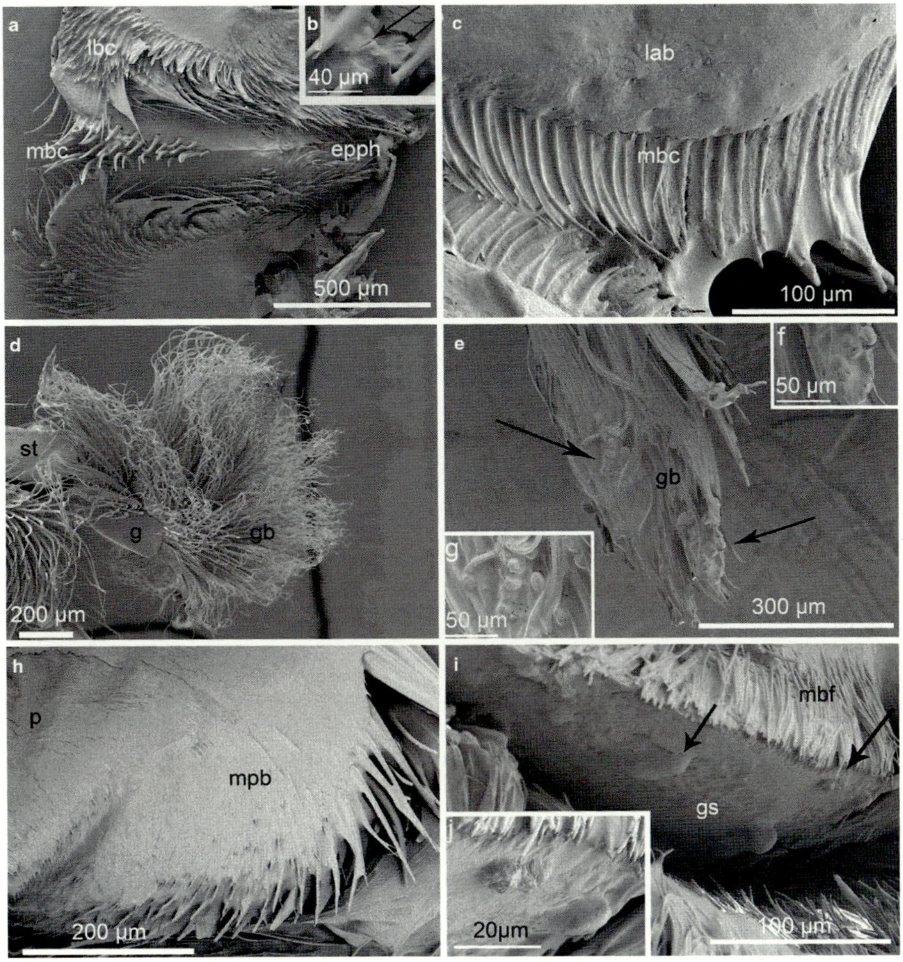

Fig. 2 Mouthparts of *Cetonia aurata*; Cryo-SEM micrographs (**a–c**, **e–j**) and SEM micrographs (**d**). **a** Ventral view of labrum with lateral bristle crests, median bristle crest and epipharynx. All bristles are covered with a fluid layer. **b** Detail of median bristle crest with pollen grains embedded in liquid layer (*arrows*). **c** Labium: dorsal view of moist median bristle crest covering the median depression. **d** Maxilla: wavy bristles of the galea forming a fan-like structure. **e** Galea tip, bristles forming a wet brush. Multiple pollen grains adhere to the liquid layer (*arrows*). **f**, **g** Pollen grains adhering to the moist galea tip. **h** Mandible: prostheca bristles with liquid layer, forming a pollen trapping and transport device. **i** Mandible: mola with moist grinding surface and wet bristle fringe. Single pollen grains are visible (*arrows*). **j** Detail view of the mola surface with a crushed and an intact pollen grain. epph, epipharynx; g, galea; gb, galea brush; gs, grinding surface; lab, labium; lbc, lateral bristle crest; mbc, median bristle crest; mbs, moist bristle fringe; mpb, moist prostheca bristles; p, prostheca; st, stipes

prostheca are entirely covered with liquid and thus adhere together, forming a large and dorsally vaulted region (Fig. 2h). Due to the convex shape it was concluded that this part of the mandible works like a pollen collecting receptacle when moving inside the cibarium. These structures clearly convey pollen to the mola region. This mandibular region with a sculptured chewing surface is covered with a liquid layer. Both crushed and intact pollen grains were detected to adhere to the fluid (Fig. 2i, j). The mola are surrounded by a fringe

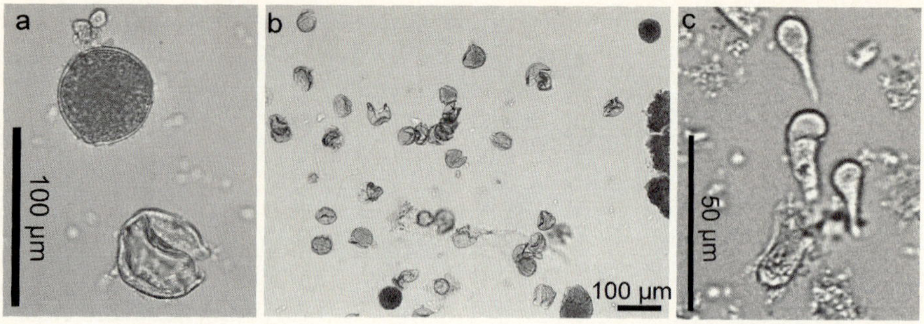

Fig. 3 *C. aurata*: Light-microscopical micrographs of the pollen grains in the gut. **a** Intact and crushed pollen grains from foregut in detail. **b** Numerous crushed and single intact pollen grains in the foregut. **c** Pollen grains with pollen tubes

of short multiply branched bristles and are also covered with a liquid layer (Fig. 2i). This moist fringe surrounds the mola and encompasses the chewing surface. Thus, it is interpreted as a sealing structure to prevent pollen grains from slipping off the chewing surface during the grinding process.

The strongly sclerotised labium forms the floor of the cibarium. Its buccal surface bears bristle crests similar to those on the labrum (Fig. 1c). Cryo-SEM micrographs show that the median crest of the labium is covered with a liquid filling the spaces between the distally orientated bristles. In this manner, the median depression of the labium is completely sealed off (Fig. 2c). Additionally the median bristle crest presumably serves to pass along the pollen grains.

Gut content

Numerous crushed pollen grains and a few intact ones were found in the examined gut regions (Fig. 3a, b). The anterior midgut contained about 2% of intact pollen and in the hindgut about 1% of the pollen was undamaged. In the hind gut of one specimen many pollen grains had a developed pollen tube (Fig. 3c).

Discussion

Feeding mechanism

Micrographs of the pollen feeding rose beetle, *Cetonia aurata*, obtained with the aid of the Cryo-SEM technique revealed that the buccal surfaces of the mouthparts were covered with a liquid coating. We conclude that this beetle uses the liquid layer to improve adhesion of pollen grains to the mouthparts. In particular, the long wavy bristles which form the galeal brushes most likely aid in retaining the fluid at the galeal tip. The moistened galeal brushes serve as effective pollen removal devices. Pollen is removed from the anthers probably not by the sweeping motion of the galeal tips, as proposed by Bürgis (1986a, b), rather by daubing the anthers with the moistened galeal brushes. This conclusion is based on the observation that the brushes and cuticle of the shock frozen mouthparts showed evidence of wetness, however this has not been experimentally verified.

The origin of the liquid is uncertain since salivary glands are allegedly absent in most Coleoptera (Beutel 2003) and are only known in representatives of few families in Coleoptera, such as Rhysodidae and Coccinellidae (Beutel 2003). However, the observed lubrication of the mouthparts could have been produced by the mandibular glands which are present in Coleoptera (Chapman 1998). Data on head glands in representatives of Cetoniinae are absent in the literature. The buccal liquid is possibly of extracorporeal origin. It may consist of nectar regurgitated from the foregut. Other specialised flower visiting insects like *Apis mellifera* regurgitate nectar to enhance the stickiness of pollen kit (Hesse 1980). This substance plays an important role in the adherence of pollen grains to each other and to the pollen collecting structures of bees.

The mouthparts of *C. aurata* have completely lost tooth-like structures that in other species of the subfamily Cetoniinae occur on the distal parts of mandible (Fuchs 1974) and maxilla (Bürgis 1989). While the maxillae primarily represent the pollen uptake organs, the mandibles serve two functions. First, the prostheca functions as a pollen conveying structure by sweeping the pollen between the mola surfaces. This hypothesis is supported by the fact that scattered pollen grains are visible in the SEM micrographs of this region. Second, the mandibles grind pollen; the molar parts work as a pollen mill, crushing pollen grains before they are swallowed. The moist fringe prevents pollen from slipping off the grinding surface. Both labrum and labium with their rows of moistened lateral bristles operate together to seal the oral cavity. Additionally, the moistened vertically orientated median crests and probably also the lateral bristle rows further convey pollen.

Mouthparts in flower visiting beetles

Representatives of protea beetle (Cetoniini) and monkey beetles (Hopliini) are reputed to be the pollinators for several South African plants (Johnson and Nicolson 2000). Their mouthparts are similar to *C. aurata*, i.e., equipped with numerous rows of bristles and the dense galeal brushes. The mandibles and maxillae are smooth plates lacking cutting edges and exhibit a similar semi-circular movement when taking food from flowers. Furthermore, these beetles are able to take up liquids by licking motions of the galeal brushes. The liquid is then presumably squeezed out when the labrum and labium are pressed together (Johnson and Nicolson 2000). *C. aurata* is also recorded to ingest nectar, plant saps and liquids from rotten fruits (Bürgis 1989). Based on the morphological similarity of its mouthparts to protea beetles and monkey beetles, it can be assumed that the mouthparts work in a comparable way to imbibe liquids. Whether these flower visiting Coleoptera also employ a buccal fluid when feeding on pollen is not known.

In pollen feeding beetles variously shaped galea bristles were described (Fuchs 1974). For example a knob-like relief or a corrugate surface are well suited to engage structured pollen grains and enhance the ability of pollen attachment. In Addition branched bristles and enlarged surfaces help convoy pollen towards the mandibles (Fig. 4). The galea bristles of *C. aurata* however are only undulate but function as adhesion structures in conjunction with fluid.

Beetles continue to be rather poorly studied as flower visitors compared to Hymenoptera and Lepidoptera (Krenn et al. 2005). As a result they are rarely considered to be specialised flower visitors or pollinators. The opinion is common that they do more harm to the flowers than benefit to the plant. This does not apply to *C. aurata* since its mouthparts lack cutting edges which could damage vital parts of flowers. Furthermore, species of the related Hopliini, are important pollinators of plants from families Asteraceae and Mesembryanthemaceae in South Africa (Picker and Midgley 1996). We expect that there is a

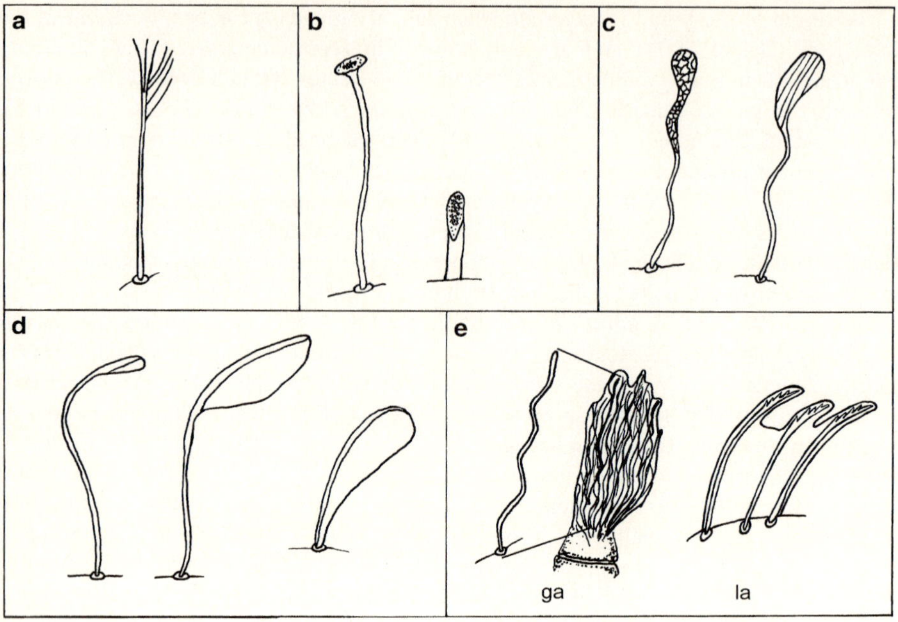

Fig. 4 Various types of specialised maxilla bristles of pollen feeding beetles. **a** Branched bristles. **b** Trumpet and spoon shaped bristles of *Malachius bipustulatus* (Malachiidae). **c** Adhesion surfaces with sculptured cuticle. **d** Variously broadened tips without sculptured surfaces. **e** Wavy bristles and serrated bristles covered with fluid in *Cetonia aurata* (Scarabaeidae). ga, galea; la, lacinia (modified after Schicha 1967 and Fuchs 1974)

wide repertoire of mouthpart adaptations to floral feeding in beetles which serve as a pollinating vector.

One of the well-investigated example of a pollen feeding beetle is *Malachius bipustulatus* (Malachiidae). Although its maxillae have special bristles for pollen uptake, they function differently from those in *C. aurata*. While the galeal surface is covered with numerous trumpet shaped bristles, the lacinia has spoon shaped bristles (Fig. 4b) which are also found laterally on the labium. Pollen grains adhere to slightly concave tips of the trumpet shaped setae and, during the retraction of the maxillae, the spoon shaped bristles of the labium sweep away the pollen. Finally, sweeping movements of the laciniae convey the pollen grains to the mandibles where they are crushed. The prostheca prevents pollen grains from escaping the cibarium and works as a guiding structure for the powdery meal (Schicha 1967). Apparently, *M. bipustulatus* relies on the specialised configuration of bristle tips instead of the rough surface or the liquid layer to convey pollen through the cibarium. Similar configurations have convergently evolved on the ventral surfaces of the tarsal pads of Coleoptera and Diptera specialised for adhesion to flat surfaces (Bauchhenss 1979; Stork 1980; Gorb 2001).

Pollen and spore feeding behaviour of Staphylinoidea was studied in detail by Betz et al. (2003). Sporophagy is a common feeding preference with particular demands on mouthpart morphology that are functionally similar to those of pollen feeding. Special adaptations for both kinds of particle feeding include brush, rake or comb like structures on the maxillae for sweeping particles, as well as setose structures on the mandible for further conveyance of the food. The mola is often reinforced with grinding structures, and the prostheca shows a microstructural diversification for sweeping particulate food material off the maxillae. The cibarial roof, epi- and hypopharynx show various microtrichia and bristle-troughs, for guiding the food through the cibarium (Betz et al. 2003).

Pollen digestion

Various mechanisms of pollen digestion have evolved in Coleoptera. Our study of *C. aurata* revealed that the mola of the mandibles are able to crush pollen grains. This is supported by observations of crushed pollen grains on the mola in Cryo-SEM preparations and the fact that crushed pollen grains are detectable in all gut sections. Thus the pollen is mechanically opened by action of the mandibles. Osmotic damaging of pollen occurs in the guts of South African protea beetles and monkey beetles whereby the pollen wall is weakened enabling digestive enzymes to enter via the pores (Johnson and Nicolson 2000). Since a small portion of undamaged pollen grains was found in all gut regions of *C. aurata*, we conclude that no further chemical process was necessary to destroy pollen exine in the digestive tract.

Another technique of pollen destruction was described in the pollen feeding beetle *Cyclocephala amazona* (Scarabaeidae) which ingests multicellular trichomes from the inflorescence axis of *Bactris gasipaes*, the beach palm of Central America, before consuming pollen. The thick walled and lignified trichomes offer no nutritional value, but work as gastroliths to crush the strong pollen walls in the beetle's gut (Rickson et al. 1990).

Pollen grains in which a pollen tube has developed were found in the hindgut of one specimen of *C. aurata.* It is reported that nectar in the gut of some insects, such as certain hoverflies (Haslett 1983) and bees (Peng et al. 1985; Dobson and Peng 1997), can initiate pseudo-germination of pollen grains. The initiation of short pollen tubes is followed by an exudation of digestible protoplasmic content from germination pores in the wall (Johnson and Nicolson 2000). However we assume that a certain amount of pollen grains have already begun to germinate before uptake.

Cryo-SEM methods

Many surface functions rely on the combination of microstructure and fluids. Therefore direct detailed visualisation of the fluid–solid surface interaction at high resolution may provide new insights into working principles of biological functional systems. The Cryo-SEM method has been previously applied for visualisation footprints of flies *Calliphora vicina* (Gorb 2006) and beetles *Gastrophysa viridula* (Schuppert and Gorb 2006), epicuticular grease on the elytra of the beetle *Leptinotarsa decemlineata* (Voigt et al. 2008) and water droplets on various surfaces of carnivore plants (Gorb et al. 2007).

The Cryo-SEM approach yields not only qualitative but also quantitative information. Of course, as every microscopy method, the Cryo-SEM approach is not absolutely free of artefacts. Typical Cryo-SEM artefacts are contamination of the surface by ice crystals, foaming of the fluid mixtures during freezing, partial sublimation of water from mixtures. Some of these can be avoided by short sublimation of the sample in the preparation chamber and more rapid freezing procedure by the use of liquid nitrogen at its melting point in vacuum. Thus, the Cryo-SEM method must be applied in combination with other microscopy techniques, including conventional SEM, phase contrast light microscopy and various preparation techniques. The application of new microscopic and experimental methods is essential for a deeper understanding of such pollen feeding insects. We have demonstrated here that the combination of Cryo-SEM and conventional SEM procedures is a promising approach to obtain detailed information about mouthparts in fresh and dried conditions and to draw conclusions on the feeding mechanisms of flower visiting insects.

Acknowledgements The authors would like to thank Dr. W. Klepal and D. Gruber from the Institution of Cell Imaging and Ultrastructure Research at the University of Vienna for providing access to the SEM-unit and for their assistance in scanning electron microscopy. We also want to thank John Plant for providing assistance with the English text, as well as the two anonymous reviewers who helped to improve the manuscript. This work, as part of the European Science Foundation EUROCORES Programme FANAS was supported from funds by the German Science Foundation DFG (contract No GO 995/4-1) and the EC Sixt Framework Programme (contract No ERAS-CT-2003-980409) to S.N.G.

References

Bauchhenss E (1979) Die Pulvillen von *Calliphora erythrocephala* Meig. (Diptera, Brachycera) als Adhäsionsorgane. Zoomorphology 93:99–123. doi:10.1007/BF00994125

Betz O, Thayer TK, Newton AF (2003) Comparative morphology and evolutionary pathways of the mouthparts in spore-feeding Staphylinoidea (Coleoptera). Acta Zool 84:179–238. doi:10.1046/j.1463-6395.2003.00147.x

Beutel RG (2003) 26. Ordnung Coleoptera, Käfer. In: Dathe HH (Hrsg.) Lehrbuch der speziellen Zoologie begründet von Alfred Kaestner Band I: Wirbellose Tiere 5. Teil: Insecta, 2. Aufl. Spektrum Akademischer Verlag, Heidelberg, Berlin, pp 426–526

Bürgis H (1986a) Blütenfressende Blatthornkäfer (Anthophaga), Der Rosenkäfer *Cetonia aurata* und seine Verwandten. 1. Kopf und Mundwerkzeuge. Mikrokosmos 75(7):207–211

Bürgis H (1986b) Blütenfressende Blatthornkäfer (Anthophaga), Der Rosenkäfer *Cetonia aurata* und seine Verwandten. 2. Das Zusammenspiel der Mundwerkzeuge. Mikrokosmos 75(11):327–330

Bürgis H (1987) Blütenfressende Blatthornkäfer (Anthophaga), Der Rosenkäfer *Cetonia aurata* und seine Verwandten. 3. Der Kaubezirk. Mikrokosmos 76(12):346–366

Bürgis H (1989) Blütenfressende Blatthornkäfer (Anthophaga), Der Rosenkäfer *Cetonia aurata* und seine Verwandten. 4. Die Nahrung des Rosenkäfers *Cetonia*. Mikrokosmos 78(2):50–53

Chapman RE (1998) The insects. Structure and function, 4th edn. Cambridge University Press, Cambridge

Dobson HEM, Peng Y (1997) Digestion of pollen components by larvae of the flower-specialist bee *Chelostoma florisomne* (Hymenoptera: Megachilidae). J Insect Physiol 43:89–100. doi:10.1016/S0022-1910(96)00024-8

Fuchs GV (1974) Die Gewinnung von Pollen und Nektar bei Käfern. Nat Mus 104:45–54

Gorb SN (2001) Attachment devices of insect Cuticle. Kluwer Academic Publishers, Dordrecht

Gorb SN (2006) Fly microdroplets viewed big: a Cryo-SEM approach. Microsc Today September 2006:38–39

Gorb SN, Voigt D, Gorb EV (2007) Visualisation of small fluid droplets on biological and artificial surfaces using the Cryo-SEM approach. In: Méndez-Vilas A, Díaz J (eds) Modern research and educational topics in microscopy, vol 2. Formatex, Badajoz, pp 812–819

Haslett JR (1983) A photographic account of pollen digestion by adult hoverflies. Physiol Entomol 8:167–171. doi:10.1111/j.1365-3032.1983.tb00345.x

Hesse M (1980) Zur Frage der Anheftung des Pollens an blütenbesuchende Insekten mittels Pollenkitt und Viscinfäden. Plant Syst Evol Suppl 133:135–148

Johnson SA, Nicolson SW (2000) Pollen digestion by flower-feeding Scarabaeidae: protea beetles (Cetoniini) and monkey beetles (Hopliini). J Insect Physiol 47:725–733. doi:10.1016/S0022-1910(00)00166-9

Krenn HW, Plant JD, Szucsich NU (2005) Mouthparts of flower-visiting insects. Arthropod Struct Dev 34:1–40. doi:10.1016/j.asd.2004.10.002

Nel A, Scholtz CH (1990) Comparative morphology of the mouthparts of Scarabaedoidea (Coleoptera). Entomol Mem 80:1–84

Peng Y, Nars ME, Marston JM, Fang Y (1985) The digestion of dandelion pollen by adult worker honeybees. Physiol Entomol 10:75–82. doi:10.1111/j.1365-3032.1985.tb00021.x

Picker MD, Midgley JJ (1996) Pollination by monkey beetles (Coleoptera: Scarabaeidae: Hopliini): flower and colour preferences. Afr Entomol 4:7–14

Rickson FR, Cresti M, Beach JH (1990) Plant cells which aid in pollen digestion within a beetle's gut. Oecologia 82:424–426. doi:10.1007/BF00317493

Schicha E (1967) Morphologie und Funktion der Malachiidenmundwerkzeuge unter besonderer Berücksichtigung von *Malachius bipustulatus* L. (Coleopt., Malacodermata). Z Morphol Oekol Tiere 60:376–433. doi:10.1007/BF00424639

Schremmer F (1961) Morphologische Anpassung von Tieren—insbesondere Insekten—an die Gewinnung von Blumennahrung. Verh Dtsch Zool Ges (Zool Anz, Suppl) 25:375–401

Schuppert J, Gorb S (2006) The wet step: visualisation of the liquid bridges in the attachment pads of the beetle *Gastrophysa viridula*. Society for Experimental Biology Main Meeting, University of Kent, Canterbury, 2–7 April 2006. Comp Biochem Physiol A 143(Suppl 1):Abstract No. A7.23

Stanley RG, Linskens HF (1974) Pollen: biology biochemistry management, vol VI. Springer Verlag, Heidelberg, p 307

Stork NE (1980) A scanning electron microscope study of tarsal adhesive setae in the Coleoptera. Zool J Linn Soc 68:173–306. doi:10.1111/j.1096-3642.1980.tb01121.x

Voigt D, Schuppert JM, Dattinger S, Gorb SN (2008) Sexual dimorphism in the attachment ability of the Colorado potato beetle *Leptinotarsa decemlineata* (Coleoptera: Chrysomelidae) to rough substrates. J Insect Physiol. doi:10.1016/j.jinsphys.2008.02.006

J Comp Physiol A
DOI 10.1007/s00359-006-0156-z

ORIGINAL PAPER

Material structure, stiffness, and adhesion: why attachment pads of the grasshopper (*Tettigonia viridissima*) adhere more strongly than those of the locust (*Locusta migratoria*) (Insecta: Orthoptera)

Pablo Perez Goodwyn · Andrei Peressadko · Heinz Schwarz · Victoria Kastner · Stanislav Gorb

Received: 28 July 2005 / Revised: 12 May 2006 / Accepted: 15 May 2006

Abstract The morphology, ultrastrucure, effective elastic modulus, and adhesive properties of two different smooth-type attachment pads were studied in two orthopteran species. *Tettigonia viridissima* (Ensifera) and *Locusta migratoria* (Caelifera) have a similar structural organization of their attachment pads. They both possess a flexible exocuticle, where the cuticular fibrils are fused into relatively large rods oriented at an angle to the surface. The compliant material of the pad contributes to the contact formation with the substrate. However, the pad material structure was found to be different in these two species. *L. migratoria* pads bear a thick sub-superficial layer, as well as a higher density of rods. The indentation experiments showed a higher effective elastic modulus and a lower work of adhesion for *L. migratoria* pads. When the indentations were made at different depths, a higher effective elastic modulus was revealed at lower indentation depths in both species. This effect is explained by the higher stiffness of the superficial pad layer. The

P. Perez Goodwyn · A. Peressadko · S. Gorb (✉)
Evolutionary Biomaterials Group, Department Arzt, Max Planck Institute for Metals Research, Heisenbergstr. 3, 70569 Stuttgart, Germany
e-mail: s.gorb@mf.mpg.de

H. Schwarz · V. Kastner · S. Gorb
Max Planck Institut für Entwicklungsbiologie, Spemannstr. 35, 72076 Tubingen, Germany
e-mail: heinz.schwarz@tuebingen.mpg.de

Present Address:
P. Perez Goodwyn
Laboratory of Insect Ecology, Graduate School of Agriculture, Kyoto University, Kyoto 606-8502, Japan
e-mail: pablogoodwyn@yahoo.com.ar

Present Address:
A. Peressadko
Nanotribology Laboratory for Information Storage and MEMS/NEMS(NLIM), Ohio State University, Columbus, OH, USA
e-mail: peresadko.1@osu.edu

obtained results demonstrate a clear correlation between density of the fibres, thickness of the superficial layer, compliance of the pad, and its adhesive properties. Such material structures and properties may be dependent on the preferred environment of each species.

Keywords Cuticle · Ultrastructure · Mechanical properties · Adhesion · Attachment

Introduction

Attachment to various substrata plays an important role during walking, jumping, and climbing in most insect species. Locomotion on rough vertical and overhanging substrata can be achieved by using claws and/or spines that interlock with the peaks and troughs of the uneven surface profile (Frantsevich and Gorb 2004). When the roughness of the substrate is comparable to or lower than the claw tip diameter, sufficient attachment cannot be achieved with the use of these leg structures (Dai et al. 2002). For reliable attachment to smooth substrata, the majority of insect groups have evolved so called attachment pads. There are many literature sources relating to these attachment devices, whose diversity is enormous (Gorb 2001). However, from the morpho-functional point of view, the devices can be divided in two major groups: hairy and smooth.

Hairy pads consist of arrays of specialized outgrowths with a high aspect ratio (5–80). They are able to form independent contacts with the substratum at numerous points (Beutel and Gorb 2001; Gorb and Beutel 2001). Smooth pads do not bear such structures on their surface, but have a very specialised material design. Regardless of such structural differences, both types of attachment systems may replicate the unevenness of the substrate at different scales. This leads to an increase in the real contact area and in contact perimeter and thus achieves a stronger adhesion and friction forces between the pad and substrate (Gorb et al. 2000; Jiao et al. 2000; Scherge and Gorb 2000; Niederegger et al. 2002). Both systems have independently appeared several times in insect evolution (Beutel and Gorb 2001). The smooth-type attachment pad is present in representatives of Orthoptera, Hymenoptera, Blattaria, Hemiptera, and in some other orders. In representatives of Orthoptera, the pads are called arolia and euplantulae and possess an extremely flexible cuticle. In *Tettigonia viridissima*, the pad surface is not perfectly smooth, but subdivided into tightly packed hexagonal outgrowths a few microns wide (Gorb et al. 2000), but representatives of some other groups possess rather smooth surfaces (Gorb 2001).

Insect cuticle is a layered fibrous composite material, in which chitin fibres are embedded in a protein matrix. The mechanical properties of the cuticle depend on the relative proportions of fibres and matrix and on the degree of the cross-linking of the components (Neville 1975; Hepburn 1985). The cuticle of smooth attachment pads is not organized in layers as is the cuticle of the rest of the body. It consists of rod-like structures formed by tightly packed fibres oriented at an angle to the surface (Slifer 1950; Gorb et al. 2000). Previous studies have shown that such pads have a very low effective elastic modulus and visco-elastic properties (Jiao et al. 2000; Gorb et al. 2001).

The present study was undertaken to compare the structure, stiffness, and adhesive properties of attachment pads in representatives of two different lineages of Orthoptera: Ensifera and Caelifera. The chosen species (*Locusta migratoria* and *T. viridissima*), have different ecological niches. *L. migratoria* lives in relatively dry environments where water is a constraint, while *T. viridissima* lives in humid temperate grasslands. The following questions were asked: (1) Is the ultrastructure of the pads of the two species different from each other? (2) Are the mechanical properties of the pads similar in the two species, and how do they correlate with the structure? (3) What is the relationship between the structure, stiffness, and adhesive properties of the species studied?

Materials and methods

Animals

Locusta migratoria specimens were either provided by the Insectarium from the Wilhelma Zoo (Stuttgart, Germany) or obtained from pet shops. *T. viridissima* specimens were collected in the field in the surroundings of Stuttgart (Germany) in July–August 2004. The specimens of both species were kept in cages with food (various grass species collected around the laboratory) and water. Cages with *L. migratoria* were heated (ca. 30°C), while those with *T. viridissima* were kept at room temperature. Individuals of the second species were used for the experiments on the day of capture.

Cryo-sectioning and light microscopy

The tarsi were severed from the bodies of both species, embedded in water-soluble mounting medium (Tissue-Tek) and frozen at −30°C in a Cryomicrotome (Leica). The 8–10 μm-thick sections were prepared using a steel knife, transferred to Superfrost® glass slides, thawed at room temperature, mounted in the water-soluble mounting medium (Moviol), and viewed in a light microscope (Zeiss Axioplan). To visualize resilin-bearing parts of the pads, the sections were additionally examined using fluorescence microscopy (Zeiss Axioplan microscope) and photographed (Sony Power HAD) (Gorb 1999). Specifically, sections were observed in the UV band (excitation 340–380 nm, emission 420 nm; Andersen and Weis-Fogh 1964), as well as green (excitation 512–546 nm, emission 600–640 nm) and red (excitation 710–775 nm, emission 810–890 nm) bands. Micrographs obtained at different wavelengths were superimposed on a single layer in Adobe Photoshop 7.0.

Scanning electron microscopy

The freeze-substitution technique of Gorb et al. (2000) was used for these experiments. The tarsi were cut off the bodies of both species and shock-frozen in the liquid nitrogen. Parts of them were fractured in the liquid nitrogen using a razor blade. All samples were transferred to 0.5% osmium tetroxide solution in absolute acetone at −80°C for 48 h, washed several times in absolute acetone at −20°C, transferred to absolute ethanol, and critical-point dried. All preparations were mounted on holders, sputter-coated with gold–palladium (10 nm) and examined in a Hitachi S−800 (Tokyo, Japan) scanning electron microscope at 20 kV.

Transmission electron microscopy

Tarsi were fixed for 12 h at 4°C in 2.5% glutaraldehyde (in 0.01 M phosphate buffer at pH 7.3), and post-fixed for 1 h in 1% osmium tetroxide in phosphate buffer at 2°C. After washing, preparations were stained for 1 h at 4°C in 0.1% aqueous uranyl-acetate solution, washed, dehydrated, and embedded in a low viscosity resin (Spurr 1969). Ultrathin sections were picked up on copper grids coated with formvar film. Sections were stained with uranyl acetate and lead citrate, and observations were made with a Philips CM10 transmission electron microscope.

Indentation experiments

Sample preparation

The hind tibia with tarsus was cut off the body of an anaesthetized *L. migratoria*. A drop of melted wax was placed on the dissected end in order to avoid rapid desiccation of the sample. The tarsomeres were glued using melted wax to a glass or aluminium sample holder, with the ventral side upwards (Fig. 1). The sample was firmly attached with its dorsal side to the holder, in order to avoid any possible undesired movement during the indentation. The whole freshly killed specimen of *T. viridissima* was mounted, and the tarsomeres were glued with wax to the sample holders in the same manner as explained for *L. migratoria*. This was done because euplantulae of cut-off leg portions of *T. viridissima* collapsed shortly after cutting from the body. In both cases, the experiments were finished within 1 h, in order to avoid the effect of the material desiccation. The second tarsomere of hind legs of *L. migratoria*, and the fourth one of *T. viridissima* were used for indentation experiments (Fig. 1b, d).

The experiments were carried out at room temperature (24–26°C, 40–60% RH). In order to measure the elastic properties of the attachment pads, micro- and nano-indentation experiments were carried out (Gorb et al. 2000; Kohane et al. 2003; Enders et al. 2004). The attachment pad was continuously loaded to a certain applied force and then unloaded.

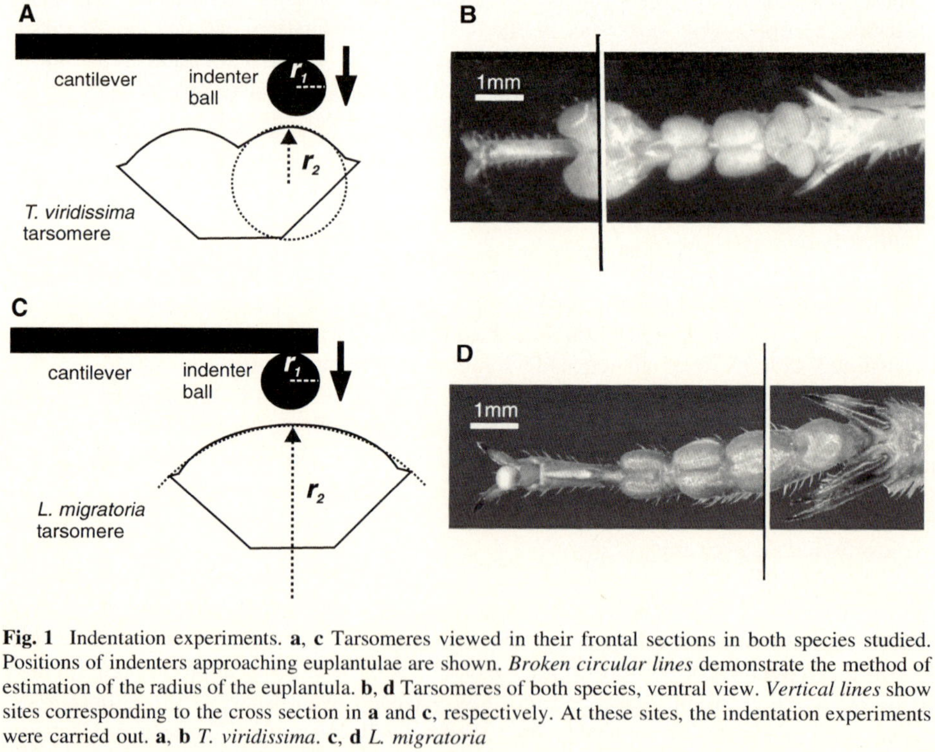

Fig. 1 Indentation experiments. **a**, **c** Tarsomeres viewed in their frontal sections in both species studied. Positions of indenters approaching euplantulae are shown. *Broken circular lines* demonstrate the method of estimation of the radius of the euplantula. **b**, **d** Tarsomeres of both species, ventral view. *Vertical lines* show sites corresponding to the cross section in **a** and **c**, respectively. At these sites, the indentation experiments were carried out. **a**, **b** *T. viridissima*. **c**, **d** *L. migratoria*

The applied load and depth of indentation were continuously recorded. A force–displacement curve was generated from each indentation. The curve provides information about the mechanical response of the cuticle to the load. From this curve, the effective elastic modulus can be calculated. Two different indenter devices were used: the force tester Basalt 01 (Tetra GmbH, Illmenau, Germany) (Jiao et al. 2000) and a XP nano-indenter (MTS Systems, USA) (Enders et al. 2004).

Micro-indentation experiments

The force tester Basalt 01 was applied for micro-indentation experiments. It includes three main parts: a platform, a glass spring with the indenter tip attached, and a fibre-optic sensor (described in detail in Scherge and Gorb 2000).

Three different indenter tips were used: two glass beads with radii 32 and 132 μm, respectively, and a flat glass surface. Indentations were made to two different depths (32 ± 8.7 and 45 ± 7.2 μm), by applying 300–1,000 μN loads. The sample sizes were as follows: *L. migratoria* 9 tarsi, and *T. viridissima* 5 tarsi; $n = 18$–20 indentations per tarsus.

Nano-indentation experiments

For comparison, an XP nano-indenter was used (MTS Systems). This system uses electromagnetic actuation to generate progressive movement of the tip, and is described in detail in Enders et al. (2004).

The sample preparation was the same as described for the force tester Basalt 01. Five tarsi per species were measured with $n = 10$ indentations per tarsus. The indenter tip used was a custom sapphire bead with a 247-μm radius. Both the indentation depth and load were kept in the same range as with the force tester Basalt 01, in order to allow the comparison of results obtained by two different measuring devices (average depth: 14.6 ± 3.0 μm, average load at maximum indentation depth: 1,000.0 ± 0.2 μN).

Data processing

Using Origin 7.03 software, the obtained load–displacement (indentation depth) data were fitted according to the Johnson, Kendall, and Roberts (JKR) contact theory (Johnson et al. 1971). For the estimation of such variables of the JKR model as the effective elastic modulus of the pad material (E_{eff}) and the work of adhesion (WOA) (W), the deformation of the glass or sapphire indenters was ignored due to the extreme difference of their effective elastic moduli compared to those of the biological materials under consideration. The effective elastic modulus was calculated from the fitting of the load curve, and the WOA was estimated from the unload curve. The WOA is the energy per unit of area (J/m^2) needed to separate two bodies in contact. It was also calculated by deriving from the pull-off force as

$$F = \frac{3}{2}(R\pi W) \tag{1}$$

$$W = \frac{F}{(3/2R\pi)}, \tag{2}$$

where R is the reduced radius. The reduced radius of the contacting spheres R_1 and R_2 was calculated as:

$$R = \frac{(R_1 \times R_2)}{(R_1 + R_2)}. \quad (3)$$

Effective elastic modulus was calculated as:

$$E_{eff} = \left(\frac{1 - v_1^2}{E_1} + \frac{1 - v_2^2}{E_2}\right), \quad (4)$$

where E_1 and v_1, and E_2 and v_2 are the elastic modulus and Poisson's ratio for the specimen and indenter, respectively (Johnson 1985). Effective elastic modulus values, as well as WOA, were compared for different treatments by applying one-way ANOVA, if data were normally distributed, or Kruskal–Wallis ANOVA on ranks, if data were not normally distributed (Sigma-Stat 3.0, SPSS).

Estimation of the pad profile

Using a white light interferometer (Zygo NewView 5000, USA), the surface of each pad was scanned, and with the software Metropro 7.9.0 the radius of curvature was calculated from the 3D surface profile. In the case of *L. migratoria*, this calculation was rather simple, since the ventral surface of the second tarsomere is rather round (Fig. 1a). In the case of *T. viridissima*, the ventral surface of the tarsomere is clearly bilobed. Each lobe of the tarsomere was therefore considered as a hemispherical solid body (Gorb et al. 2000) (Fig. 1c).

Dessication experiments

In order to assess the dehydration rate of the cuticle of *L. migratoria* euplantulae, the tarsus was cut off the body and sealed with melted wax at the dissected end. This preparation was allowed to dry under normal laboratory conditions (24–26°C, 40–60% RH) on a high precision balance (Mettler-Toledo Delta Range AG204). The weight was measured every 20 min for 24 h. Control measurement was carried out with a cut-off portion of the tibia. Three samples from three different animals were measured for both parts of legs (tarsus and tibia). The data were compared with those for *T. viridissima*, previously measured under similar laboratory conditions (Jiao et al. 2000).

Results

Dessication dynamics

The tarsi of both species studied lost their water content much faster than the tibiae. However, the water loss rate was relatively higher for both parts of the leg of *T. viridissima* which inhabits humid areas compared with *L. migratoria* which lives in more arid habitats (Fig. 2).

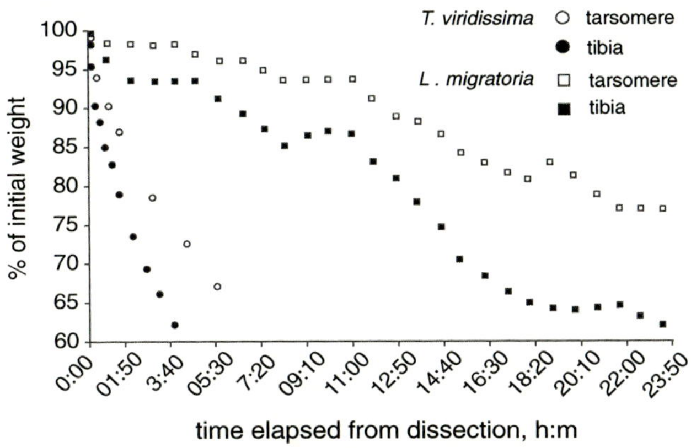

Fig. 2 Desiccation dynamics of tarsi and pieces of tibiae cut off the body. Data for *T. viridissima* are taken from Jiao et al. (2000)

Morphology

Tettigonia viridissima

The exocuticle is 45–50 µm thick (Fig. 3a, b). The surface is subdivided into hexagonal structures (Fig. 4a, b). Primary filaments (rods), branching into the secondary filaments, are present in the exocuticle. Filaments are oriented at some angle to the surface. The primary filaments are much thinner (1.0–1.3 µm) and less densely packed than those in *L. migratoria* with 4.0–4.5 µm space between them (Fig. 3a, b). At relatively regular intervals (5–10 µm), cross-links interconnecting fibres are present. In the branching zone, at a depth of 10–22 µm, the primary filaments branch into secondary ones terminating at the epicuticle (Fig. 3a, c). Since the distance between primary rods is approximately the same as between the centres of the hexagons, it seems that each single primary rod, branching into numerous secondary rods, supports each hexagon. The dense superficial layer, present in *L. migratoria*, was not found in *T. viridissima*. There is only a thin superficial layer of the epicuticle (0.2–0.3 µm thick) (Fig. 3c). A relatively thick layer (20 µm), fluorescing blue at UV wavelength, approximately corresponds to the layer of secondary rods (Fig. 3b).

Locusta migratoria

The organization of the pad cuticle in *L. migratoria* is rather different from that of *T. viridissima*. The exocuticle is approximately 60 µm thick, consisting of long densely packed filaments (rods), oriented at an angle to the surface (Fig. 3d, e). The filaments are composed of several fibres bundled together. Between the filaments, numerous interconnecting cross-links are present, so that the material resembles a foam-like structure. The filaments are 2.5–3.5 µm wide extending through the entire exocuticle. The space between rods is comparable to their width (3–4 µm). The exocuticle can be subdivided into three zones according to its ultrastructural organization. The deepest zone, starting from the endocuticle, is 35–40 µm thick and represents the layer of filaments, where single filaments are clearly visible (Fig. 3d–f). Fluorescent microscopy reveals also the presence of resilin, a rubber-like protein fluorescing blue at UV wavelength (Fig. 3e). The intermediate

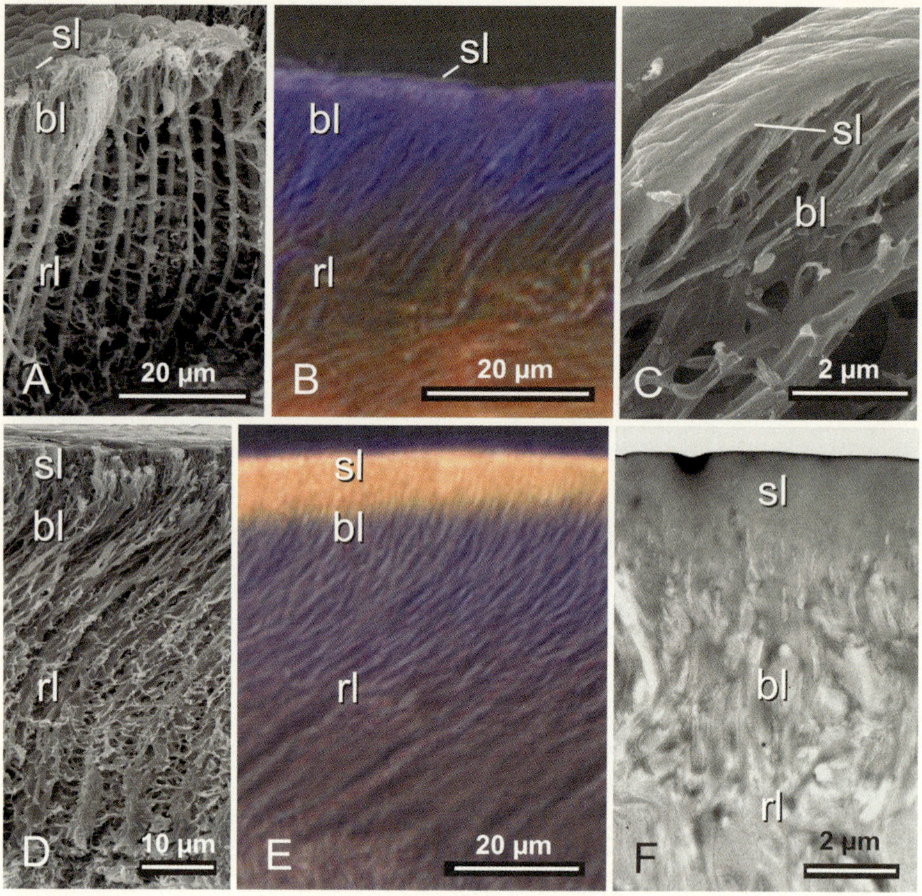

Fig. 3 Material structure of the cuticle of euplantulae. **a**, **c**, **d** Scanning electron microscopy. Freezing-fractured, substituted, dehydrated, and critical-point dried preparations. **b**, **e** Fluorescence microscopy. **f** Transmission electron microscopy. **a**–**c** *T. viridissima*. **d**–**f** *L. migratoria*. *sl* superficial layer, *bl* layer of branching rods, *rl* layer of primary rods

zone is the layer of branching filaments. In the vicinity of the surface, filaments divide into thinner fibre bundles. The branching is observed approximately at a depth of 20–25 μm. This layer shows fluorescence at all three tested wavelengths (Fig. 3e). The most superficial layer is 2–4 μm thick. Here the filaments are fused together forming a dense layer seen as an electron-dense structure in transmission electron microscopy (Fig. 3f). The surface is smooth, when observed in SEM or with a white light interferometer (Fig. 4c, d).

Effective elastic modulus of the pad

The pads of *L. migratoria* have higher values of the effective elastic modulus (E_{eff}) at the shallower indentation depth of 32.0 ± 8.7 μm than at the greater indentation depth, regardless of the indenter tip used (Fig. 5b). Two depths comparisons for the flat indenter, for 32 μm radius indenter ($P < 0.01$, *t*-test), and for 132 μm radius indenter were significantly different ($P < 0.01$, Mann–Whitney rank sum test). The experiments with different

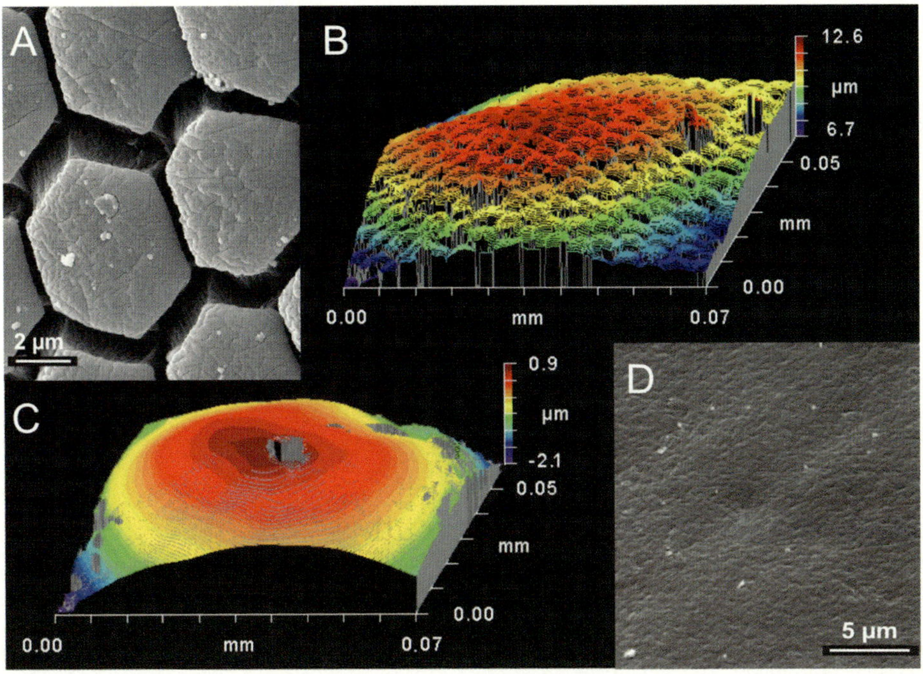

Fig. 4 Surface of the attachment pads. **a**, **d** Scanning electron micrographs, critical-point dried preparations. **b**, **c** 3D images of fresh euplantulae obtained with the white light interferometer. **a**, **b** *T. viridissima*. **c**, **d** *L. migratoria*

indenter tips at same indentation depth demonstrated no significant difference between effective elastic moduli ($P > 0.05$, ANOVA on ranks).

Similar results were obtained from *T. viridissima*. The shallower indentations (32 ± 8.7 μm) yielded higher E_{eff} ($P \leq 0.001$, Mann–Whitney rank sum test), but similar values for the other indenters (two depths comparisons for the flat indenter: $P = 0.11$, *t*-test, and for 132 μm radius indenter: $P = 0.26$, Mann–Whitney rank sum test) (Fig. 5a). When comparing E_{eff} values obtained with different indenter tips at the same depth, those with the 32 μm radius tip were significantly higher than the others (Kruskal–Wallis ANOVA on ranks, $P \leq 0.001$, Dunn's post-comparisons, $P < 0.05$). This result was obtained only for shallower indentations, but not for deeper ones (Kruskal–Wallis ANOVA on ranks, $P \leq 0.001$, Dunn's post-comparisons, $P > 0.1$) (Fig. 5a).

E_{eff} values obtained, using the XP nano-indenter, were qualitatively similar to those obtained with the force tester Basalt 01. E_{eff} in *L. migratoria* was clearly higher than that of *T. viridissima* ($P < 0.0001$, *t*-test). E_{eff} in *T. viridissima* was rather similar for both measuring devices. However, E_{eff} values in *L. migratoria* were approximately twice as high as in experiments with the micro-indenter Basalt 01 (Fig. 6).

Work of adhesion

Estimates of the WOA was different depending on the method of estimation (Fig. 7a). The values obtained directly from the unload curve fitting according to the JKR model showed higher WOA values in *T. viridissima* than in *L. migratoria*. In both species there was a

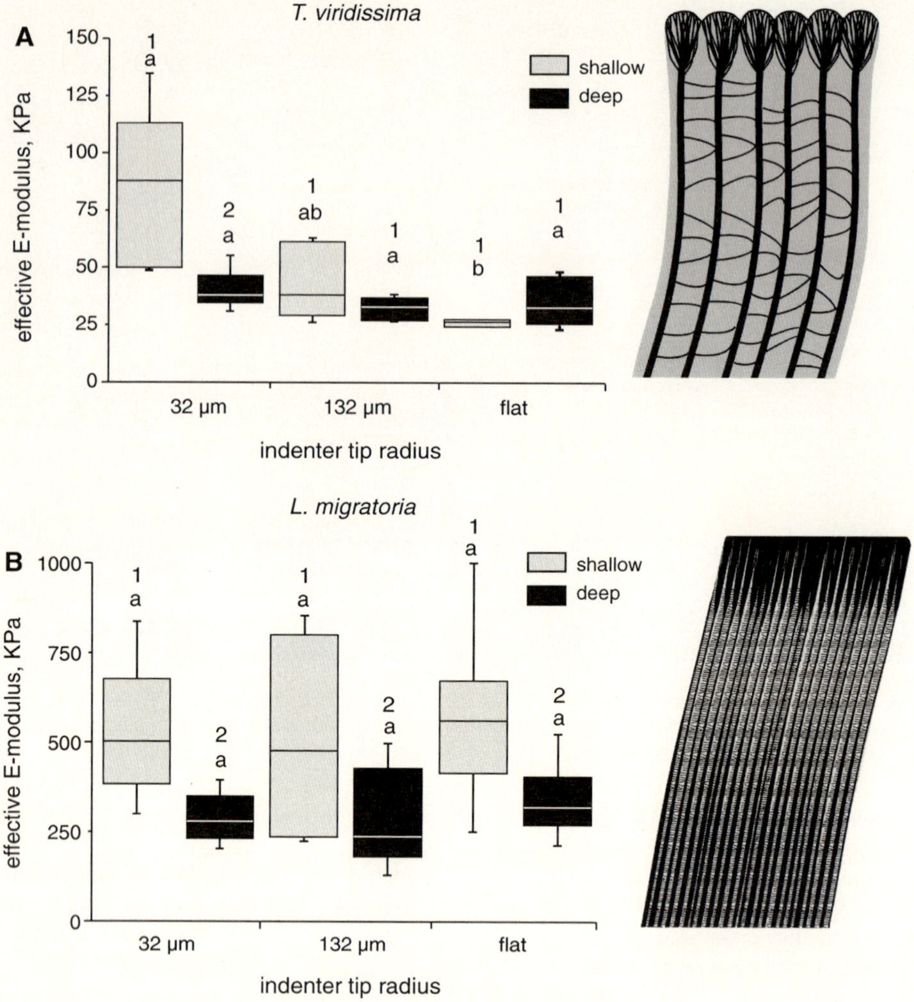

Fig. 5 Effective elastic modulus (E_{eff}) obtained with the force tester Basalt 01 using different indenter tips (32 μm radius, 132 μm radius, *flat*). Indentation depths discriminated as "shallow" (32 ± 8.7 μm) and "deep" (45 ± 7.2 μm). **a** *T. viridissima*. **b** *L. migratoria*. The ends of the boxes define the 25th and 75th percentiles, with a line at the median and error bars defining the 10th and 90th percentiles. *1* and *2* indicate statistically different samples according to the comparison of data obtained with same indenter at different depths; *a* and *b* indicate statistically different samples according to the comparison of data obtained with different indenters at same depth. Samples indicated with *different letters* are significantly different. Samples having *two letters* are not significantly different from samples indicated with just one of those letters. For *L. migratoria*, there is no statistical difference between data obtained with different indenters at same depth. *Insets* schematically demonstrate internal structures of pad materials in both species

certain trend to a lower WOA, when the indenter tip was larger (Fig. 7b). However, WOA values calculated directly from the pull-off force (Eq. 2) are very different. Values obtained for *L. migratoria* were higher than those for *T. viridissima* (Fig. 7c). The maximum difference was of one order of magnitude, obtained for the 32 μm radius indenter. A somewhat smaller difference was obtained with the 132 μm radius indenter, while WOA

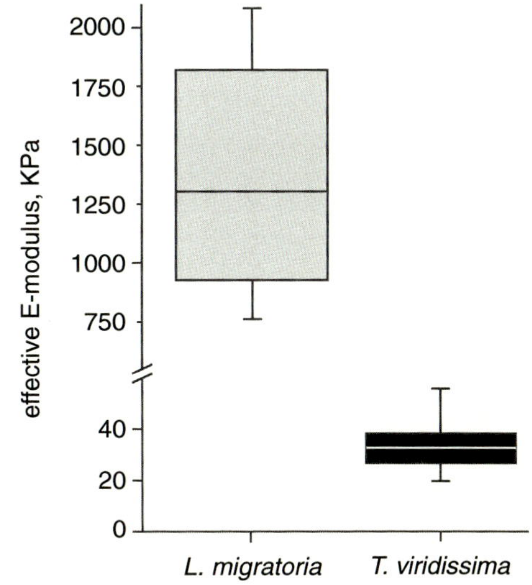

Fig. 6 Effective elastic modulus (E_{eff}) obtained using an XP nano-indenter with the tip radius of 250 μm for attachment pads of *T. viridissima* and *L. migratoria*

values for the two species were not significantly different when the flat indenter was used. There was a clear trend in both species, regarding the relationship between the tip diameter and WOA. In *T. viridissima*, the indenter diameter is positively correlated with the WOA values, however in *L. migratoria* this correlation was negative (Fig. 7c).

The indentation depth had no significant effect on the WOA obtained for either species, irrespective of the method used for estimating it.

Discussion

Insect cuticle is a layered composite material, with anisotropic material properties (Neville 1975). Cuticle shows an extremely wide range of properties (review by Vincent and Wegst 2004) and serves multiple mechanical functions (Gorb 2001). Smooth attachment pads responsible for contact formation on diverse substrata have a remarkable inner structure and high degree of deformability (Jiao et al. 2000; Gorb et al. 2000; Federle and Endlein 2004). In the present study, we used a comparative approach to understand the functional significance of different pad organization in representatives of Orthopera belonging to two different clades.

To obtain the effective elastic moduli of the pads, we used two different types of indentation devices: the micro-indenter Basalt 01 (Gorb et al. 2000) and MTS XP nano-indenter (Enders et al. 2004). We have chosen the JKR contact model to estimate the E_{eff}, because it considers the surface attraction effects between two solids (Johnson 1985), which is not the case for the Hertz model (Hertz 1881). The E_{eff} values obtained in this study with different indentation methods and with the use of JKR model for E_{eff} estimation are in the same range previously obtained for *T. viridissima* (Gorb et al. 2000).

Work of adhesion

The discrepancy between the WOA values for the two data processing methods chosen may be due to the fact that both the pull-off force of a visco-elastic material (Gorb et al.

Fig. 7 Work of adhesion (*WOA*) of *T. viridissima* and *L. migratoria* attachment pads. **a** Force–distance ▶ diagram showing the two ways of calculation of the work of adhesion (for further explanations see Sect. "Materials and methods"). **b** Work of adhesion calculated from the JKR model fitting of the unload curve. **c** Work of adhesion calculated directly from the pull-off force. The ends of the boxes define the 25th and 75th percentiles, with a line at the median and error bars defining the 10th and 90th percentiles. *a–e* indicate statistically different samples according to the comparison of data obtained for different species and different indenters. Samples indicated with *different letters* are significantly different. Samples having *two letters* are not significantly different from samples indicated with just one of those letters, but different from samples indicated with other letters

2000) as well as the action of secreted fluids (Vötsch et al. 2002) are not considered in the JKR model. When the model fits the unload part of the force–distance curve to the point, where it crosses the force axis (Fig. 7a), the WOA can be estimated under consideration that the pad behaves ideally elastically and contains no fluid layer at its surface (Fig. 7b). The fitting of the unload curve expects a pull-off force at the point P_1 (Fig. 7a), which is used then for calculation of the WOA according to Eq. 1. However, the experimentally measured pull-off force of the attachment pad (P_2) is usually much higher than that estimated from the JKR model (P1). This discrepancy between P_1 and P_2, would explain the overestimation of the WOA values calculated from the measured pull-off force, as well as the stability of the WOA value obtained from adhesion force estimated by JKR-fitting.

Pad functional morphology

In *T. viridissima*, there are three distinct layers of the pad material (Slifer 1950; Gorb et al. 2000). There seem to be two levels of deformation in the euplantula when it contacts the substratum (Gorb et al. 2000). The superficial level replicates the surface roughness due to deformation of secondary filaments located in several first micrometres of the cuticle. The deeper level involves bending of the primary filaments located deeper.

Superficial deformations were observed to the depth of 5 μm, which corresponds to one depth of single hexagonal structure of the pad and to the half of thickness of the layer of secondary filaments (Gorb et al. 2000). The hexagons of the *T. viridissima* surface can be considered as independent contact elements. The multiplying of the contact elements leads to an increase of the ability of replication of the uneven surface profile, and to the enhancement of adhesion force (considering JKR equation fitting derived WOA, see the former paragraph) (Autumn et al. 2002; Arzt et al. 2003; Persson 2003).

The fluid within the pad between the filaments contributes to the visco-elastic behaviour of the pad, because the fluid is able to flow through the gaps between the rods, when the pad deforms during contact formation and breakage (Gorb et al. 2000). The interconnecting filaments presumably prevent rod buckling, when these are bending during contact. The pad presumably works as a damper during jump or landing at high speed of the deformation, as well as a means of replicating a complex substratum profile at slow deformation speeds. Also, the rod-like organization of the pad architecture would allow an independent local load distribution over the area of contact between pad and substrate. This would aid in an enhancement of the adaptation of the pad to imperfections of natural substrata. Presumably, the pad structure of *L. migratoria* works in a similar way to that of *T. viridissima*. The foam-like organization should have a similar ability to conform to the substrate profile during contact formation.

However, there seem to be some major differences in the organization of the pad structure of the two species studied. The higher density of rods and interconnecting

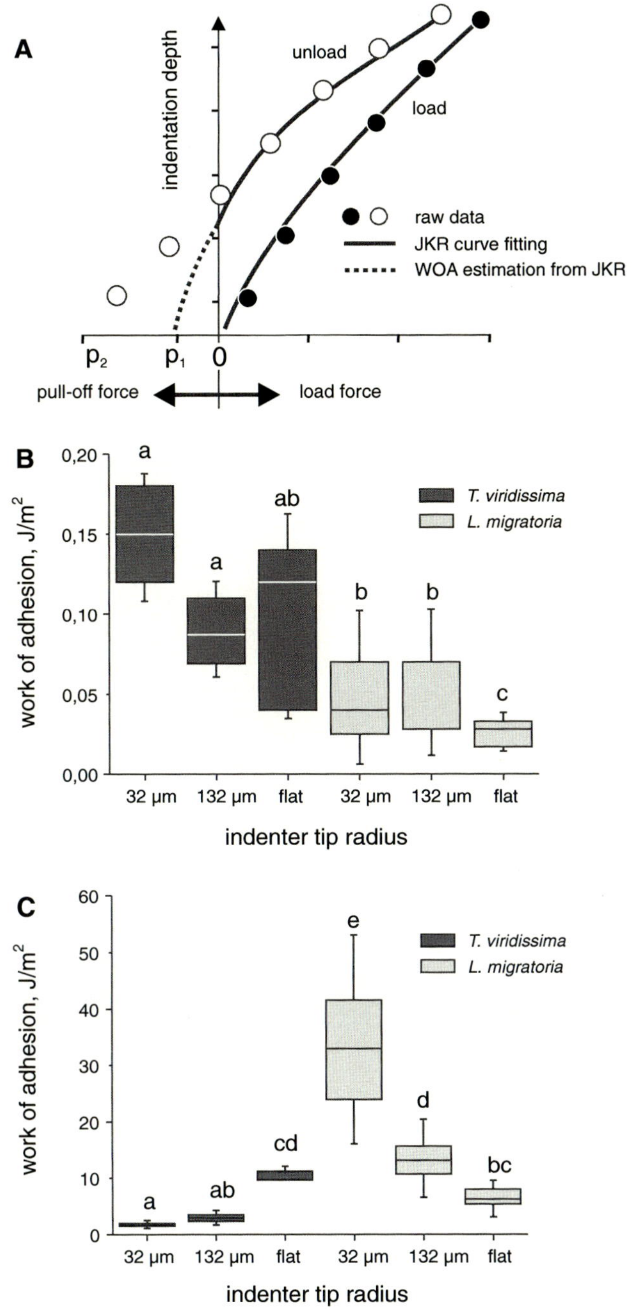

filaments are responsible for a higher E_{eff} in the pad of *L. migratoria*. Moreover, the natural substrata for *T. viridissima*, which lives in a temperate climate, are mostly plant surfaces. *L. migratoria*, which lives in desert areas, walks, jumps, and lands on various surfaces, including stones, sand, and probably hot surfaces. In this species, both the presence of a thick superficial layer and dense distribution of rods in the pad architecture, suggest a protective function of this pad organization against this harsh environment. It has been recently demonstrated that a thin highly deformable layer would be better at conforming to substrate asperities compared to bulk material with the same material properties (Persson and Gorb 2003; Carbone et al. 2004). This is only true, however, as long as the wavelength of the surface roughness is greater than the layer thickness. The thickness of the stiffer surface layer (superficial layer plus branching layer) of *L. migratoria* is about 20 μm. The layer is much stiffer and supported by the rods layer, containing resilin and fluid. Presumably the much higher E_{eff} measured with the nano-indentation technique in *L. migratoria* could be due to this layer. With nano-indentation the volume involved in the deformation is much smaller than with the micro-indentation technique, because of the shallower indentation. In the case of nano-indentation of *L. migratoria* pads only the upper stiffer layer would be involved in the deformation. With micro-indentation the upper stiffer layer is bent as a thin stiff sheet, and the deformation is affecting the underlying rod layer, thus resulting in lower values of E_{eff}. The pad, consisting of a soft core covered by a stiffer layer, combines conformability to the surface roughness and resistance to the environment.

The remarkably low desiccation rate of *L. migratoria* pads compared to that of *T. viridissima* might be a possible trade-off between attachment efficiency and resistance to a dry environment. The pads of *T. viridissima* are softer and possess a higher WOA, but are too sensitive to live in a dry environment. Since the pads of *L. migratoria* are presumed to have a dehydration-resistant superficial layer, they are stiffer and thus less effective in adapting to imperfections of the substrata.

This study shows that the same basic architecture of the pads can be tuned in the evolution of species to different needs and constraints. The structural principle, based on branching rods, contributes to holding the shape of the pad. This principle, combined with the presence of a relatively stiff superficial layer, provides the adaptability of the attachment pad to unevenness of natural substrata. In species living in arid environments, the superficial layer must protect animals from desiccation. However, the adhesive properties will be reduced at the same time, because of the reduced ability to form close contact with rough substrata.

Acknowledgments The authors are grateful to Ms Koch (Wilhelma Zoo Insectarium, Stuttgart, Germany) for supply of insects for this study. Dr S. Enders and Dr N. Barbakadse (Max Planck Institute for Metals Research, Stuttgart, Germany) introduced and guided the first author in the nano-indentation technique. This study was partly supported by the project BioFuture 0311851 (Federal Ministry of Education and Science, BMBF, Germany) to SNG.

References

Andersen SO, Weis-Fogh T (1964) Resilin, a rubber-like protein in arthropod cuticle. Adv Insect Physiol 2:1–65

Arzt E, Gorb SN, Spolenak R (2003) From micro to nano contacts in biological attachment devices. PNAS 100:10603–10606

Autumn K, Sitti M, Peattie A, Hansen W, Sponberg S, Liang YA, Kenny T, Fearing R, Israelachvili J, Full RJ (2002) Evidence for van der Waals adhesion in gecko setae. PNAS 99:12252–12256

Beutel RG, Gorb SN (2001) Ultrastructure of attachment specializations of hexapods (Arthropoda): evolutionary patterns inferred from a revised ordinal phylogeny. J Zoolog Syst Evol Res 39:177–207

Carbone G, Mangialardi L, Persson BNJ (2004) Adhesion between a thin elastic plate and a hard randomly rough substrate. Phys Rev B 70:125407

Dai Z, Gorb SN, Schwarz U (2002) Roughness-dependent friction force of the tarsal claw system in the beetle *Pachnoda marginata* (Coleoptera, Scarabaeidae). J Exp Biol 205:2479–2488

Enders S, Barbakadse N, Gorb SN, Arzt E (2004) Exploring biological surfaces by nanoindentation. J Mater Res 19:880–887
Federle W, Endlein T (2004) Locomotion and adhesion: dynamic control of adhesive surface contact in ants. Arthropod Struct Dev 33:67–75
Frantsevich L, Gorb S (2004) Structure and mechanics of the tarsal chain in the hornet, *Vespa crabro* (Hymenoptera: Vespidae): implications on the attachment mechanism. Arthropod Struct Dev 33:77–89
Gorb SN (1999) Serial elastic elements in the damselfly wing: mobile vein joints contain resilin. Naturwissenschaften 86:552–555
Gorb SN (2001) Attachment devices of insect cuticle. Kluwer Academic Publishers, Dordrecht, The Netherlands
Gorb SN, Beutel RG (2001) Evolution of locomotory attachment pads of hexapods. Naturwissenschaften 88:530–534
Gorb SN, Jiao Y, Scherge M (2000) Ultrastructural and mechanical properties of attachment pads in *Tettigonia viridissima* (Orthoptera, Tettigoniidae). J Comp Physiol A 186:821–831
Gorb S, Gorb E, Kastner V (2001) Scale effects on the attachment pads and friction forces in syrphid flies (Diptera, Syrphidae). J Exp Biol 204:1421–1431
Hepburn HR (1985) Structure of the integument. In: Kerkut GA, Gilbert LI (eds) Comprehensive insect physiology, biochemistry, and pharmacology. Pergamon, Oxford, pp. 1–58
Hertz H (1881) Über den Kontakt elastischer Körper. J Reine Angew Math 92:156–171
Jiao Y, Gorb SN, Scherge M (2000) Adhesion measured on the attachment pads of *Tettigonia viridissima* (Orthoptera, Insecta). J Exp Biol 203:1887–1895
Johnson KL (1985) Contact mechanics. Cambridge University Press, Cambridge
Johnson KL, Kendall K, Roberts AD (1971) Surface energy and the contact of elastic solids. Proc R Soc Lond A 324:301–313
Kohane M, Daugela A, Kutomi H, Charlson L, Wyrobek A, Wyrobek J (2003) Nanoscale in vivo evaluation of the stiffness of *Drosophila melanogaster* integument during development. J Biomed Mater Res A 66:633–642
Neville AC (1975) Biology of arthropod cuticle. Springer, Berlin Heidelberg New York
Niederegger S, Gorb SN, Jiao Y (2002) Contact behaviour of tenant setae in attachment pads of the blowfly *Calliphora vicina* (Diptera, Calliphoridae). J Comp Physiol A 187:961–970
Persson BNJ (2003) On the mechanism of adhesion in biological systems. J Chem Phys 118:7614–7621
Persson BNJ, Gorb S (2003) The effect of surface roughness on the adhesion of elastic plates with application to biological systems. J Chem Phys 119:11437–11444
Scherge M, Gorb SN (2000) Biological micro- and nanotribology. Nature's solutions. Springer, Berlin Heidelberg New York
Slifer EH (1950) Vulnerable areas on the surface of the tarsus and pretarsus of the grasshopper (Acrididae, Orthoptera) with special reference to the arolium. Ann Entomol Soc Am 43:173–188
Spurr AR (1969) A low-viscosity epoxy resin embedding medium for electron microscopy. J Ultrastruct Res 26:31–43
Vincent JFV, Wegst UGK (2004) Design and mechanical properties of insect cuticle. Arthropod Struct Dev 33:189–199
Vötsch W, Nicholson G, Müller R, Stierhof Y-D, Gorb S, Schwarz U (2002) Chemical composition of the attachment pad secretion of the locust *Locusta migratoria*. Insect Biochem Mol Biol 32:1605–1613

Arthropod-Plant Interactions
DOI 10.1007/s11829-008-9049-0

ORIGINAL PAPER

Attachment force of the beetle *Cryptolaemus montrouzieri* (Coleoptera, Coccinellidae) on leaflet surfaces of mutants of the pea *Pisum sativum* (Fabaceae) with regular and reduced wax coverage

Elena Gorb · Dagmar Voigt · Sanford D. Eigenbrode · Stanislav Gorb

Received: 1 April 2008 / Accepted: 22 October 2008

Abstract This study represents an investigation of surface-related plant–insect interactions. Surface micro-morphology of leaflets in pea (*Pisum sativum*) with wild-type crystalline surface waxes (waxy) and with reduced crystalline surface waxes (glossy) caused by a mutation (*wel*) were studied using various microscopy techniques. The free surface energy of these plant surfaces was estimated using contact angles of droplets of three different liquids. The morphological study of the attachment system in the ladybird beetle *Cryptolaemus montrouzieri* was combined with measurements of attachment (traction) forces, generated by beetles on these plant substrates. Differences were found in wax crystal shape, dimensions, and density between the adaxial and abaxial surfaces of waxy and glossy plants. The crystalline wax was not completely eliminated in the glossy plant: it was only slightly reduced on the adaxial side and underwent greater changes on the

Handling editor: Lars Chittka.

E. Gorb (✉) · D. Voigt
Evolutionary Biomaterials Group, Department of Thin Films and Biological Systems, Max Planck Institute for Metals Research, Heisenbergstr. 3, 70569 Stuttgart, Germany
e-mail: o.gorb@mf.mpg.de

D. Voigt
e-mail: voigt@mf.mpg.de

S. D. Eigenbrode
Department of Plant, Soil and Entomological Sciences, College of Agricultural and Life Sciences, University of Idaho, Moscow, ID 83844-2339, USA
e-mail: sanforde@uidaho.edu

S. Gorb
Department of Zoology: Functional Morphology and Biomechanics, Zoological Institute, University of Kiel, Am Botanischen Garten 1-9, 24098 Kiel, Germany
e-mail: s.gorb@mf.mpg.de; sgorb@zoologie.uni-kiel.de

abaxial side. The free surface energy for both surfaces of both pea types was rather low with strongly predominating dispersion component. Insects generated low traction forces on all intact plant surfaces studied, except the abaxial surface of the glossy plant, on which the force was greater. After being treated with chloroform, all the surfaces allowed much higher traction forces. It is demonstrated that the difference in the crystal length and density of the epicuticular wax coverage within the observed range did not influence wettability of surfaces, but affected insect attachment. The reduction in insect attachment force on plant surfaces, covered with the crystalline wax, is explained by the decrease of the real contact area between setal tips of beetles and the substrate.

Keywords Adhesion · Coccinellidae · Coleoptera · Contact angle · Epicuticular crystalline waxes · Fabaceae · Free surface energy · Mutant · Traction force

Introduction

Epicuticular waxes, covering primary aerial surfaces in leaves, flowers, fruits and non-woody stems in terrestrial plants, appear predominantly as a thin smooth amorphous film or layer. Additionally, microscopically small crystals, protruding from the film, cause the pruinosity or powdery appearance of the surface in many plant species (Barthlott et al. 1998). Such pruinose surfaces contribute prominently to the mechanical strength of the tissue, to the protection against harmful radiation, to the limitation of uncontrolled water loss or leaching from interior and foliar uptake, to the control of water status, to the cueing of host-pathogens/insects recognition and epidermal cell development, to the reduction of contamination and pathogen attack, and to the promoting or preventing insect attachment and locomotion (see reviews by Barthlott 1990 and Bargel et al. 2006).

Regarding the latter function, it has been previously repeatedly reported that plant surfaces, bearing crystalline wax coverage, reduce the attachment of insects (Stork 1980, 1986; Edwards 1982; Bodnaryk 1992; Eigenbrode 1996; Eigenbrode and Kabalo 1999; Eigenbrode et al. 2000; Federle et al. 2000; Eigenbrode and Jetter 2002; Gaume et al. 2002, 2004; Gorb and Gorb 2002; Gorb et al. 2005; Voigt et al. 2007). In some of these studies, conducted with different insect and plant species, it has been experimentally shown that insects were able to attach successfully to smooth substrates, without wax or with removed wax, but failed on pruinose surfaces.

Recent studies, carried out with mutants of *Pisum sativum* L. (Fabaceae), exhibiting essential differences in the wax coverage, confirmed anti-adhesive properties of the wax bloom (Eigenbrode et al. 1998; 2008; Rutledge and Eigenbrode 2003). Mutants with a highly reduced wax coverage demonstrated considerable enhancement of the insect attachment force: a multi-fold (from almost 2 to >30) force increase on glossy plants compared to waxy ones was measured, depending on the insect species and method of force measurement. Force values obtained in different studies with the same *P. sativum* mutants differed considerably and only the adaxial leaflet surfaces were examined.

The aim of this study was to compare attachment ability of beetles *Cryptolaemus montrouzieri* Mulsant to different leaflet surfaces in the *P. sativum* with well developed or reduced crystalline wax in order to understand these surface-related plant–insect interactions in further detail. A comprehensive investigation was undertaken to examine the leaflet surfaces, including surface micro-morphology and physicochemical properties, and the morphology of the beetle attachment system, and to use this information to interpret differences in attachment (traction) forces generated by the insects on four plant substrates.

Materials and methods

Plant and insect species

The field or garden pea *P. sativum* is an annual herb with pinnately compound leaves, which in leafy types consist of two stipules at the base of the leaf, one or more pairs of opposite, broad and ovate petiolulate leaflets, several pairs of tendrils and a single or compound terminal tendril (Aichele and Schwegler 2000). The plant originated from the Mediterranean area and Middle East (wild genotypes *P. humile* J.I. and *P. elatius* L., Waines 1975) and is cultivated worldwide mostly as a human and animal food, and as a source of pharmacologicals (Kunkel 1984; Cooper and Johnson 1998; McGuffin et al. 2000).

Two sister lines of *P. sativum* differing in expression of the wax mutation *wel* (Marx 1969) were developed from accession PI W6-15368 (Marx 406). They differ in the epicuticular wax coverage in both (1) amount of wax material and (2) wax micromorphology and composition, as described by Eigenbrode et al. (1997). The line 406 N (later called waxy), which is homozygous for the wild-type allele *Wel*, has a well-developed wax bloom on all photosynthetically active surfaces. The line 406G (later called glossy) is homozygous for *wel* and shows substantial wax reduction. Both lines are also homozygous for the recessive mutation *tl* (acacia leaf), which turns all tendrils into leaflets and because of this provides larger surface area for foraging insects compared to standard peas, where some leaflets appear as tendrils (Eigenbrode et al. 2008).

Seeds, treated with the commercial rate of Captan 400 fungicide (Gustafson Ltd., Plano, USA), were sown in small, 9 cm in diameter, plastic pots (3 plants per pot) in a commercial potting soil ASB-Greenworld Premium, NPK: 14-16-18 (ASB-Gruenland Helmut Aurenz GmbH, Lugwigsburg, Germany). Plants were grown in a room greenhouse at 20–25°C, 40–45% humidity, a 12L:12D photoregime and bottom-watered daily. Plants having from 3 to 4 nodes were used in the study.

The Australian ladybird beetle *C. montrouzieri* was chosen as a model insect species for traction experiments because of its commercial availability and appropriate body size (4–5 mm in length). It originated from Australia and after being imported first to North America, is widely used for biological control against mealy bugs throughout the world (Booth and Pope 1986). Adult beetles were obtained from a commercial supplier (Katz Biotech AG, Welzheim, Germany) and kept in a ventilated cage at 22–24°C, 40–60% humidity. Insects were fed with a solution of honey in tap water.

Structural studies

The micromorphology of the leaflet surfaces in *P. sativum* was studied using scanning electron microscopy (SEM). Leaflets, cut off of the proximal part of the compound leaf using a razor blade, were air dried, glued with polyvinyl alcohol Tissue-Teck® O.C.T.™ Compound (Sakura Finetek Europe B.V., Zoeterwoude, The Netherlands) to holders, sputter-coated with gold-palladium (3–6 nm) and examined in a SEM Hitachi S-4800 (Hitachi High-Technologies Corporation, Tokyo, Japan) at 3 kV accelerating voltage. Samples were observed either intact or after treatment with chloroform for 20 s at 60°C. Type of wax projections was identified according to the generally accepted classification of plant epicuticular waxes (Barthlott et al. 1998). Wax crystal dimensions and density were quantified from digital images using SigmaScan Pro 5 software (SPSS Inc., Chicago, USA).

To study attachment devices in *C. montrouzieri*, beetles were air-dried, mounted dorsally on holders, sputter-coated with gold-palladium (3–6 nm) and examined in the SEM at 3 kV.

Contact angle measurements and estimation of the free surface energy

The free surface energy and its dispersion and polar components were estimated by applying a high-speed optical contact angle measuring device OCAH 200 (DataPhysics Instruments GmbH, Filderstadt, Germany; for detailed device description see Maier 2002). To calculate the free surface energy, contact angles of a series of well-characterised liquids (water, density = 1.000 kg/m^3; diiodomethane, dens. = 3.325 kg/m^3; ethylene glycol, dens. = 1.113 kg/m^3) were used. Leaflets were cut off of the proximal part of the compound leaf using a razor blade and then attached with a double-sided tape tesa®fotostrip (Tesa AG, Hamburg, Germany) to a glass slide. Static contact angles of liquids on surfaces were evaluated using the sessile drop method (drop volume 2 μl) with circle- or ellipse-fitting. The free surface energy and its components were calculated according to the universal Owens–Wendt–Kaelble method (Owens and Wendt 1969). For both, the waxy and glossy mutants, the free surface energy of the adaxial and abaxial surfaces of 3 leaflets belonging to 3 plants was calculated. For each plant, 4–5 measurements of the contact angle of each liquid on each surface were conducted. Clean flat glass plate was used as a reference surface (1 sample, 5 measurements per liquid). In total, 193 contact angle measurements were performed.

Traction experiments with insects

Traction experiments with tethered walking beetles *C. montrouzieri* were carried out to measure insect attachment forces on different surfaces. Force tests were performed with a load cell force transducer (10 g capacity, Biopac Systems Ltd., Santa Barbara, CA, USA; Gorb and Popov 2002). Both males and females were used in experiments. To make experimental insects incapable of flying, elytra were glued together with a small droplet of molten wax prior to experiments. The beetle was attached to the force sensor by means of a hair (10–15 cm long), glued to the dorsal surface of the insect thorax with a droplet of molten wax. Leaflets (1.5–2.0 cm in diameter) were cut off of the proximal part of the leaf using a razor blade and then, intact or treated, attached with double-sided adhesive tape to a horizontal glass plate. Five types of substrates were tested: (1) intact adaxial and abaxial leaflet surfaces of the waxy plant, (2) intact adaxial and abaxial leaflet surfaces of the glossy plant, (3) adaxial and abaxial leaflet surfaces of waxy plants, treated with hot (60°C) chloroform for 20 s, (4) adaxial and abaxial leaflet surfaces of glossy plants, treated with hot chloroform, and (5) a glass plate as a control. With each individual beetle, two force tests were carried out: first on the glass substrate, then on one of the plant surfaces. The force, generated by the insect walking horizontally on test substrates, was measured. Force–time curves, where the beetle stretched the hair for ca. 1 min, were used to estimate the maximal traction force (Gorb et al. 2005). Tests were carried out at the room temperature of 23–25°C and 40–45% of relative humidity. For each plant surface type, experiments with 9 individual insects were conducted. Three leaflets from three plants per surface were used. In all, 72 insects were tested and 144 force measurements were performed.

Results

Leaflet surface micromorphology

There is a difference in the crystalline epicuticular wax coverage between the adaxial and abaxial leaflet surfaces in both waxy and glossy plants, and between two mutants (Table 1).

The adaxial leaflet surface in waxy plants bears a network of interconnected numerous membranous platelets, occurring in high density and covering the surface uniformly and completely (Fig. 1a, b). Although crystals vary greatly in size and shape, they are rather small and have predominantly irregular shapes with indented, often fringed outside edges (Table 1). The platelets are connected with the underlying surface through edges. Crystal planes are oriented randomly according to their longitudinal axes, but perpendicular or slightly sloped relatively to the plane of the leaflet.

In glossy plants, the wax coverage present on the adaxial side is similar to the waxy pea. However, wax crystal density on the adaxial surface of the glossy mutant is reduced by a factor of three compared to the waxy plants, and the surface is not completely covered with the crystalline wax (Table 1; Fig. 1c). The membranous platelets have similar shapes as in waxy peas, but are longer and narrower (Fig. 1d). They emerge often as aggregated crystals (rosettes) and are connected with the underlying surface through a larger edge. The crystals show no specific orientation, and their planes are either perpendicular to or more often have acute angles with the leaflet surface.

The abaxial surface of waxy plants is completely covered with the dense crystalline wax coverage, composed of crystals of three types: large elongated filaments, ribbons, and transversely ridged rodlets (Table 1; Fig. 1e). Crystals vary essentially in size and shape, emerge solitarily, are connected with the underlying surface through a small edge, show no preferred orientation and are somewhat crowded to the surface (Fig. 1f). Ribbons have slightly irregular edges and possess widened, sometimes spatulae-like tips with rather corrugated surface. Rodlets are usually bar- or ribbon-shaped, with very uneven sides and edges.

Although the density of wax crystals on the abaxial side of the glossy mutant is much higher than on all other surfaces studied here, this surface is not completely covered with the crystalline wax (Table 1; Fig. 1g, h). Crystals of two types, short ribbons and scales,

Table 1 Wax crystal type, dimensions and density on the adaxial (*ad*) and abaxial (*ab*) leaflet surfaces in waxy and glossy plants of *Pisum sativum*

Surface	Crystal type	Length (μm)			Width (μm)			Density (μm^{-2})		
		ave	*SD*	*n*	*ave*	*SD*	*n*	*ave*	*SD*	*n*
Waxy ad	Platelets	1.54	0.31	100	0.80	0.27	100	4.38	0.92	5
Waxy ab	Ribbons	2.68	0.67	50	0.49	0.16	50	1.79	0.31	5
	Filaments	2.19	0.65	50	0.14	0.04	50			
	Rodlets	2.60	1.04	40	0.61	0.22	40			
Glossy ad	Platelets	1.20	0.44	100	0.57	0.23	100	1.84	0.74	5
Glossy ab	Ribbons	1.14	0.36	50	0.27	0.13	50	5.97	0.90	5
	Scales	0.41	0.19	50	0.28	0.14	50			

ave mean value, *n* sample size, *SD* standard deviation

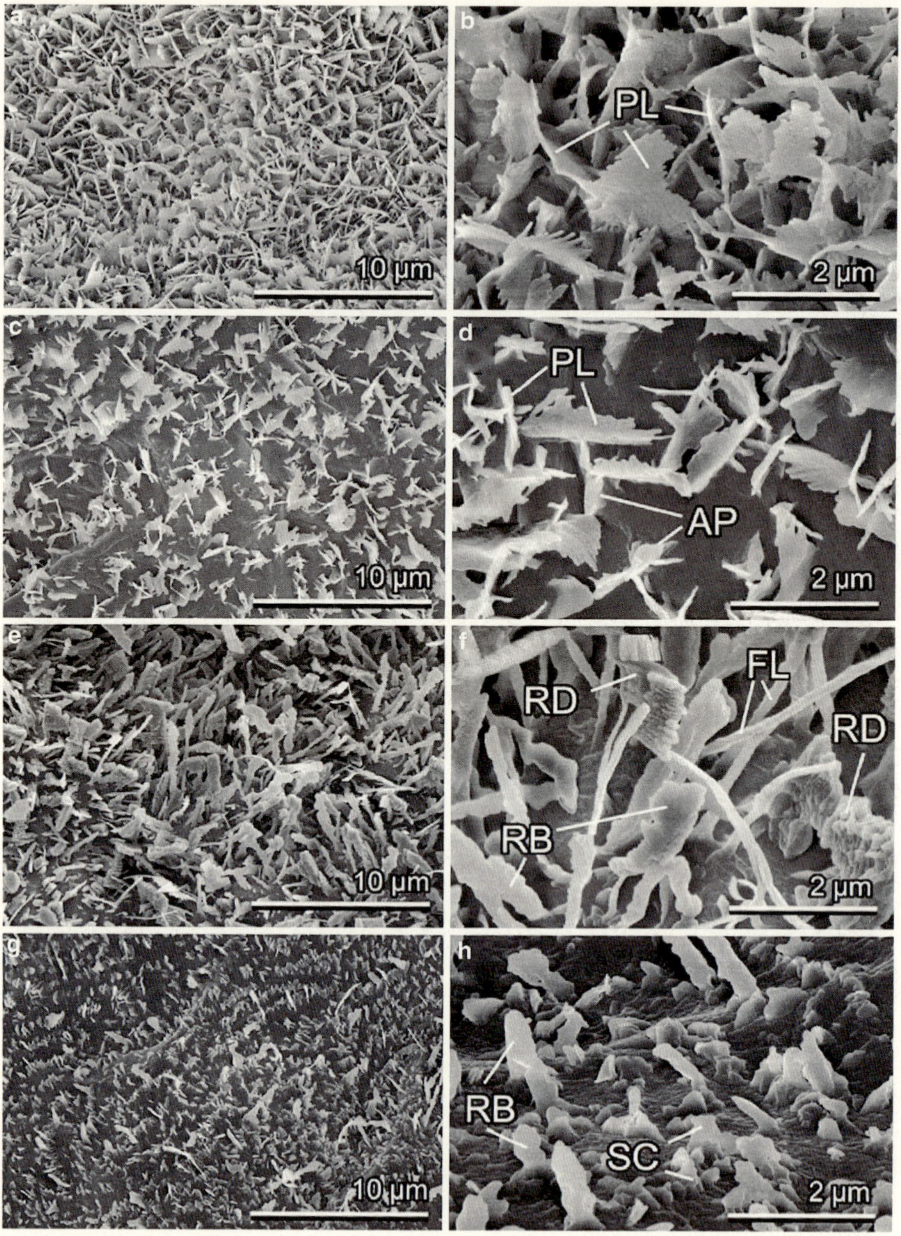

Fig. 1 SEM micrographs of the crystalline epicuticular wax coverage on the adaxial (**a**–**d**) and abaxial (**e**–**h**) leaflet surfaces in waxy (**a**, **b**, **e**, **f**) and glossy plants (**c**, **d**, **g**, **h**) of *Pisum sativum*. AP, aggregated platelets; FL, filaments; PL, platelets; RB, ribbons; RD, rodlets; SC, scales

are very variable in both size and shape being essentially smaller and having more regular edges than crystals on the abaxial surface of the glossy mutant. As well as latter, these crystals originate solitarily and have random orientation relatively to their longitudinal axes, but are connected with the surface through small (ribbons) or large (scales) sides and may be either erect or crowded to the substrate.

After treatment with warm chloroform, both leaflet sides of both mutants were completely free of wax crystals (Fig. 2a, c, e, g) and therefore, the microroughness was

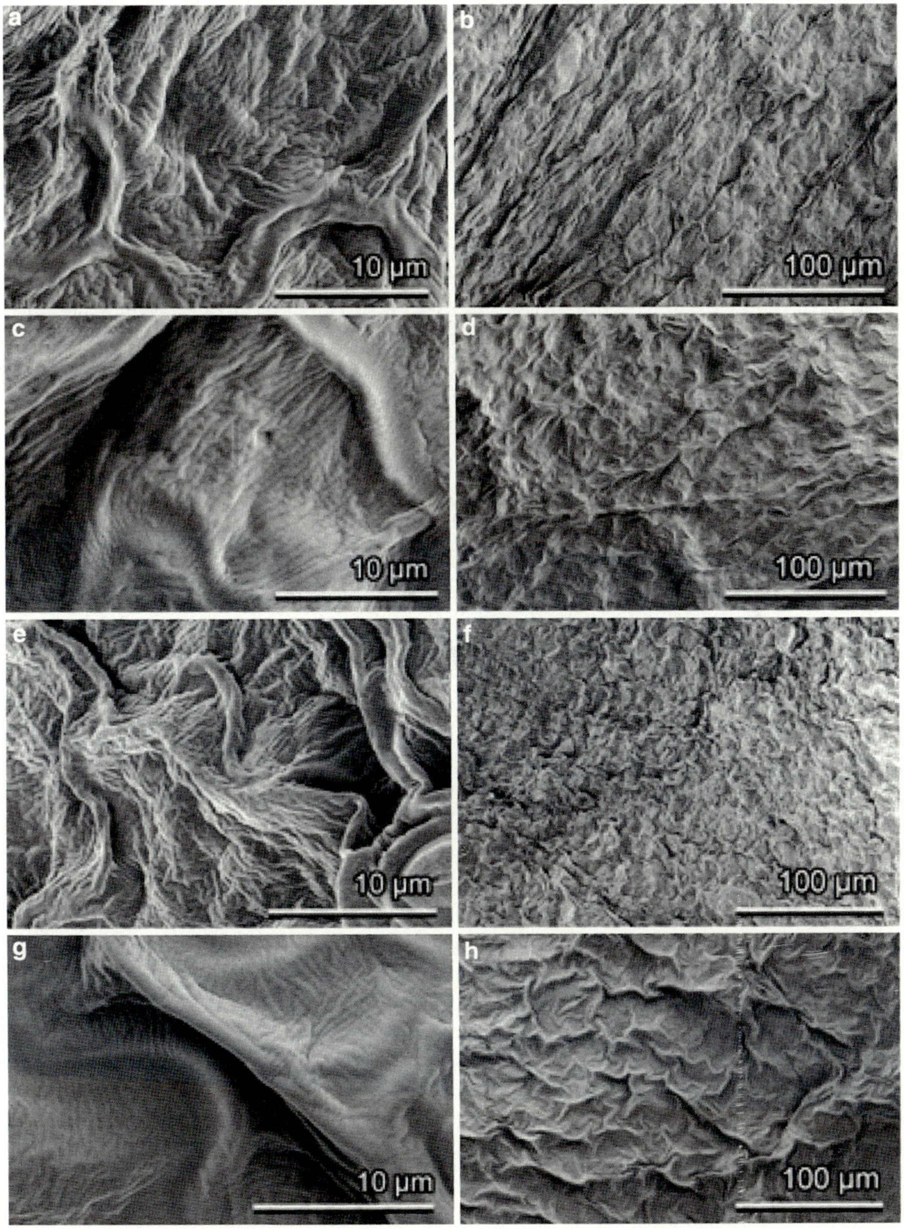

Fig. 2 SEM micrographs of the adaxial (**a–d**) and abaxial (**e–h**) leaflet surfaces in waxy (**a**, **b**, **e**, **f**) and glossy plants (**c**, **d**, **g**, **h**) of *Pisum sativum* after treatment with warm chloroform

removed. However, treated surfaces became corrugated, with a noticeably increased large-scale roughness (Fig. 2b, d, f, h).

Physicochemical properties of leaflet surfaces

In all plants studied, both adaxial and abaxial leaflet surfaces were unwettable with both polar liquids, such as water (surface tension = 72.1 mN/m, dispersion component = 19.9 mN/m, polar component = 52.2 mN/m, Busscher et al. 1984) and ethylene glycol (surf. tension = 48.0 mN/m, disp. comp. = 29.0 mN/m, pol. comp. = 19.0 mN/m, Erbil 1997) (Fig. 3a). Contact angles of water were the highest compared to other liquids and seemed to be rather similar on all surfaces. However, the statistic difference was found in contact angles of water between plant surfaces studied (Kruskal–Wallis one way ANOVA on ranks, $H_{3,58} = 9.728$, $P = 0.021$), where the adaxial side of the waxy mutant and the abaxial side of the glossy mutant was the only surface pair with significantly different angles (Dunn's method carried out after Kruskal–Wallis one way ANOVA on ranks, d.f. = 28, diff. of ranks = 17.367, $Q = 2.769$, $P < 0.05$). Contact angles of ethylene glycol were significantly lower in glossy plants compared to waxy ones (one way ANOVA, $F_{3,59} = 24.100$, $P < 0.001$). In both waxy and glossy mutants, no significant difference in contact angles of ethylene glycol was detected between the abaxial and adaxial leaflet sides (Table 2). Contact angles of non-polar diiodomethane (surf. tension = 50.0 mN/m, disp. comp. = 47.4 mN/m, pol. comp. = 2.6 mN/m, Busscher et al. 1984) were lower than those of both water and ethylene glycol. The abaxial surface of glossy plants was wetted with diiodomethane (contact angle = 75.3 ± 6.4°, mean ± SD, $n = 15$ drops, $N = 3$ plants). The leaflet surfaces studied showed highly significant difference in contact angles of latter liquid (one way ANOVA, $F_{3,58} = 111.810$, $P < 0.001$). The abaxial side of the waxy mutant and the adaxial side of the glossy mutant was the only surface pair, where contact angles of diiodomethane did not differ significantly (Table 3).

Considering morphometrical parameters of the waxy coverage of the leaflet surfaces (average size and density of wax crystals), there was no significant correlation between contact angles and any morphometrical parameters tested (cont. angle of water versus cryst. length: $R^2 = 0.118$, $P = 0.144$; cont. angle of water versus cryst. dens.: $R^2 = 0.526$, $P = 0.144$; cont. angle of ethylene glycol versus cryst. length: $R^2 = 0.488$, $P < 0.001$; cont. angle of ethylene glycol versus cryst. dens.: $R^2 = 0.082$, $P = 0.186$; cont. angle of diiodomethane versus cryst. length: $R^2 = 0.121$, $P = 0.011$; cont. angle of diiodomethane versus cryst. dens.: $R^2 = 0.023$, $P = 0.159$).

Leaflet surfaces in both plant mutants showed rather low values of the free surface energy with the prevailing dispersion component (Fig. 3b). The lowest surface energy with the highest dispersion component was observed for the adaxial surface of waxy plants, whereas in the abaxial leaflet side of the glossy mutant, the surface energy was several times higher with markedly higher polar component. As for the abaxial surface of waxy plants and the adaxial side in glossy plants, there was no considerable difference besides slightly higher values of the total surface energy in latter surfaces.

Morphology of the beetle attachment system

The tarsus of *C. montrouzieri* consists of three segments (Fig. 4a, e). The distal tarsomere (T3) bears paired claws, curved ventrally, each with a tooth. Attachment pads are of the hairy type (Beutel and Gorb 2001; Gorb 2001). The ventral side of the two first proximal tarsomeres (T1 and T2) is densely covered by tiny tenent setae (Fig. 4b, f). The setae of the

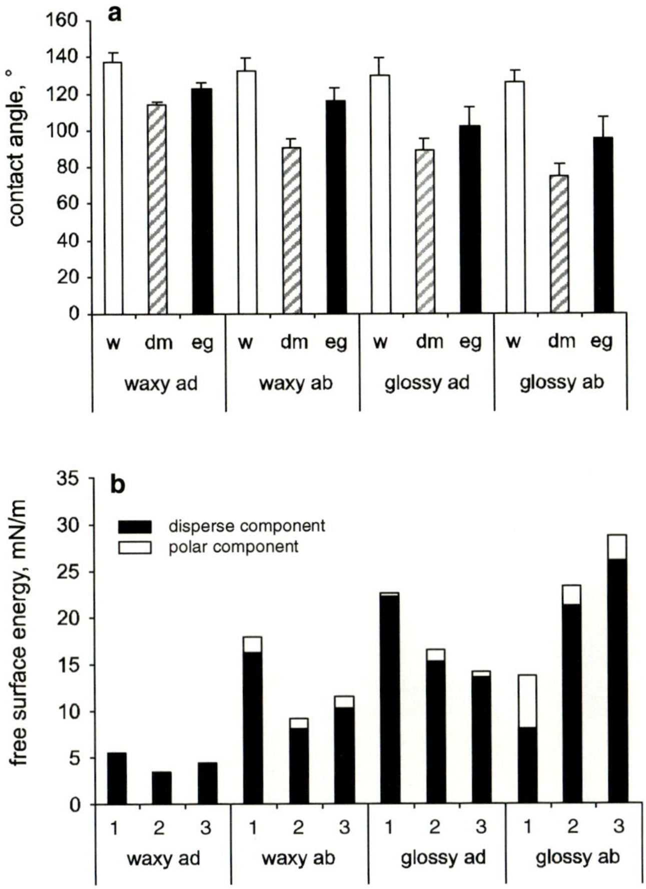

Fig. 3 Contact angles of water, diiodomethane, and ethylene glycol (**a**) and values of the free surface energy (**b**) in different leaflet surfaces of *Pisum sativum*. dm, diiodomethane; eg, ethylene glycol; glossy ab, abaxial surface of the glossy mutant; glossy ad, adaxial surface of the glossy mutant; w, water; waxy ab, abaxial surface of the waxy plant; waxy ad, adaxial surface of the waxy plant; 1, 2, 3, three different plants. For statistics see Tables 2 and 3

Table 2 Results of pairwise multiple comparisons of means (Holm-Sidak method) carried out after one way ANOVA for the contact angles of ethylene glycol on adaxial (*ad*) and abaxial (*ab*) leaflet surfaces in waxy and glossy plants of *Pisum sativum*

Comparison	DM	p	t	$P < 0.05$
Waxy ad versus glossy ab	26.680	15	7.576	Yes
Waxy ad versus glossy ad	20.347	15	5.778	Yes
Waxy ab versus glossy ab	19.927	15	5.658	Yes
Waxy ab versus glossy ad	13.593	15	3.860	Yes
Waxy ad versus waxy ab	6.753	15	1.918	No
Glossy ad versus glossy ab	6.333	15	1.798	No

DM difference of means, *no* no significant difference, *p* number of means spanned in the comparison, *P* probability value, *t* test statistics, *yes* significant difference

Table 3 Results of pairwise multiple comparisons of means (Holm-Sidak method) carried out after one way ANOVA for the contact angles of diiodomethane on adaxial (*ad*) and abaxial (*ab*) leaflet surfaces in waxy and glossy plants of *Pisum sativum*

Comparison	DM	*p*	*t*	$P < 0.05$
Waxy ad versus glossy ab	38.393	15	18.069	Yes
Waxy ad versus glossy ad	24.220	15	11.398	Yes
Waxy ad versus waxy ab	23.051	15	10.660	Yes
Waxy ab versus glossy ab	15.342	15	7.095	Yes
Glossy ad versus glossy ab	14.173	15	6.670	Yes
Waxy ab versus glossy ad	1.169	15	0.540	No

DM difference of means, *no* no significant difference, *p* number of means spanned in the comparison, *P* probability value, *t* test statistics, *yes* significant difference

first tarsomere T1 are more or less uniform, with straight or slightly curved pointed tips (Fig. 4c, d). In the second tarsomere T2, setae of the two types were found: (1) with pointed tips and (2) with flattened and widened tips called spatulae (Fig. 4g, h). No sexual dimorphism in the morphology of the attachment system in *C. montrouzieri* was detected.

Traction forces of beetles on pea leaflet surfaces

Although the maximal traction force generated on glass by different insect individuals (0.5–11.8 mN) did not vary significantly in different experiments (Kruskal–Wallis one way ANOVA on ranks, $H_{7,71} = 9.597$, $P = 0.213$), we used data, normalised to values obtained on glass, for comparison of different surfaces. For each individual, the force, obtained on a test plant surface, was compared to that on glass (considered as 100%).

For all intact plant surfaces tested, with the only exception of the abaxial side in the glossy mutant, the values of the maximal traction force of beetles differed significantly from those, produced on the glass plate (Kruskal–Wallis one way ANOVA on ranks, $H_{4,44} = 27.954$, $P < 0.001$; Table 4; Fig. 5a). The comparison of different intact plant surfaces (one way ANOVA, $F_{3,35} = 4.681$, $P = 0.008$) showed that on waxy plants, insects produced similar forces regardless of a leaflet surface, while higher force was measured on the abaxial side in the glossy pea (Table 5). There was no significant difference between adaxial surfaces in waxy and glossy plants, whereas on the abaxial side, the force was stronger in the glossy mutant.

Considering morphometrical parameters of the waxy coverage of the leaflet surfaces (average size and density of wax crystals) and surface wettability (contact angles of water, ethylene glycol and diiodomethane), a significant correlation was found for the traction force with both the wax crystal length and density (Fig. 6). There was no significant correlation between the force and contact angles of any liquid (force versus cont. angle of water: $R^2 = 0.202$, $P = 0.145$; force versus cont. angle of ethylene glycol: $R^2 = 0.544$, $P = 0.014$; force versus cont. angle of diiodomethane: $R^2 = 0.376$, $P = 0.043$).

Treatment of plant surfaces with chloroform resulted in a substantial increase of traction forces on both leaflet sides in both waxy and glossy plants, compared to those on intact surfaces (Table 6; Fig. 5). Compared with glass (Kruskal–Wallis one way ANOVA on ranks, $H_{4,44} = 12.283$, $P = 0.015$; Fig. 5b), the abaxial leaflet side of the waxy pea was the only substrate, where the force was significantly higher (Tukey test carried out after Kruskal–Wallis one way ANOVA on ranks, d.f. = 17, diff. on ranks = 168.000,

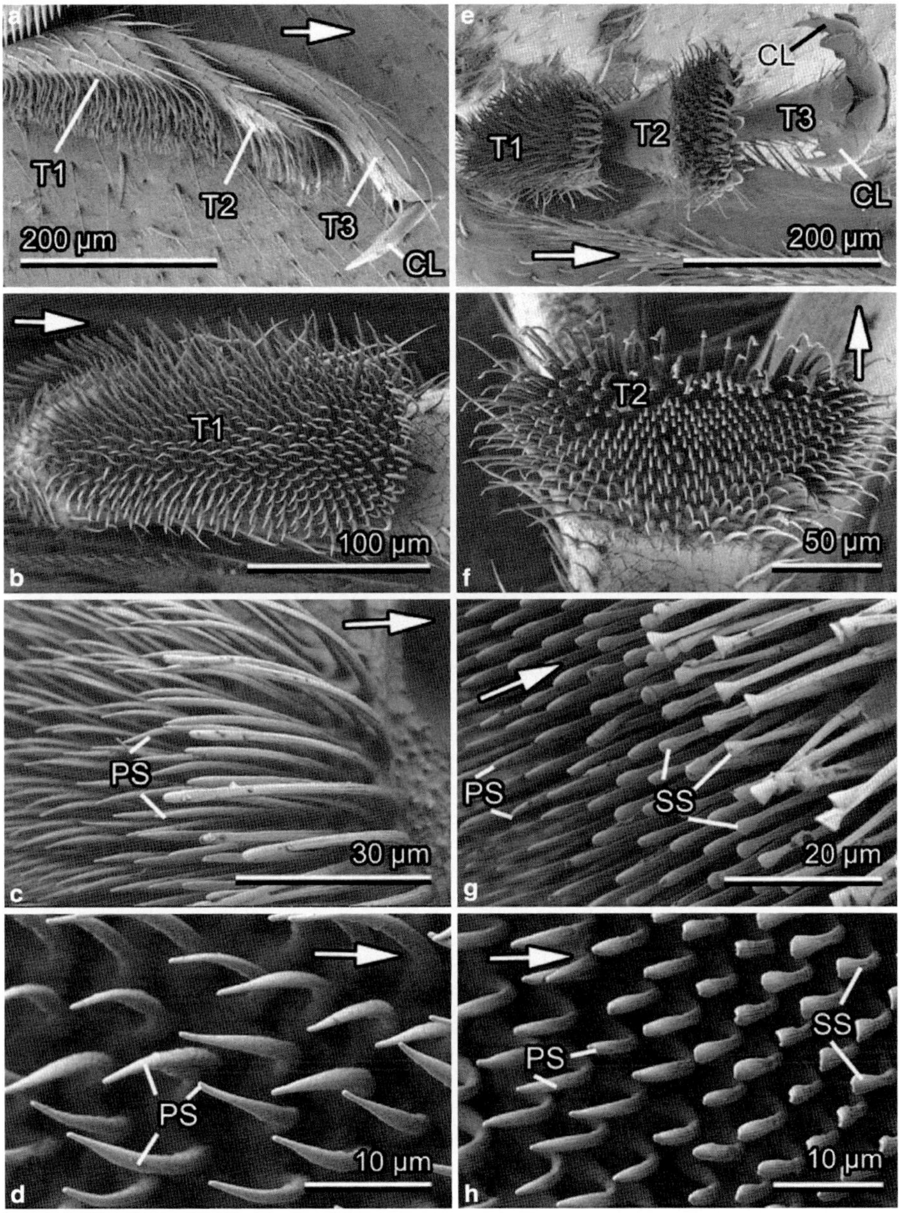

Fig. 4 Attachment system of the beetle *Cryptolaemus montrouzieri* (female, SEM). (**a**) Lateral and (**e**) ventral view of the tarsus. (**b–d**) The first proximal tarsomere. (**f–h**) The second proximal tarsomere. CL, claws; PS, setae with pointed tips; SS, setae with flattened and widened tips, spatulae; T1–T3, tarsomeres 1–3. Arrows show distal direction

$q = 4.264$, $P < 0.05$), and no difference was found between other treated plant surfaces and glass. The comparison of treated leaflet surfaces showed no statistically significant difference in traction forces of insects between plant substrates tested (one way ANOVA, $F_{3,35} = 1.675$, $P = 0.192$).

Table 4 Results of pairwise multiple comparisons of means (Tukey test) carried out after Kruskal–Wallis one way ANOVA on ranks for the maximal traction forces, produced by beetles *Cryptolaemus montrouzieri* on glass and intact adaxial (*ad*) and abaxial (*ab*) leaflet surfaces in waxy and glossy plants of *Pisum sativum*

Comparison	DR	*p*	*q*	$P < 0.05$
Glass versus waxy ab	276.000	9	5.898	Yes
Glass versus waxy ad	234.000	9	4.958	Yes
Glass versus glossy ad	233.000	9	4.937	Yes
Glass versus glossy ab	134.000	9	2.839	No

DR difference of ranks, *no* no significant difference, *p* number of means spanned in the comparison, *P* probability value, *q* test statistics, *yes* significant difference

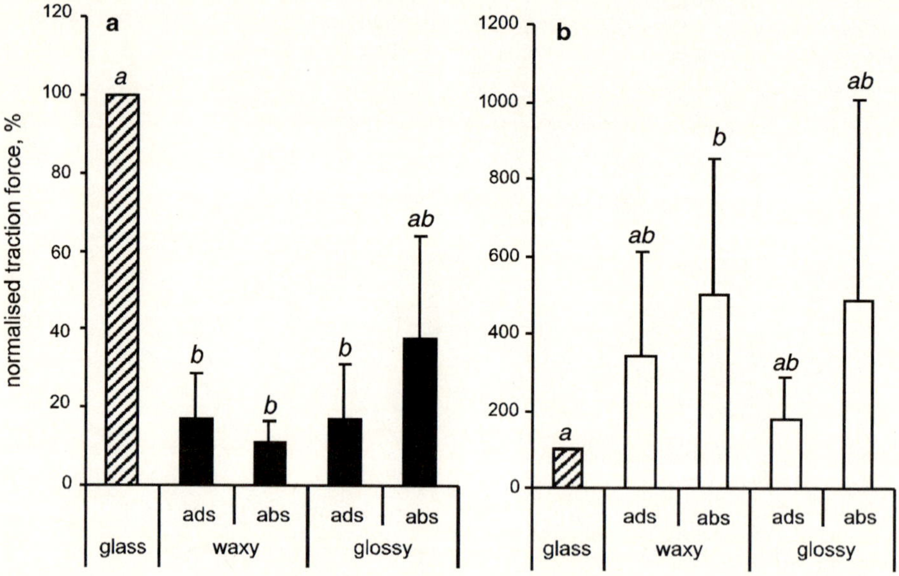

Fig. 5 Maximal traction force, generated by beetles *Cryptolaemus montrouzieri* (males and females pooled together) on glass compared to intact (**a**) and treated (**b**) leaflet surfaces of *Pisum sativum* (normalised data, i.e. in comparison to measurements on glass). abs, abaxial surface; ads, adaxial surface. Different letters show significant differences between traction force values on different substrates (for statistics see text and Tables 4 and 5)

Table 5 Results of pairwise multiple comparisons of means (Holm-Sidak method) carried out after one way ANOVA for the maximal traction forces, produced by beetles *Cryptolaemus montrouzieri* on different intact surfaces of *Pisum sativum*: the adaxial (*ad*) and abaxial (*ab*) leaflet surfaces in waxy and glossy plants

Comparison	DM	*p*	*t*	$P < 0.05$
Glossy ab versus waxy ab	27.001	9	3.504	Yes
Glossy ab versus glossy ad	20.593	9	2.672	Yes
Waxy ad versus waxy ab	6.109	9	0.793	No
Glossy ad versus waxy ad	0.300	9	0.039	No

DM difference of means, *no* no significant difference, *p* number of means spanned in the comparison, *P* probability value, *t* test statistics, *yes* significant difference

Fig. 6 The maximal traction force, generated by beetles *Cryptolaemus montrouzieri* on leaflet surfaces of *Pisum sativum* versus the wax crystal length (**a**) and density (**b**). Lines indicate linear regressions

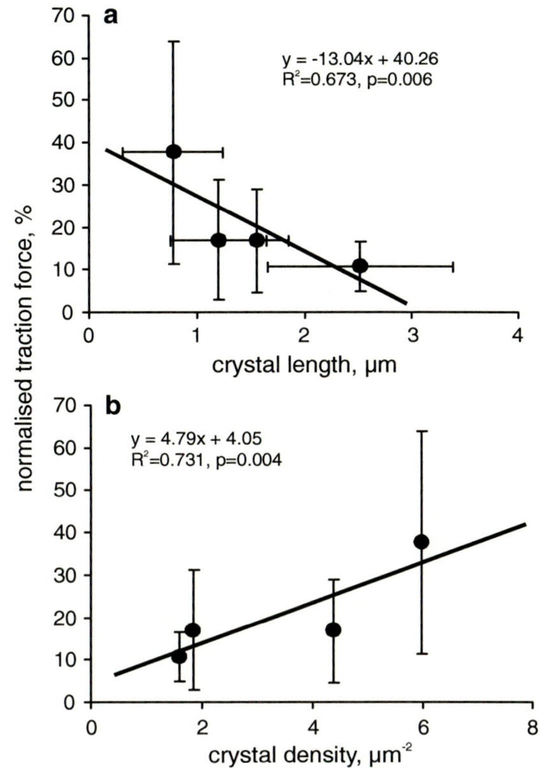

Table 6 Results of pairwise multiple comparisons of means (Tukey test) carried out after Kruskal–Wallis one way ANOVA on ranks for the maximal traction forces, produced by beetles *Cryptolaemus montrouzieri* on intact (*int*) and treated with chloroform (*tr*) leaflet surfaces of *Pisum sativum*: the adaxial (*ad*) and abaxial (*ab*) surfaces in waxy and glossy plants

Comparison	DR	*p*	*q*	$P < 0.05$
Kruskal–Wallis one way ANOVA on ranks $H_{3,35} = 26.560$, $P < 0.001$				
tr Waxy ab versus int waxy ab	179.000	9	5.663	Yes
tr Waxy ad versus int waxy ad	145.000	9	3.244	Yes
Kruskal–Wallis one way ANOVA on ranks $H_{3,35} = 25.675$, $P < 0.001$				
tr Glossy ad versus int glossy ad	167.000	9	5.284	Yes
tr Glossy ab versus int glossy ab	141.000	9	4.461	Yes

DR difference of ranks, *H* test (Kruskal–Wallis one way ANOVA on ranks) statistics, *p* number of means spanned in the comparison, *P* probability value, *q* Tukey test statistics, *yes* significant difference

Discussion

Ultrastructure and properties of the wax coverage

It is established that the wax structure can differ between the leaf surfaces of the same plant individual (Juniper 1959; Juniper and Cox 1973). For *P. sativum*, it has been previously repeatedly reported that both leaflet surfaces are densely covered with crystalline epicuticular waxes having markedly divergent micro-morphology and chemical composition: alcohol-rich plates/platelets on the adaxial surface and alkane-rich ribbons or filaments on the abaxial side (Halloway et al. 1977; Hunt and Baker 1982; Gniwotta et al. 2005; Jeffree 2006). Additionaly, few cylindrical rodlets (cultivars Avanta and Lincoln) and short crystals with different shapes of cross sections and flat upper surface (cultivar Maiperle) have been found on the abaxial surfaces (Gniwotta et al. 2005). Our SEM study of this leaflet surface in the waxy plants reveals the presence of both ribbons and rather large amount of filaments, which were not absolutely upright, as it has been reported previously, but somewhat crowded to the surface. Transversely ridged rodlets were found for the first time for this leaflet side.

In low-wax pea plants, *wel* mutation (wax eliminator) has been previously described to be unique among wax mutations in peas, since it reduced strongly the wax bloom on all aerial plant surfaces (Marx 1969). Glossy plants had shinier, darker green leaves (White and Eigenbrode 2000a), significantly lower (up to approximately 80%) wax load (Eigenbrode et al. 1997, 2008; White and Eigenbrode 2000a, b; Rutledge and Eigenbrode 2003), decreased amount of large crystals (Eigenbrode et al. 1997; 1998; White and Eigenbrode 2000a, 2000b) and higher density of minute crystals on both leaf sides (Eigenbrode et al. 1997, 1998). No ribbons, but only 1 μm needles and granules were found on the abaxial surface (Eigenbrode et al. 1997). Wax in glossy plants was less extractable (White and Eigenbrode 2000a) and showed changed composition with decreased proportions of alcohols and alkanes (Eigenbrode et al. 1997, 1998). This literature contains some contradictions regarding the crystal density and wax micro-structure of the *wel* mutation: reduced density of small crystals has been reported for the abaxial surface (Eigenbrode et al. 1998). Glossy plants (*wel*) have been described as possessing amorphous wax in one study (Rutledge and Eigenbrode 2003). Our study showed the presence of crystalline waxes on both leaflet sides in the glossy mutant. We also found that micro-structure of the wax coverage in the adaxial surface did not differ very much between waxy and glossy plants. Rather, in the glossy mutant, crystals were of the same type (platelets), slightly smaller and much more sparsely distributed compared to those of the waxy pea. In contrast, we observed considerable difference between waxy and glossy mutants in the wax coverage of the abaxial leaflet side. No large crystals, but only short ribbons and rather small scales occurred in a great number on the abaxial surface of glossy plants (see Table 1). The reduction in wax crystals we detected on the glossy mutant line 406G (*wel* homozygote) is not as severe as described previously for this line (Eigenbrode et al. 1997, 1998). The seed stock used for our study was identical with that used in the prior work, and the growing conditions were similar. The slightly younger stage tested here (3–4 node) versus the older plants used in the prior studies, or an undetermined environmental factor, could account for this apparent discrepancy.

Although superhydrophobic properties of plant surfaces covered with the crystalline wax have been described previously (e.g., Holloway 1969; Juniper and Cox 1973), the contact angle of water on pea leaf surface was measured only recently (Gniwotta et al. 2005). It was found that in three cultivars of *P. sativum*, contact angles were similar (144.7° ± 4.9° for the adaxial side and 141.2° ± 4.3° for the abaxial side, mean ± SD), since the epicuticular waxes of these cultivars caused similar roughness and polarity of surfaces. The free surface energy of leaflet surfaces of *P. sativum* was estimated in this study for the first time by the use of three different liquids. Waxy pea plants were found to have slightly higher contact angles of water than glossy plants, but these angles did not significantly differ between the adaxial and abaxial sides. Glossy plants (especially their abaxial side) were wetted somewhat more strongly by all three liquids tested than were waxy plants.

Although the wax crystal size and density in glossy mutants differ from those in waxy plants, there was no correlation between these two morphometrical parameters of the crystalline wax coverage and contact angles of liquids used. This means that within the range found on these plants, differences in the wax coverage morphology did not explain differences in surface wettability.

Wax-mediated beetle attachment

A great number of insects have evolved specialised adhesive devices, smooth flexible pads or hairy structures, in order to attach to smooth substrates (Gorb 2001; Beutel and Gorb 2001). Compliant pad material in the first case and fibrillar surface structures in the second one ensure sufficiently large contact area between pads and substrate and by this, a high attachment force. Excellent attachment ability on various smooth artificial and natural surfaces has been previously reported for many insects (Way and Murdie 1965; Stork 1980; Edwards 1982; Eigenbrode 1996; Eigenbrode et al. 1999; Federle et al. 2000; Gorb 2001; Gorb and Gorb 2002, 2006a). In waxy plant substrates, the deteriorative effect of the crystalline wax coverage on insect attachment has been experimentally shown by many authors (Stork 1980, 1986; Edwards 1982; Bodnaryk 1992; Eigenbrode 1996; Eigenbrode and Kabalo 1999; Eigenbrode et al. 2000; Federle et al. 2000; Eigenbrode and Jetter 2002; Gaume et al. 2002, 2004; Gorb and Gorb 2002; Gorb et al. 2005; Voigt et al. 2007). It has been found for several insect species and a series of plant species that insects can attach well to smooth substrates, without wax or with removed wax bloom, whereas they fail on waxy surfaces.

Although there is a lot of information on attachment of different insects to *P. sativum* in the literature, the attachment force was measured only in a very few studies (Eigenbrode et al. 1998, 2008; Rutledge and Eigenbrode 2003). Different techniques for force measurement were used, and results reported are rather variable in terms of force values. Force tests on the abaxial surface of waxy (*Wel*) and glossy (*wel*) plants with coccinellid predatory beetles *Hippodamia convergens* Guerin-Menéville, performed by using a force tenaxometer, have shown that adhesion was >30-fold stronger in the glossy pea (6.448 mN) than in the waxy one (0.197 mN) (Eigenbrode et al. 1998). Centrifugal experiments with the same insect species and both leaf surfaces of the same plant mutants have revealed significant, but only up to 2-fold, enhancement of attachment force on glossy plants (0.182 ± 0.026 μN, mean ± SD) compared with waxy ones (0.109 ± 0.017 μN, mean ± SD), and no difference between leaf sides has been detected (Rutledge and Eigenbrode 2003). Tests with adults of five coccinellid beetle species (*H. convergens, Coccinella transversoguttata* Fald., *C. septempunctata* L., *Adalia bipunctata* L., and *Harmonia axyridis* Pallas) on the adaxial side in the same pea mutants by applying of the centrifuge method have shown that each insect species achieved lower attachment forces on waxy peas than on glossy plants (Eigenbrode et al. 2008). Depending on the beetle species, the force was 5–15 mN and <5 mN, respectively.

Our results of traction experiments with *C. montrouzieri* showed that the attachment force (in range of several mN) was significantly lower on all pea surfaces studied than that on smooth glass. The only exception was the abaxial leaflet side of the glossy mutant, which permitted significantly higher attachment force than the other plant surfaces. We found that the crystal wax length correlated with beetle attachment force. Additionally, there was a significant correlation between the force and crystal density. Both these morphometrical parameters of the crystalline wax coverage contribute to the surface roughness. The influence of the surface roughness on insect attachment has been tested experimentally only in a few previous studies (Gorb 2001; Peressadko and Gorb 2004; Gorb and Gorb 2006a; Voigt et al. 2008). It has been found that attachment forces were much higher on smooth and rough surfaces compared to micro-rough samples (0.3 and 1.0 μm asperity size), since spatula-like terminal elements of tarsal setae of the tested insects are able to generate sufficient contact with smooth substrates or smooth islands within rough substrates. In the present study, a considerable decrease of the wax crystal length (on the abaxial side of glossy plants) led to the reduction of the surface roughness amplitude, and this evidently caused an increase of the traction force (Fig. 7).

On the other hand, our results may be also explained by wax crystals contaminating the tarsi of the insects (Stork 1980; Eigenbrode et al. 1996; Gorb and Gorb 2002, 2006c; Gorb et al. 2005). Although we have not examined pads for contamination here, it has been previously demonstrated for a series of plant species that crystals with high aspect ratio are more breakable and have a considerably stronger ability to contaminate insect adhesive pads and hinder essentially the adhesion (Gorb et al. 2005; Gorb and Gorb 2006c).

Physicochemical properties of plant surfaces (wettability by water and the free surface energy) in terms of their influence on insect attachment have been recently discussed in some studies (Riedel et al. 2003; Bohn and Federle 2004; Gorb et al. 2004; Gorb and Gorb 2006a, b). It has been reported that hydrophobic substrates reduced adhesion (Riedel et al. 2003; Gorb et al. 2004; Gorb and Gorb 2006a, b). However, we found here no significant

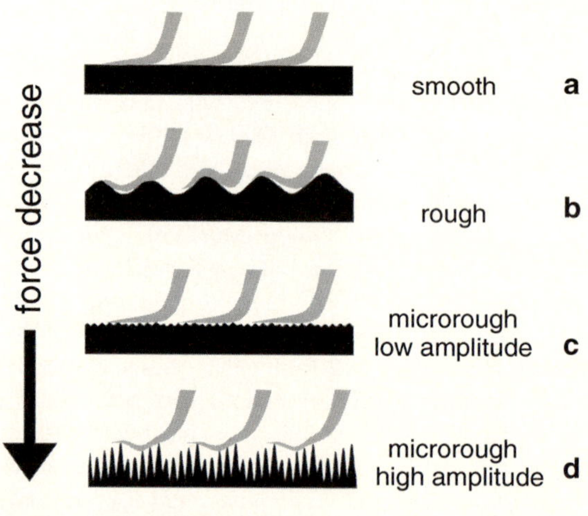

Fig. 7 Diagram explaining results obtained in traction tests. Since force, generated by insects, depends on the real contact area between setal tips of their adhesive organs and substrate, the substrate geometry is a crucial parameter influencing insect attachment. **a** On the smooth substrate (glass), setal tips have the maximum contact and therefore, maximal attachment force. **b** On the rough substrates, where dimensions of asperities are comparable with the size of setal tips, the real contact area is slightly reduced; however, tips can adjust to the substrate profile. **c** When dimensions of asperities are much smaller than the size of setal tips, the real contact area is strongly reduced, because terminal elements of setae can not adjust to the minute asperities. **d** In the case of small dimensions of the asperities and their large amplitude (the case similar to long wax crystals of the pea plant), the real contact area is even more strongly reduced, which results in the strongest decrease of the attachment force. The cases **a**, **b** and **c** were experimentally proven in the previous studies (Gorb 2001; Peressadko and Gorb 2004; Gorb and Gorb 2006a; Voigt et al. 2008). The cases **a**, **c** and **d** correspond to the present study

correlation between the traction force of *C. montrouzieri* and contact angles of liquids on *P. sativum* leaflet surfaces.

The correlation analyses (Fig. 6) demonstrated dependence of traction forces on the crystal length and density. Crystals of different length and different crystal densities did not strongly affect the surface wettability, but significantly influenced insect attachment. This means that the surface roughness affects the force reduction not necessarily through the wettability reduction, especially if we take into account purely lipide-like nature of a beetle pad secretion (Ishii 1987; Kosaki and Yamaoka 1996; Eisner and Aneshansley 2000). It is clear that here we deal with a complex contact mechanics phenomenon and observe an integration of overlaying effects of various parameters, but data obtained let us assume that the attachment force depends primarily on the surface profile and ability of insects to establish real contact with the substrate and secondarily on the surface wettability.

Interestingly, the comparison of abaxial and adaxial leaflet surfaces showed comparable force values in waxy plants, whereas a strong difference between these surfaces was found in glossy plants. Thus, based on our data, we can conclude that not only the presence of the crystalline wax, but its morphology influences insect attachment force.

As for the plant substrates, treated with chloroform, removing the wax from the surface, we obtained much higher forces (of several dozens mN) than on a glass, independently of the leaflet side and plant mutant. This effect can be explained by removal of the microroughness caused by wax crystals and probably by a strongly increased surface roughness, enabling beetles to use their claws for mechanical interlocking with the surface (Dai et al. 2002; Gorb et al. 2005; Voigt et al. 2007). In addition, surfaces with large roughness provide enough smooth sites for proper contact formation by single spatulae of adhesive hairs.

Conclusions

Our data on the ultrastructure and physicochemical properties of the adaxial and abaxial leaflet surfaces in waxy and glossy plants of *P. sativum* and the experimental results obtained in traction tests with beetles *C. montrouzieri* show that insect attachment does not depend solely on plant genetic background.

The epicuticular crystalline wax was not completely eliminated in the glossy mutant: it was only slightly reduced on the adaxial side and underwent greater changes on the abaxial side. We demonstrated here that the difference in the crystal length and density of the wax coverage (within the range found on these plants) did not influence wettability of surfaces, but affected insect attachment.

Acknowledgements Sincere thanks to V. Kastner (MPI for Developmental Biology, Tübingen, Germany) for assistance with traction tests. This work was partly supported by Landesstiftung Baden-Württemberg, Germany (Project "Plants as Source for Biomimetics of Anti-Adhesive Surfaces").

References

Aichele D, Schwegler HW (2000) Die Bluetenpflanzen Mitteleuropas. Bd. 2. Franckh-Kosmos Verlags, Stuttgart, 544 pp

Bargel H, Koch K, Cerman Z, Neinhuis C (2006) Structure–function relationship of the plant cuticle and cuticular waxes – a smart material? Funct Plant Biol 33:893–910. doi:10.1071/FP06139

Barthlott W (1990) Scanning electron microscopy of the epidermal surfaces in plants. In: Claugher D (ed) Scanning electron microscopy in taxonomy and functional morphology. Clarendon Press, Oxford, pp 69–94

Barthlott W, Neinhuis C, Cutler D, Ditsch F, Meusel I, Theisen I, Wilhelmi H (1998) Classification and terminology of plant epicuticular waxes. Bot J Linn Soc 126:237–260

Beutel R, Gorb SN (2001) Ultrastructure of attachment specialization of hexapods (Arthropoda): evolutionary patterns inferred from a revised ordinal phylogeny. J Zool Syst Evol Res 39:177–207. doi:10.1046/j.1439-0469.2001.00155.x

Bodnaryk RP (1992) Leaf epicuticular wax, an antixenotic factor in Brassicaceae that affects the rate and pattern of feeding in flea beetles, *Phyllotreta cruciferae* (Goeze). Can J Plant Sci 72:1295–1303

Bohn HF, Federle W (2004) Insect aquaplaning: *Nepenthes* pitcher plants capture prey with the peristome, a fully wettable water-lubricated anisotropic surface. Proc Natl Acad Sci USA 101:14138–14143. doi:10.1073/pnas.0405885101

Booth RG, Pope RD (1986) A review of the genus *Cryptolaemus* (Coleoptera: Coccinellidae) with particular reference to the species resembling *C. montrouzieri* Mulsant. Bull Entomol Res 76:701–717

Busscher HJ, Vanpert AWJ, Deboer P, Arends J (1984) The effect of the surface roughening of polymers on measured contact angle of liquids. Coll Surf 9:319–331. doi:10.1016/0166-6622(84)80175-4

Cooper MR, Johnson AW (1998) Poisonous plants and fungi in Britain: Animal and human poisoning. The Stationery Office, London, 398 pp

Dai Z, Gorb S, Schwarz U (2002) Roughness-dependent friction force of the tarsal claw system in the beetle *Pachnoda marginata* (Coleotera, Scarabaeidae). J Exp Biol 205:2479–2488

Edwards PB (1982) Do waxes of juvenile *Eucalyptus* leaves provide protection from grazing insects? Aust J Ecol 7:347–352. doi:10.1111/j.1442-9993.1982.tb01309.x

Eigenbrode SD (1996) Plant surface waxes and insect behaviour. In: Kerstiens G (ed) Plant cuticles – an integral functional approach. BIOS, Oxford, pp 201–222

Eigenbrode SD, Jetter R (2002) Attachment to plant surfaces by insect predators. Integr Comp Biol 42:1091–1099. doi:10.1093/icb/42.6.1091

Eigenbrode SD, Kabalo NN (1999) Effects of *Brassica oleracea* waxblooms on predation and attachment by *Hippodomia convergens*. Entomol Exp Appl 91:125–130. doi:10.1023/A:1003613710304

Eigenbrode SD, Castagnola T, Roux M-B, Steljes L (1996) Mobility of three generalist predators is greater on cabbage with glossy leaf wax than on cabbage with a wax bloom. Entomol Exp Appl 81:335–343. doi:10.1007/BF00187043

Eigenbrode SD, White C, Rohde M, Simon CJ (1997) Epicuticular wax phenotype of *wel* mutation and its effect on pea aphid populations in the greenhouse and in the field. Pisum Genet 29:13–17

Eigenbrode SD, White C, Rohde M, Simon CJ (1998) Behavior and effectiveness of adult *Hippodamia convergens* (Coleoptera: Coccinellidae) as a predator of *Acyrthsiphon pisum* (Homoptera: Aphididae) on a wax mutant of *Pisum sativum*. Environ Entomol 91:902–909

Eigenbrode SD, Kabalo NN, Stoner KA (1999) Predation, behavior, and attachment by *Chrysoperla plorabunda* larvae on *Brassica oleracea* with different surface waxblooms. Entomol Exp Appl 90:225–235. doi:10.1023/A:1026416831201

Eigenbrode SD, Rayor L, Chow J, Latty P (2000) Effects of wax bloom variation in *Brassica oleracea* on foraging by a vespid wasp. Entomol Exp Appl 97:161–166. doi:10.1023/A:1004051823556

Eigenbrode SD, Snyder WE, Clevenger G, Ding H, Gorb S (2008) Variable attachment to plant surface waxes by predatory insects. In: Gorb SN (ed) Functional surfaces in biology. Springer-Verlag (in press)

Eisner T, Aneshansley DJ (2000) Defence by foot adhesion in a beetle (*Hemisphaerota cyanea*). Proc Natl Acad Sci USA 97:6568–6573. doi:10.1073/pnas.97.12.6568

Erbil HY (1997) Surface tension of polymers. In: Birdi KS (ed) CRC Handbook of surfaces and colloid chemistry. CRC Press Inc., Boca Raton, pp 265–312

Federle W, Rohrseitz K, Hölldobler B (2000) Attachment forces of ants measured with a centrifuge: better "wax-runners" have a poorer attachment to a smooth surface. J Exp Biol 203:505–512

Gaume L, Gorb S, Rowe N (2002) Function of epidermal surfaces in the trapping efficiency of *Nepenthes alata* pitchers. New Phytol 156:479–489. doi:10.1046/j.1469-8137.2002.00530.x

Gaume L, Perret P, Gorb E, Gorb S, Labat J-J, Rowe N (2004) How do plant waxes cause flies to slide? Experimental tests of wax-based trapping mechanisms in the three pitfall carnivorous plants. Arthropod Struct Dev 33:103–111. doi:10.1016/j.asd.2003.11.005

Gniwotta F, Vogg G, Gartmann V, Carver TLW, Riederer M, Jetter R (2005) What do microbes encounter at the plant surface? Chemical composition of pea leaf cuticular waxes. Plant Physiol 139:519–530. doi:10.1104/pp.104.053579

Gorb SN (2001) Attachment devices of insect cuticle. Kluwer Academic Publishers, Dordrecht, 305 pp

Gorb EV, Gorb SN (2002) Attachment ability of the beetle *Chrysolina fastuosa* on various plant surfaces. Entomol Exp Appl 105:13–28. doi:10.1023/A:1021777517981

Gorb E, Gorb S (2006a) Combination of the surface profile and chemistry reduces the attachment of the beetle *Gastrophysa viridula* on the *Rumex obtusifolius* leaf surface. In: Salmén L (ed) Proceedings of the 5th plant biomechanics conference, Stockholm, Sweden, August 28–September 1, 2006. Stockholm, vol 2, pp 537–542

Gorb EV, Gorb SN (2006b) Physicochemical properties of functional surfaces in pitchers of the carnivorous plant *Nepenthes alata* Blanco (Nepenthaceae). Plant Biol 8:841–848. doi:10.1055/s-2006-923929

Gorb EV, Gorb SN (2006c) Do plant waxes make insect attachment structures dirty? Experimental evidences for the contamination-hypothesis. In: Herrel A, Speck T, Rowe NP (eds) Ecology and biomechanics: a mechanical approach to the ecology of animals and plants. CRC Press Inc., Boca Raton, pp 147–162

Gorb SN, Popov VL (2002) Probabilistic fasteners with parabolic elements: biological system, artificial model and theoretical considerations. Philos Trans R Soc Lond A 360:211–225. doi:10.1098/rsta.2001.0926

Gorb E, Kastner V, Peressadko A, Arzt E, Gaume L, Rowe N, Gorb S (2004) Structure and properties of the glandular surface in the digestive zone of the pitcher in the carnivorous plant *Nepenthes ventrata* and its role in insect trapping and retention. J Exp Biol 207:2947–2963. doi:10.1242/jeb.01128

Gorb E, Haas K, Henrich A, Enders S, Barbakadze N, Gorb S (2005) Composite structure of the crystalline epicuticular wax layer of the slippery zone in the pitchers of the carnivorous plant *Nepenthes alata* and its effect on insect attachment. J Exp Biol 208:4651–4662. doi:10.1242/jeb.01939

Halloway PJ, Hunt GM, Baker EA, Macey MJK (1977) Chemical composition and ultrastructure of the epicuticular wax in four mutants of *Pisum sativum* (L). Chem Phys Lipids 20:141–155. doi:10.1016/0009-3084(77)90087-1

Holloway PJ (1969) The effects of superficial wax on leaf wettability. Ann Appl Biol 63:145–153. doi:10.1111/j.1744-7348.1969.tb05475.x

Hunt GM, Baker EA (1982) Developmental and environmental variations in plant epicuticular waxes: some effects on the penetration of naphthylacetic acid. In: Cutler DF, Alvin KL, Price CE (eds) The plant cuticle. Academic Press, London, pp 279–292

Ishii S (1987) Adhesion of a leaf feeding ladybird *Epilachna vigintioctomaculata* (Coleoptera: Coccinellidae) on a vertically smooth surface. Jap J Appl Entomol Zool 22:222–228

Jeffree CE (2006) The fine structure of the plant cuticle. In: Riederer M, Müller C (eds) Biology of the plant cuticle. Blackwell Publishing Ltd., Oxford, pp 11–125

Juniper BE (1959) Die Oberflächen von Pflanzen. Endeavour 18:20–25. doi:10.1016/0160-9327(59)90112-7

Juniper BE, Cox GC (1973) The anatomy of the leaf surface: the first line of defence. Pestic Sci 4:543–561. doi:10.1002/ps.2780040413

Kosaki A, Yamaoka R (1996) Chemical composition of footprints and cuticula lipids of three species of lady beetles. Jap J Appl Entomol Zool 40:47–53

Kunkel G (1984) Plants for human consumption: an annotated checklist of the edible phanerogams and ferns. Koeltz Scientific Books, Koenigstein, 393 pp

Maier G (2002) Operating manual DataPhysics OCA. Release 2.0. Filderstadt, Germany, 111 pp

Marx GA (1969) Two additional genes conditioning wax formation. Pisum Newslett 1:10–11

McGuffin M, Kartesz JT, Leung AY, Tucker AO (2000) Herbs of commerce, 2nd edn. American Herbal Products Association, Silver Spring, USA, p 442

Owens DK, Wendt RC (1969) Estimation of the surface free energy of polymers. J Appl Polym Sci 13:1741–1747. doi:10.1002/app.1969.070130815

Peressadko AG, Gorb SN (2004) Surface profile and friction force generated by insects. In: Boblan I, Bannasch R (eds) Proceedings of the first international conference Bionik 2004, Hannover, Germany. VDI Verlag, Duesseldorf, pp 257–263

Riedel M, Eichner A, Jetter R (2003) Slippery surfaces of carnivorous plants: composition of epicuticular wax crystals in *Nepenthes alata* Blanco pitchers. Planta 218:87–97. doi:10.1007/s00425-003-1075-7

Rutledge CE, Eigenbrode SD (2003) Epicuticular wax on pea plants decreases instantaneous search rate of *Hippodamia convergens* larvae and reduces attachment to leaf surfaces. Can Entomol 135:93–101

Stork NE (1980) Role of wax blooms in preventing attachment to brassicas by the mustard beetle, *Phaedon cochleariae*. Entomol Exp Appl 28:100–107. doi:10.1007/BF00422029

Stork NE (1986) The form of plant waxes: a means preventing insect attachment? In: Juniper BE, Southwood TRE (eds) Insects and the plant surface. Edward Arnold, London, pp 346–347

Voigt D, Gorb E, Gorb S (2007) Plant surface–bug interactions: *Dicyphus errans* stalking along trichomes. APIS 1:221–243

Voigt D, Schuppert JM, Dattinger S, Gorb SN (2008) Sexual dimorphism in the attachment ability of the Colorado potato beetle *Leptinotarsa decemlineata* (Coleoptera: Chrysomelidae) to rough substrates. J Insect Physiol 54:765–776. doi:10.1016/j.jinsphys.2008.02.006

Waines JG (1975) The biosystematics and domestication of peas (*Pisum* L.). Bull Torrey Bot Club 102:385–395. doi:10.2307/2484765

Way MJ, Murdie G (1965) An example of varietal resistance of Brussel sprouts. Ann Appl Biol 56:326–328. doi:10.1111/j.1744-7348.1965.tb01245.x

White C, Eigenbrode SD (2000a) Effects of surface wax variation in *Pisum sativum* on herbivorous and entomophagous insects in the field. Environ Entomol 20:773–780

White C, Eigenbrode SD (2000b) Leaf surface waxbloom in *Pisum sativum* influences predation and intra-guild interactions involving two predator species. Oecologia 124:252–259. doi:10.1007/s004420000374

DOI 10.1007/978-3-540-85049-6_1

Visualization of Epicuticular Grease on the Covering Wings in the Colorado Potato Beetle: A Scanning Probe Approach

D. Voigt · H. Peisker · S. Gorb

Abstract Insects and spiders are supposed to release a greasy layer on their body surface, which may be involved in chemical and physical interactions between the organisms and their environment. In mating events, males frequently adhere to the female's dorsal body site by means of their feet, whereas grease should play an important role at the feet-attachment substrate interface. The properties and thickness of epicuticular grease have been diversely reported, but no definite visualizations and measurements have been previously carried out. Using the Colorado Potato beetle as a model species, we visualized the epicuticular grease on covering wings and characterized its adhesive properties. In this study, three different AFM modes (contact, tapping, and phase contrast) were applied. Obtained data were compared with the results of the Cryo-SEM. The grease layer thickness is about 8 nm on elevated sites of the epicuticle. A strong adhesion on the beetle epicuticle due to the presence of the grease layer was measured. The influence of a semi-fluid greasy layer on male adhesion to female's wings during copulation is discussed.

Springer

Keywords Adhesion · AFM · Chrysomelidae · Coleoptera · Covering wings · Cuticle · Elytra · Insect · *Leptinotarsa decemlineata* · SPM · Surface

Abbreviations

AFM	Atomic force microscopy
Cryo-SEM	Cryo scanning electron microscopy
GR	Epicuticular grease

Introduction

Epicuticular Grease: Introductory Remarks

The covering wings (elytra) in the Colorado potato beetle *Leptinotarsa decemlineata* appear shiny, smooth, water-repellent and slightly slippery. These properties are due to the presence of epicuticle, the outermost layer of the insect integument, covered by a wax-like lipid surface layer called grease. Surface waxes have been previously reported in a variety of conditions, from liquid viscous coatings to crystalline structures in the form of plates, rods and filaments from many insects (adults and larvae) and arachnids [1]. Beament [2] and Wigglesworth [3] considered an outer thin layer of solid wax, whereas Lewis [4] rather assumed a lipophilous oil film on the epicuticle surface to be widespread throughout insects. According to Noble-Nesbitt [5], epicuticular lipids are found to be either waxes or mobile greases at ambient temperatures. Outermost greasy layers on the epicuticle have been reported in cock roaches [2, 6, 7], beetles [8], flies [9], true bugs [10], crickets [11], locusts [12], ticks [8, 13, 14], and spiders [15, 16].

The consistence of grease has been controversially discussed. Beament [2] extracted hard wax grease from the fresh cuticle of *Periplaneta* cockroaches, later described as mobile grease extracted from the same species [6]. In spite of observations that wax secretion films in cockroach grease remain stable [17], the cockroach's secretions spread over their entire surface by the pressure caused by this polar material and strongly reducing agent [13]. It may be a relatively fluid lipid layer hardening due to chemical reactions [7, 18].

The thickness of the superficial greasy wax layer covering the epicuticle is of molecular dimensions [17] varying from less than a few nm to several μm [11]. The bulk of lipids on an insect's surface form a layer of 0.1–1.0-μm thickness, probably in less well-oriented layers permeating the cement layer of the epicuticle [19]. But a precise definition of grease thickness has only been reported for the hunting spider *Cupiennius salei* Keys (Ctenidae), where the thickness of the surface viscous layer was measured within 20–40 nm [16].

The greasy material on the cuticle surface is considered to play a fundamental role in preventing water loss [e.g. 20, 21]. Furthermore, cuticular lipids may serve as chemical cues used by insects for signalling in olfactory communication [22].

Thus, previous research on grease mainly considered chemical and physicochemical approaches. Chloroform extractions and chemical analyses have indicated the presence of lipids in the epicuticular layers of several arthropod species [2, 7, 12, 15, 23]. In the bug *Rhodnius*, a viscous semi-liquid protein layer rich in polyphenols was found [17]. In *L. decemlineata*, a protective, yellow, oily fluid of strong odour and acrid taste, secreted from hypodermal glands located along the margins of the dark stripes on the elytron's surface, has been described [24].

Both fluid and solid coverage, located on the surface, may promote or prevent adhesive interactions at the interface between attachment devices of insects and the surface [25].

Males of the Colorado potato beetle *L. decemlineata* firmly adhere to the female surface during copulation using adhesive tarsal setae. Tarsal secretion has been previously supposed to be an adhesive agent in leaf beetles (Chrysomelidae) [26, 27] and for this reason one may conjecture that the grease of the female elytra may play an important role in male attachment on the female surface during copulation.

Covering Wings and Mating Behaviour

The female covering wings (elytra) of *L. decemlineata* are a common attachment substrate for male feet [24, 28–31]. The copulation process consists of several phases [28–32]. The male mounts the female laterally and backwards, at first wobbling on the convex female elytra (Fig. 1A, B). The male repeatedly touches the smooth surface of the elytra with shearing movements of his feet. After reaching an optimal position, the male may stay motionless for several hours, adhering to the female's elytra with tarsi of foreand mid-legs. With the hind-legs, the male is mostly grasping terminal sclerites of the female's abdomen without touching the surface of the elytra. The male's feet, in contact with the female's surface, seem to be attached with the entire area of the attachment system, consisting of tenent setae located on the ventral surface of tarsomeres (Fig. 1D). Observation of the copulation posture lets us assume a role of the female elytra epicuticular grease on male adhesion (Fig. 1C).

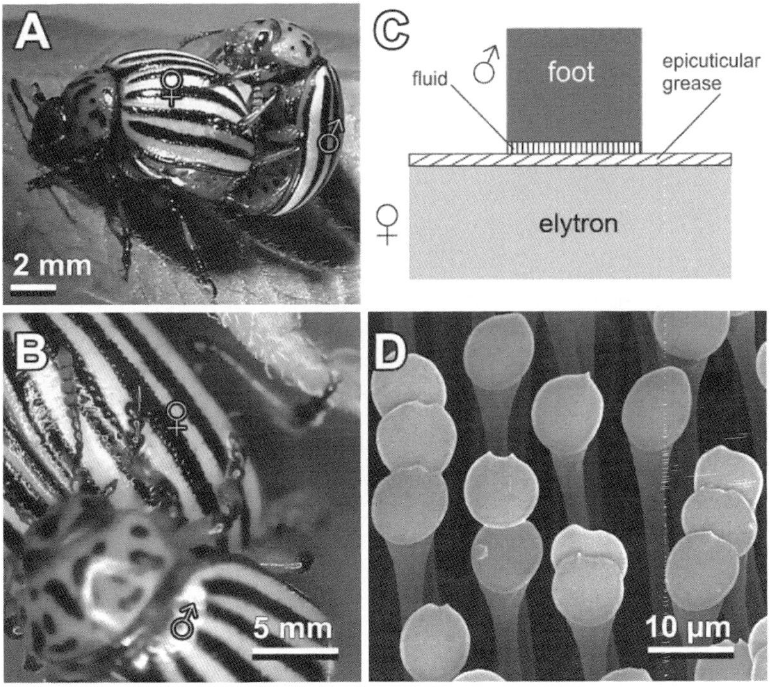

Fig. 1 Role of attachment in the mating process of the beetle *Leptinotarsa decemlineata*. **A, B** Typical mating posture demonstrating position of male attachment pads on the female surface. **C** Very simplified hypothetical diagram of the contact between the male tenent seta (foot) and female covering wing (elytron). Two fluid layers are presumably present between the cuticle of the seta and elytron. **D** Male tenent setae, Cryo-SEM image

Leptinotarsa decemlineata possess oval, triangular, convex elytra with a pale yellow to flavous coloured base [33]. Elytra give the insect a hemispherical appearance. Elytra are about 7.8 mm long and 6.8 mm wide [33]. The convex, anterior margin of the elytron tapers downward abruptly into the apophysis, whereas the two other sides of each elytron taper gradually into a point [34]. Each elytron has five black stripes (striae, vittae) extending from its base to its apex. The sutural margin of the elytron is black, the first vitta nearly joins it at the apex. Vittae 2 and 3 join at the apex. Punctation is course, patterned in the form of irregular double rows along the vittae. With two rows of punctuation between each two veins, and four in the subcostal space, they correspond exactly in position to the edges of the stripes [24]. In addition to these larger punctuations, smaller ones are also distributed without any discernible scheme of arrangement. Compound hypodermal glands are connected to the pores of the larger punctuation [24].

To understand mechanical interactions between male attachment devices and female elytra, detailed information on layer thickness, release, distribution, and physical properties of the grease are required. Previous morphological and ultrastructural studies using conventional electron microscopy methods (SEM, TEM) resulted in the removal of surface films, because samples were either dried and/or washed in organic solvents, such as ethanol, acetone, and propylene oxide, according to the conventional preparation procedures [15].

The present study was undertaken to characterize the thickness of the epicuticular grease layer and its adhesive properties in beetle elytra. For this purpose, atomic force microscopy (AFM) was applied, which has been previously demonstrated to be an excellent approach for geometrical characterization of insect and spider cuticles and for estimation of their adhesive properties on a local scale [16, 35, 36]. We have compared AFM data of the present study with previous results obtained in cryo SEM [30, 31].

Methods

Insects and Sample Preparation

The beetles, *Leptinotarsa decemlineata* Say (Coleoptera, Chrysomelidae), were taken from a stock culture (25 C, 60% RH, 16-h photoperiod) at Bayer Cropscience AG (Monheim, Germany).

Elytra were removed from anaesthetized alive females using micro-scissors and tweezers. For measurements, two types of samples were used: (1) fresh elytra without treatment and (2) fresh elytra washed in 50-ml chloroform shaking for 1 min in a vibrator, rinsed with aqua Millipore and dried using compressed air. Samples of the second type were treated with chloroform in order to remove grease and obtain a clean control surface without surface coverage. Four pieces (5 mm^2) were cut out from the dorsal part of a single elytron, using a razor blade, and immediately mounted on a glass slide using double-sided adhesive tape. Prior to measurements, samples where kept in clean Petri dishes with a small piece of moist filter paper to prevent contamination and desiccation. In all, five female beetle individuals (ten elytra) were studied.

Scanning Probe Microscopy

We used a NanoWizard® atomic force microscope (JPK Instruments AG, Berlin, Germany) mounted on an inverted microscope Zeiss Axiovert 135 (Carl Zeiss

MicroImaging GmbH, Göttingen, Germany). NanoWizard® image acquisition software 3.1.13 (JPK Instruments AG, Berlin, Germany) was used to obtain AFM images and NanoWizard® image processing software 3.1.6 was used to process images and calculate all necessary geometrical parameters of the surface. The most-common operating modes in the AFM are tapping mode and contact mode. The latter provides a better resolution, whereas, in the tapping mode, both the normal force and shear force applied to the sample are minimized. Visualization of the epicuticular grease in both modes provided us with complementary information. The force applied to the surface, was set to a minimum to ensure the integrity of the sample in both operating modes.

Tapping Mode

Scans were carried out in air at a scan rate of 1 Hz and a resolution of 512 × 512 pixels using standard, non-contact, high frequency cantilevers with reflex coating (NST-NCHF-R, Nascatec Technologies GmbH, Stuttgart, Germany). AFM images were taken from four pieces of the same elytron, two fresh untreated ones and two chloroform treated ones (see Insects and Sample Preparation). Five scans (20 × 20 μm) per specimen were carried out. A total of 20 images per sample were taken.

Contact Mode

In contrast to the tapping mode, the contact mode always exerts a much larger lateral force, making it more difficult to properly visualize sensitive surfaces like insect epicuticular grease. However, load force must be sufficient to properly resolve the surface topography. The optimal balance between these two parameters was determined by scanning the same surface repeatedly while optimizing the setpoint, scan rate, and gains of the feedback loop.

Scans with a size of 20 × 20 μm were obtained at 1-Hz line rate and a resolution of 512 × 512 pixels using standard contact mode cantilevers (NST-CM, Nascatec Technologies GmbH, Stuttgart, Germany). A total of 20 images per sample were taken (four specimens used, five images per specimen).

Adhesion Measurements

The attraction and repulsion forces acting on the standard contact-mode cantilever tip (NST-CM, Nascatec Technologies GmbH, Stuttgart, Germany) were measured in pull-off experiments. The spring constant of each cantilever was determined according to the thermal noise technique [36, 37]. The idea of the technique is that while the tip is far from the sample, its motion is due only to thermal fluctuations, and measurements of this motion at frequencies near the resonant frequency of the spring allow estimating the spring constant [37].

Adhesive properties of fresh samples of beetle elytra (see Insects and Sample Preparation) were compared with those of a silanized Si-wafer (114° contact angle of aqua Millipore). The Si-wafer was used as a defined hydrophobic surface with the surface roughness comparable to that of single polygonal cells observed on the dorsal surface of the elytron (unpublished data). To exclude the influence of sites with stronger adhesion located in the vicinity of pores due to the accumulated secretion, force-distance curves were taken on the tops of pre-scanned, convex elytron cells. The pre-scanning procedure was carried out to identify the proper site at the top of the hexagonally shaped cell for taking force-distance curves.

In total, 127 force-distance curves were collected on both types of surfaces. The adhesion force was evaluated from the curves with NanoWizard® image processing software 3.1.6 (JPK Instruments AG, Berlin, Germany). The Mann–Whitney Rank Sum test was used to estimate statistic differences in the adhesion force acting on the cantilever tip in contact with the beetle elytra and with the silanized Si-wafer (SigmaStat 3.1.1® software, Systat Software, Inc., Richmond, California, USA).

Cryo-SEM

Cryo-SEM allows observation of fresh specimens and provides a unique possibility to prove our data obtained from AFM scans. Female elytra surfaces were analyzed using a cryo-SEM Hitachi S 4800 (Hitachi High-Technologies Corp., Tokyo, Japan) equipped with a Gatan ALTO 2500 cryo-preparation system (Gatan Inc., Abingdon, UK). Previously, this method has been successfully applied for visualization of tarsal adhesive fluid in flies [38]. Pieces of elytra from live, anaesthetized beetles were cut off with a fine knife, mounted on metal holders, frozen on a cryo-stage at −140°C, sputter-coated with gold-palladium (3 nm) in the preparation chamber, and examined in the SEM at −120°C and with an accelerating voltage of 5 kV. Also, elytra treated with chloroform were examined for comparison.

Results

Elytra Topography and Grease Visualization

Cryo-SEM images of fresh, shock-frozen elytra demonstrated large pores (diameter 75.1 ± 6.98 μm, mean ± SD, n = 10) located at margins of dark stripes (Fig. 2). Additionally, there are some mid-sized pores unevenly distributed over the surface (diameter 29.5 ± 3.32 μm, mean ± SD, n = 10). The elytron surface consists of a hexagonal

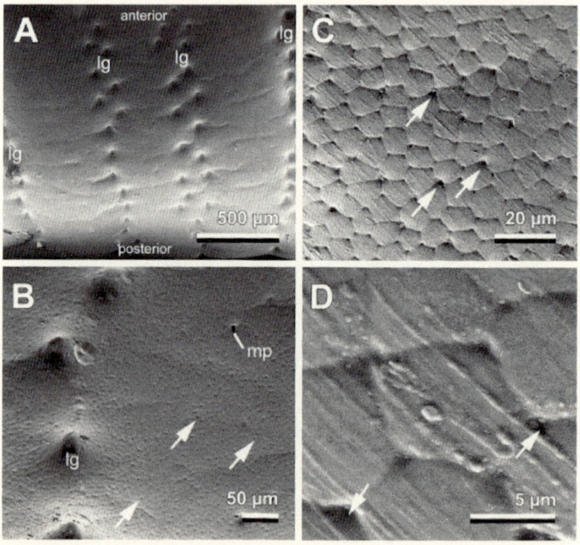

Fig. 2 Cryo SEM micrographs of the elytra surface in a fresh female *Leptinotarsa decemlineata*. **A** Overview of the posterior elytra surface showing irregular double rows of pores corresponding to large compound hypodermal glands. **B** Detail of **A**, a row of large pores on the left side, hexagonal surface pattern, regularly dispersed small pores (*arrows*) and a single mid-sized pore is visible. **C** Detail of **B**, hexagonal-shaped cells and small pores (*arrows*). **D** Small pores (*arrows*) and single cells of the hexagonal surface pattern. Grease smeared over the surface is visible.
lg, large glands; mp mid-sized pores

shaped, cell-like pattern. Smaller pores (diameter 3.9 ± 0.02 μm, mean ± SD, n = 10) are often situated at the corners of hexagons. Several tiny pores of 0.2 ± 0.02-μm diameter (mean ± SD, n = 10) were found on the top of cells. The pores observed are presumably responsible for delivering secretory substances to the cuticle surface. At higher magnification, an amorphous layer smeared over the surface can be detected (Fig. 2D). This layer often covers small pores and gaps between hexagons and thus makes the surface uniformly smooth.

The gathered AFM tapping-mode height images of fresh samples provided 3D information about hexagonal cells at the nanoscale (Fig. 3). However, no grease was visible.

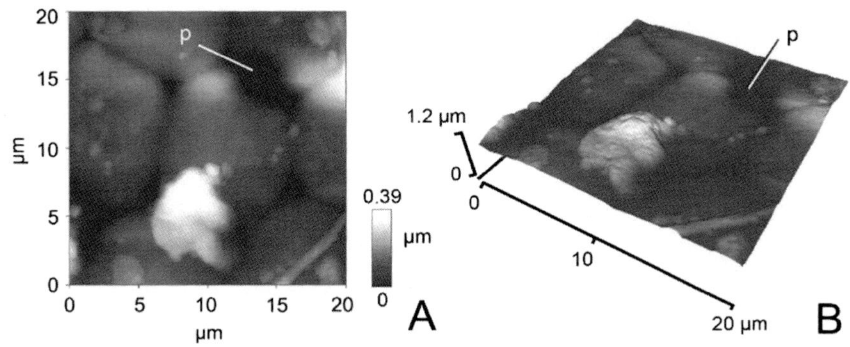

Fig. 3 AFM tapping-mode images of a fresh female elytron of *Leptinotarsa decemlineata*. **A** Height image. **B** 3D image. p, pores

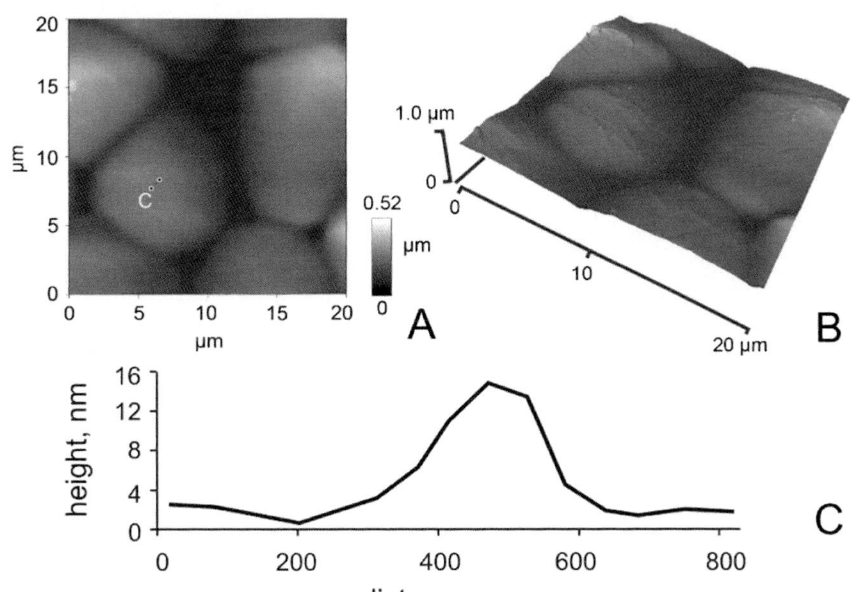

Fig. 4 AFM tapping-mode images of a chloroform-treated female elytron of *Leptinotarsa decemlineata* showing residues of grease. **A** Height image. **B** 3D image of the surface. **C** Profile diagram of a representative cross section along the *line* indicated between two dots in **A**

On surfaces washed with chloroform, irregular runlet-like patches of presumably waxy residues were found in AFM tapping-mode images (Fig. 4). These residues demonstrate that chloroform can dissolve the grease, but after evaporation of the solvent, they form a specific pattern on the surface. The typical width of such runlets in a cross section was determined to be in the range of 250 nm, and their height about 12 nm.

In order to obtain a more detailed view of fresh surfaces, phase images were taken and compared to height images (Fig. 5). Height images alone did not allow a definition of the grease distribution over the surface (Fig. 5A), whereas phase images (Fig. 5B, C) showed distinct differences in material properties and/or geometry indicated by a phase shift. The grease often accumulated in the vicinity of small pores, within the pores themselves, or in the regions between cells. The grease layer was about 8 nm thick, shown by the profile crossing the boundary between a grease patch and the surface of the epicuticle (Fig. 5D).

Topographical images of the fresh elytra, obtained in the contact mode, showed a heavy, smeary, striped appearance especially in the proximity of cuticle pores (Fig. 6). The visualization of surface fine structure was disturbed in the vicinity of the smear. An attempt to decrease the force applied to the sample surface resulted in almost non-interpretable images of surface topography.

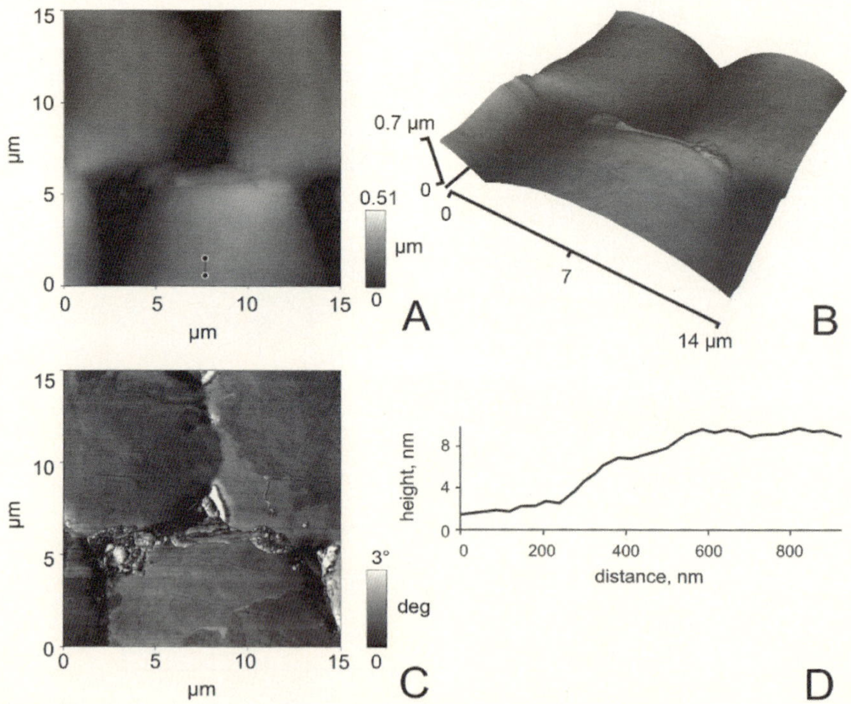

Fig. 5 AFM tapping-mode images of a fresh female elytron of *Leptinotarsa decemlineata*. **A** Height image. **B** Phase image of **A** indicating differences in surface properties/geometry by the use of phase-shift visualization. Grease distribution (*darker sites*) is clearly visible. **C** 3D phase image showing the grease boundary. **D** Profile diagram of the height image representing a cross section of the grease boundary along the *line* indicated between two dots in **A**

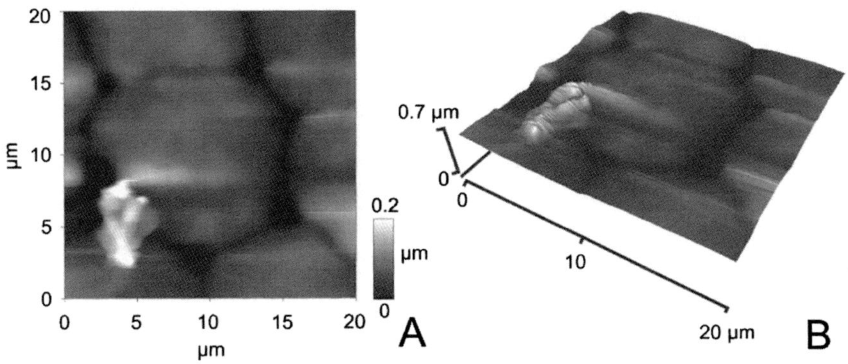

Fig. 6 AFM contact-mode scans of a fresh female elytron of *Leptinotarsa decemlineata*. **A** Height image. **B** 3D image. Images show smearing artefacts, presumably caused by pushing cuticle grease with the cantilever tip

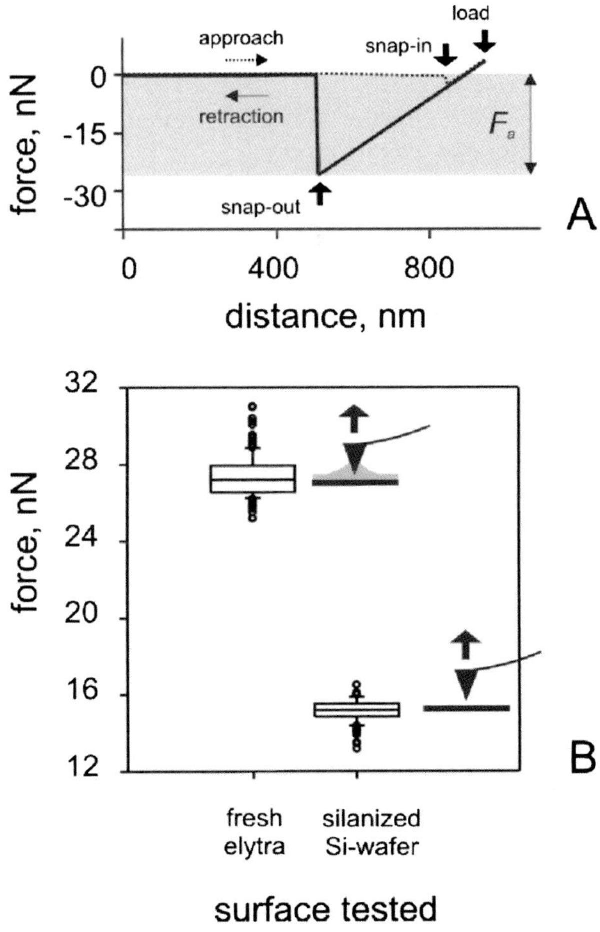

Fig. 7 Results of adhesion measurements. **A** Example of a force-distance curve obtained on a fresh female elytron of the beetle *Leptinotarsa decemlineata*. The *dotted line* represents data obtained while the cantilever approaches the surface and has to be read from left to right. The *black line* represents data acquired while the cantilever retracts from the sample showing adhesion of 27 nN. This line has to be read from right to left. The right side of the curve, where the force is positive, corresponds to the load applied to the sample (about 4 nN). The left part of the graph indicates the change in direction of the cantilever movement while not in contact. **B** Box-and-whiskers diagram of the adhesion force measured on fresh female elytra, and on a silanized Si-wafer. Statistically significant differences between the groups are revealed ($T = 24{,}257$, $p \leq 0.001$, Mann–Whitney Rank Sum test, $n = 127$ per group)

Adhesive Properties of the Elytra Surface

Having a thin layer of viscous grease on the surface of female elytra, one can expect an increase in adhesion (due to an increase of capillary forces), when male tarsal setae contact the surface. To prove this idea, we compared pull-off forces measured on fresh female elytra with those obtained on the silanized Si-wafer. A relatively low adhesion force, ranging from 14 to 16 nN, was measured presumably due to the low capillary interactions on the hydrophobic surface (Fig. 7A). On fresh elytra, a considerably stronger pull-off force was detected (up to 28 nN). A relatively sharp contact breakage event was observed when retracting the cantilever tip from the elytra surface (Fig. 7B). The contact breakage was smoother in absolutely fresh elytra and became sharper when elytra began to dry.

Discussion

The present paper is a case study demonstrating the application of the AFM for visualization of the epicuticular grease in females of the Colorado potato beetle. The location and thickness of the grease was demonstrated, showing its non-uniform distribution over the elytra surface. The grease residues also remain on the top of the hexagonal cells in the form of patches, whereas the rest runs down the curvature and accumulates within the pores. Also, grease accumulation between the cells is clearly visible.

The data, obtained with the AFM, were partly proven with the use of Cryo-SEM, which supported the presence and geometry of the grease layer. These data, providing an important extension to the previous data of chemical analyses, were obtained for the first time with each of both microscopy methods.

The combined use of both AFM scanning modes (tapping and contact) shows the semi-fluid nature of the grease layer: the contact mode resulted in smeared images of the elytron surface, corresponding to fluid properties, whereas the tapping mode provided clear images of the surface, corresponding to solid properties.

Waxy residues found on the chloroform-treated samples indicate that lipid-like cuticle materials take a different shape after solvent evaporation, but are not completely removed from the surface. That is why chloroform treatment seems not to be a sufficient method to clean the elytra surface from cuticle fluids/secretions/grease. In addition, we cannot be sure that lipids situated in the depth of the cuticle are not dissolved and deposited on the surface after fluid evaporation. Thus, the method of chloroform treatment should also be critically evaluated in other kinds of experiments, for example in chemical analyses. The proper way of material collection for analysis of the cuticle grease is the preparation of surface prints on glass with further washing out of substances from the glass surface, as previously shown for tarsal adhesive secretion in the locust [12].

The phase image, used in this study for visualization of grease, has demonstrated very good results in combination with height images obtained in the tapping mode. Phase images aided in localization of places on the smooth epicuticle covered with the grease. In spite of uncertainty of the origin of such a phase contrast (caused either by (1) surface geometry, (2) differences in local material properties, or (3) both), it seems to be a powerful tool for detection and visualization of submicron-thick layers on the smooth surface of biological objects.

Hydrocarbons are a major component of the surface lipids in *L. decemlineata* [39]. There were quantitative compositional differences in epicuticular hydrocarbons found between different sexes: females contain a higher amount of branched methylalkanes

[39, 40]. It has been previously supposed that quantitative differences observed in the hydrocarbon profile of both sexes may be used in sex recognition [39]. Indeed, behaviour experiments on mating preferences have demonstrated that compounds of the female elytra stimulate mating responses in male, whereas those found on male elytra act as mating response inhibitors [41]. Taking into account the mating posture of both sexes, one can suggest a possible role of the female epicuticular grease in male attachment on the female surface.

Since roughness of the Si-wafer was comparable with the roughness of single hexagonal cells of elytra (Ra = 7.3 ± 0.77 nm, RMS = 9.1 ± 0.77 nm; mean ± SD, n = 15, scanning area 2 × 2 μm, unpublished data), detected differences in the pull-off force were not due to the surface topography. Also, epicuticle of the tanned cuticle is rather stiff [42] compared to the spring constant of the cantilever beam ($k = 0.0627$ N/m). That is why the differences in the material compliance between epicuticle and Si-wafer cannot explain differences in adhesion. We tend to explain differences in adhesive properties between these two substrates by the presence of the fluid layer on the beetle epicuticle. However, the influence of surface energy, which is presumably lower in silanized Si-wafer compared to clean epicuticle, cannot be completely excluded. The sharp contact breakage event, while retracting the cantilever tip from the elytra surface can be explained by highly viscous properties of the grease, especially in non-fresh specimens (unpublished data).

The sharp character of pull-off events that occurred when the cantilever tip was retracted from the sample surface was typical of a silicon substrate. Considering the assumed fluid-like properties of the cuticle grease, a less abrupt return of the cantilever to the zero deflection was expected (see our previous AFM data on the adhesive fluid of fly pulvilli [43–46]). The sharp character of the pull-off part of force-distance curves on the surface of the elytra may be explained by a very small radius of the cantilever tip (less than 10 nm).

As mentioned above, having a layer of greasy semi-fluid material on the elytra surface, one can expect its influence on male adhesion during copulation. The insect adhesive system relies on a certain amount of fluid in the contact area [47–50]. An increase in the fluid layer due to the presence of the grease presumably leads to an aquaplaning effect when, instead of an increase in friction forces, a lubrication (slipping) effect occurs. Video recordings of copulations of Colorado potato beetles [32] demonstrate the difficulties male beetles have in getting a firm grip on the female elytra surfaces. Male tarsi continuously slide along the female elytra. This means that, even having the proper foot position on the female elytra, male attachment systems fail on the female surface. It is known that females in many species of the beetle family Dytiscidae are dimorphic bearing smooth or structured elytra surfaces caused by sexual selection and sexual conflict. Female elytra and male attachment systems underlie continuous adaptations and counter-adaptations determined by the reproductive mechanisms and mate choice of females in a mix of cooperative and conflict interactions with males [51]. Whereas most males are adapted to the smooth surface of elytra, only the fittest ones succeed by proving their adhesion to rough surfaces [52]. Polyandry, mating and long mate guarding events have been suggested to be costs influencing female fitness [53, 54], also known in *L. decemlineata* [55]. They may drive the evolution of female adaptations to interfering with the male attachment system. Furthermore, females of *L. decemlineata* have been recorded to mate size-assortively [24] and could show clear preference for individual males [29]. All of these female adaptations may be related to the presence and properties of the grease. If female grease thickness changes during their life cycle, and if females are able to control the grease thickness, one may hypothesize that females may potentially use the grease thickness/properties for mating control.

Conclusions

No information about the distribution and thickness of cuticle grease on the cuticle surface of the Colorado potato beetle has been found in the literature. This study demonstrates three different AFM applications (contact mode, tapping mode, and phase contrast) not only for visualization of cuticle grease in living insect cuticle, but also for probing its properties. Cryo-SEM was applied as a control method to prove data obtained with AFM. The grease layer was about 8 nm thick on elevated sites of the epicuticle. Grease accumulation was observed between the hexagonally shaped cells of the cuticle and within the pores. We have demonstrated strong adhesion on the beetle epicuticle due to the presence of the grease layer. The possible influence of the presence of a layer of greasy semi-fluid material on male adhesion during copulation is discussed.

Acknowledgments Peter Meisner, Bayer Cropscience AG, Research Insecticides/Insecticide Biology (Monheim, Germany) kindly provided the insects. Valuable discussions with Michael Varenberg (Technion, Haifa, Israel) on contact mechanics and tribology are greatly acknowledged. Cristian Löbbe (JPK, Berlin, Germany) helped with the data processing. Victoria Kastner (Tü bingen, Germany) provided linguistic corrections on an early draft of the manuscript. This study was partly funded by the Federal Ministry of Education and Research, Germany (project InspiRat 01RI0633D).

References

1. Hadley NF (1981) Cuticular lipids of terrestrial plants and arthropods: a comparison of their structure, composition, and waterproofing function. Biol Rev 56:23–47
2. Beament JWL (1945) The cuticular lipoids of insects. J Exp Biol 21:115–131
3. Wigglesworth VB (1945) The insect cuticle. Biol Rev 23:408–451
4. Lewis CT (1962) Diffusion of oil films over insects. Nature 183:904
5. Noble-Nesbitt J (1992) Cuticular permeability and its control. In: Binnington K, Retnakaran A (eds) Physiology of the insect epidermis. CSIRO Australia, pp 252–283
6. Beament JWL (1958) The effect of temperature on the waterproofing mechanism of an insect. J Exp Biol 35:494–519
7. Gilby AR, Cox ME (1963) The cuticular lipids of the cockroach *Periplaneta americana* (L.). J Insect Physiol 9:671–681
8. Richards AG (1951) The integument of arthropods. The chemical components and their properties, the anatomy and development, and the permeability. University of Minnesota Press, Minneapolis
9. Wolfe LS (1954) Studies of the development of the imaginal cuticle of *Calliphora erythrocephala*. J Cell Sci s3-95:67–78
10. Wigglesworth VB (1933) The physiology of the cuticle and of ecdysis in *Rhodnius prolixus* with special reference to the function of the oenocytes and of the dermal glands. Quart J Micr Sci 76:269
11. Hendricks GM, Hadley NF (1983) Structure of the cuticle of the common house cricket with reference to the location of lipids. Tissue Cell 15:761–779
12. Vötsch W, Nicholson G, Müller R, Stierhof Y-D, Gorb S, Schwarz U (2002) Chemical composition of the attachment pad secretion of the locust *Locusta migratoria*. Insect Biochem Mol Biol 32:1605–1613
13. Lees AD, Beament JWL (1948) An egg-waxing organ in ticks. Quart J Micr Sci 89:291–332
14. Gilby AR (1957) Studies of cuticular lipids of Arthropods. III. The chemical composition of the wax from *Boophilus microplus*. Arch Biochem Biophys 67:320–324
15. Hadley NF (1981) Fine structure of the cuticle of the black widow spider with reference to surface lipids. Tissue Cell 13:805–817

16. McConney ME, Schaber CF, Julian MD, Barth FG, Tsukruk VV (2007) Viscoelastic nanoscale properties of cuticle contribute to the high-pass properties of spider vibration receptor (*Cupiennius salei* Keys). J R Soc Interface 4:1135–1143
17. Beament JWL (1955) Wax secretion in the cockroach. J Exp Biol 32:514–538
18. Gilby AR (1962) Absence of natural volatile solvents in cockroach grease. Nature 195:729
19. Locke M (1964) The structure and formation of the integument in insects. In: Rockstein M (ed) The physiology of Insecta. Academic Press, New York, pp 123–213
20. Ramsay JA (1935) The evaporation of water from the cockroach. J Exp Biol 12:373
21. Neville AC (1975) Biology of the arthropod cuticle. Springer, Berlin
22. Espelie KE, Bernays EA, Brown JJ (1991) Plant and insect cuticular lipids serve as behavioural cues for insects. Arch Insect Biochem Physiol 17:223–233
23. Lockey KH (1988) Lipids of the insect cuticle: origin, composition and function. Comp Biochem Physiol 89B:595–645
24. Tower WL (1906) An investigation of evolution in chrysomelid beetles of the genus *Leptinotarsa*. Carnegie Institution of Washington, Publication No. 48. Papers of the station for Experimental Evolution at Cold Spring Harbor, New York, No. 4, Press of Judd & Detweiler, Inc., Washington, DC
25. Gorb E (2007) Plant surfaces preventing insect adhesion. In: Brickwedde EF, Erb R, Lefèvre J, Schwake M (eds) Bionik und Nachhaltigkeit-Lernen von der Natur, 12. Internationale Sommerakademie St. Marienthal. Erich Schmidt Verlag GmbH & Co., Berlin, Initiativen zum Umweltschutz 68:103–110
26. Stork NE (1983) The adherence of beetle tarsal setae to glass. J Nat Hist 17:583–597
27. Eisner T, Aneshansley DJ (2000) Defence by foot adhesion in a beetle (*Hemisphaerota cyanea*). Proc Nat Acad Sci 97:6568–6573
28. Thibout E (1982) Le comportement sexuel du doryphore, *Leptinotarsa decemlineata* Say et son possible controle par l'homone juvenile et les corps allates. Behaviour 80:199–217
29. Szentesi Á (1985) Behavioral aspects of female guarding and inter-male conflict in the Colorado potato beetle. Proceedings of the Symposium on the Colorado potato beetle, XVIIth international congress of Entomology, Research Bull 704:127–137
30. Voigt D, Schuppert JM, Dattinger S, Gorb SN (2008) Sexualdimorphismus der Haftfähigkeit an rauen Oberflächen bei *Leptinotarsa decemlineata* Say (Coleoptera, Chrysomelidae). Mitt Dtsch Ges Allg Angew Ent 16:431–434
31. Voigt D, Schuppert JM, Dattinger S, Gorb SN (2008) Sexual dimorphism in the attachment ability of the Colorado potato beetle *Leptinotarsa decemlineata* (Coleoptera, Chrysomelidae) to rough substrates. J Insect Physiol 54:765–776
32. Wyss U (2005) Lebensweise und Entwicklung des Kartoffelkäfers *Leptinotarsa decemlineata*. Video documentation. Institut für Phythopatologie, Christian-Albrechts-Universität zu Kiel, Germany, www.entofilm.com
33. Jaques RL (1988) The potato beetles. The genus *Leptinotarsa* in North America (Coleoptera: Chrysomelidae). Flora & Fauna Handbook, no. 3, E. J. Brill, Leiden, New York, Kobenhavn, Köln
34. Rivnay E (1928) External morphology of the Colorado beetle (*Leptinotarsa decemlineata* Say). J New York Entomol Soc 26:25–145
35. Scherge M, Gorb SN (2001) Biological microand nanotribology. Springer, Berlin Heidelberg New York
36. Langer MG, Ruppersberg JP, Gorb S (2004) Adhesion forces measured at the level of a terminal plate of the fly's seta. Proc R Soc Lond B 271:2209–2215
37. Hutter JL, Bechhoefer J (1993) Calibration of atomic-force microscope tips. Rev Sci Instrum 64:1868–1873
38. Gorb SN (2006) Fly microdroplets viewed big: a Cryo-SEM approach. Microsc Today N9:38–39
39. Dubis E, Malínski E, Dubis A, Szafranek J, Nawrot J, Poplawski J, Wróbel JT (1987) Sex-dependent composition of cuticular hydrocarbons of the Colorado beetle, *Leptinotarsa decemlineata* Say. Comp Biochem Physiol A 87:839–843
40. Nelson DR, Adams TS, Fatland CL (2003) Hydrocarbons in the surface wax of eggs and adults of the Colorado potato beetle, *Leptinotarsa decemlineata*. Comp Biochem Physiol B 134:447–466
41. Mpho M, Seabrook WD (2003) Functions of antennae and palpi in the mating behaviour of the Colorado potato beetle *Leptinotarsa decemlineata* (Coleoptera: Chrysomelidae). Bull Entomol Res 93:91–95
42. Barbakadze N, Enders S, Gorb S, Arzt E (2006) Local mechanical properties of the head articulation cuticle in the beetle *Pachnoda marginata* (Coleoptera, Scarabaeidae). J Exp Biol 209:722–730
43. Stadler H, Mondon M, Then D, Jiao Y, Gorb SN, Ziegler C (2000) Scanning force microscopy measurements of the viscosity force of fly pad secretion. In: Scanning-probe microscopes and organic materials IX, Workshop, Hannover, 9.10.-11.11.2000, Abstract Booklet, edited by Kolb H-A, Enders O, Guckenberger R, Heckl WM, Hörber JKH, Rabe JP, and Ziegler C, Hannover

44. Stadler H, Mondon M, Wallentin J, Jiao Y, Gorb S, Ziegler C. (2001) Viscosity force of the fly's pad secretion measured by atomic force microscopy. Technische Biologie und Bionik 5, ed. by Nachtigall W. Mainz: Akademie der Wissenschaften und der Literatur. Biona Report 15:340–344
45. Wallentin J, Mondon M, Stadler H, Gorb SN, Ziegler C (1999) Rasterkraftspektroskopische Untersuchung der Adhäsionseigenschaften von Fliegensekret. In: Deutsche Physikalische Gesellschaft e.V., Frühjahrstagung des Arbeitskreises Festkörperphysik bei der DPG, Regensburg, März 27–31, 2000, Abstract Booklet, Regensburg, p 807
46. Wallentin J, Mondon M, Stadler H, Gorb SN, Ziegler C (1999) The secretes of fly secretes: adhesion properties probed by force-distance curves. In: Scanning-probe microscopes and organic materials VIII, Workshop, Basel, October 4–6, 1999, Abstract Booklet, ed. by Müller DJ, and Knapp HF, Basel, 1999, p 14
47. Gorb S (2001) Attachment devices of insect cuticle. Kluwer Academic Publishers, Dordrecht, p 305
48. Ishii S (1987) Adhesion of a leaf feeding ladybird *Epilachna vigintioctomaculata* (Coleoptera: Coccinellidae) on a vertically smooth surface. Appl Entomol Zool 22:222–228
49. Walker G, Yule AB, Ratcliffe J (1985) The adhesive organ of the blowfly, *Calliphora vomitoria*: a functional approach (Diptera: Calliphoridae). J Zool Lond A 205:297–307
50. Attygalle AB, Aneshansley DJ, Meinwald J, Eisner T (2000) Defence by foot adhesion in a chrysomelid beetle (*Hemisphaerota cyanea*): characterization of the adhesive oil. Zoology 103:1–6
51. Alcock, J (2006) Animal behavior. An evolutionary approach, 8th edn. Elsevier Spektrum Akademischer Verlag, p 577
52. Härdling R, Bergsten J (2006) Nonrandom mating preserves intrasexual polymorphism and stops population differentiation in sexual conflict. Am Nat 167:401–409
53. Arnquist G, Nilsson T (2000) The evolution of polyandry: multiple mating and female fitness in insects. Anim Behav 60:145–164
54. Miller KB (2003) The phylogeny of diving beetles (Coleoptera: Dytiscidae) and the evolution of sexual conflict. Biol J Linn Soc 79:359–388
55. Orsetti DM, Rutowski RL (2003) No material benefits, and a fertilization cost, for multiple mating by female leaf beetles. Anim Behav 66:477–484

Appl Phys A (2010) 100: 63–71
DOI 10.1007/s00339-010-5794-x

Applied Physics A
Materials Science & Processing

BRIEF COMMUNICATION

Fracture behaviour of plant epicuticular wax crystals and its role in preventing insect attachment: a theoretical approach

Feodor M. Borodich · Elena V. Gorb · Stanislav N. Gorb

Received: 12 October 2009 / Accepted: 12 May 2010 / Published online: 10 June 2010

Abstract It is known that attachment ability of insects depends among others on the texture of the substrate. On plant surfaces, covered with microscopic epicuticular wax crystals, insect attachment was found to be highly reduced for many insect species. In some plants, this effect may be explained by the contamination of insect adhesive organs (pads) by wax crystals. In the present study, mechanics of the wax crystals fracture during contact formation between insect adhesive pads and plant surface is examined, in order to explain the observed contamination of pads by wax. It is shown that mechanisms of the wax crystal fracture may be rather different, depending on the slenderness ratio of the crystals. Crystals with high values of the ratio may buckle elastically or in an elastic-plastic way. For five plant species under consideration, the critical value of the slenderness ratio for elastic buckling is 26.5, while for elastic-plastic buckling it is 18.7. If the values of the slenderness ratio are lower than the critical values, one has to consider bending of the crystals under the weight of insects. Although this study considered only crystals of a tubular shape, the general approach is valid also for crystals of other shapes.

F.M. Borodich
School of Engineering, Institute of Theoretical, Applied and Computational Mechanics, Cardiff University, The Parade, Cardiff CF24 3AA, Wales, UK
e-mail: BorodichFM@cardiff.ac.uk

E.V. Gorb (✉) · S.N. Gorb
Evolutionary Biomaterials Group, Max Planck Institute for Metals Research, Heisenbergstr. 3, 70569 Stuttgart, Germany
e-mail: egorb@zoologie.uni-kiel.de
Fax: +49-431-8801389

S.N. Gorb
e-mail: sgorb@zoologie.uni-kiel.de

E.V. Gorb · S.N. Gorb
Department of Functional Morphology and Biomechanics, Zoological Institute, Christian Albrecht University of Kiel, Am Botanischen Garten 1–9, 24118 Kiel, Germany

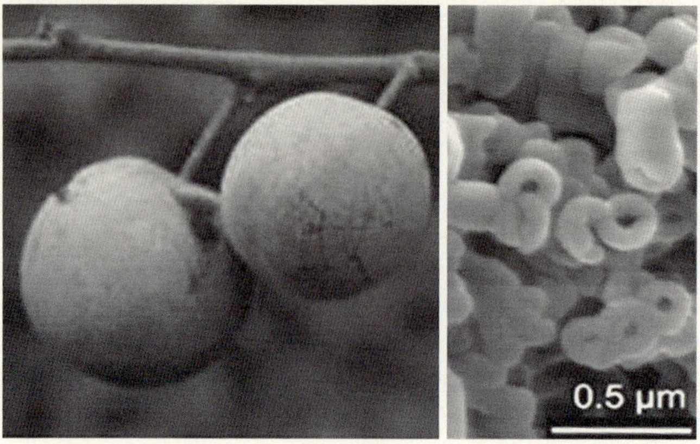

Fig. 1 European plum (*Prunus domestica*) fruits (*left*) covered by tubular wax crystals (*right*, SEM micrograph)

1 Introduction

Attachment ability of insects depends on various factors. Thus, depending on the texture of the substrate, insects use different mechanisms of attachment. For example, they can apply their claws to interlock with surface irregularities on rough (at the mesoscale) surfaces. On smooth and microrough substrates, insects rely on specialised attachment organs called adhesive pads [1]. Pad functioning is based on one of two different mechanisms, smooth flexible pads or hairy structures [2,3]. Compliant pad material in the first case and fibrillar surface structures called tenent setae in the second guarantee maximisation of the contact area and ensure high attachment force.

In the case of attachment to plant surfaces, roughness and other surface features affect the adhesion of insects. It is well known that most of aerial surfaces of land plants are covered by epicuticular waxes. The epicuticular wax is a mixture of a few major classes of alicyclic and long-chain aliphatic compounds, where the overall chain length of homologues is usually in the range of C16–C35 [4,5]. Waxes differ from one another in nature, number and relative percentage of different classes and the homologue chain lengths, and vary among plant species, with plant parts, ages, and environmental conditions [6]. Microscopically small crystals often emerge from the wax film, covering the plant cuticle, and cause the pruinosity or powdery appearance of the surface in many plant species [7–9]. Length of the wax crystals ranges from few hundred nanometres to several micrometres. Such pruinose waxy surfaces (Fig. 1) are found on stems, leaves, flowers, seeds and fruits in many plants [8].

The plant crystalline waxes perform different functions (see reviews by Barthlott [10] and Bargel et al. [11]). In particular, they can protect plants by reducing insect attachment to plant surfaces covered by wax crystals. This phenomenon is very important, because most of insect species are associated with plants, and hence, they should be able to attach successfully to plant surfaces [12].

Dealing with insect attachment to plant substrates, one has to consider a problem of interaction between insect adhesive organs and the plant surface. Contacting surfaces of many insect adhesive devices and pruinose waxy plant surfaces possess various patterns of micro- or nanostructures. The reduction of insect adhesion on the plant surfaces, covered with the crystalline epicuticular wax, may be explained by (1) the decrease of the real contact area

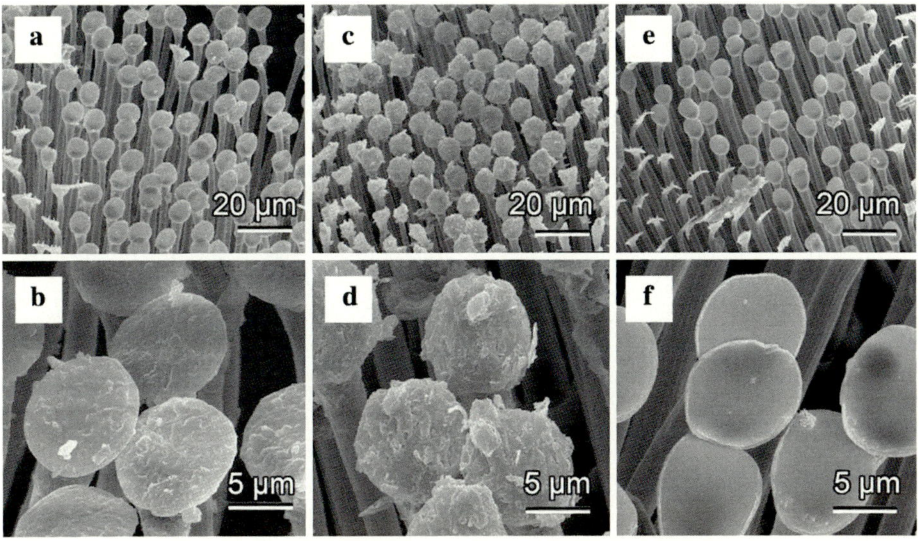

Fig. 2 SEM micrographs of the pad surface in the third tarsomere of the dead-nettle leaf beetle *Chrysolina fastuosa* after walking on waxy plant surfaces of leaves in the white goosefoot *Chenopodium album* (**a**, **b**), young stems in the maple ash *Acer negundo* (**c**, **d**), and leaves in the common barberry *Berberis vulgaris* (**e**, **f**) (after Gorb and Gorb [15])

between the substrate and insect adhesive pads caused by plant surface microroughness (roughness-hypothesis); (2) the absorption of the pad fluid from the pad surface by the structured wax coverage (fluid-absorption-hypothesis); (3) the dissolving of the wax by the pad secretion, resulting in hydroplaning and substrate slipperiness (wax-dissolving-hypothesis); and (4) the contamination of pads by wax crystals (contamination-hypothesis) [12,13].

The contamination-hypothesis was proven experimentally for several insects and a series of plant species. The effect of pad contamination by wax crystals on the reduction of insect attachment was found for the upper wax layer in the trapping organs of the carnivorous tropical pitcher plant *Nepenthes alata* [14]. The data on the contamination of adhesive pads in the dead-nettle leaf beetles *Chrysolina fastuosa* by wax crystals of 12 waxy plant surfaces showed that the plants differ essentially in their contaminating effects on insect pads [15] (Fig. 2). Differences were found in both the nature of the contamination, i.e. the structure of the contaminating material and the presence of recognisable crystals, and the degree of contamination (the portion of setal tip surface covered with contaminating particles, and the portion of setae covered with the wax). These differences were hypothesised to be caused by various micromorphologies of waxes in the plant species studied.

In this paper, we present an attempt to model mechanical behaviour of the plant wax crystals in order to explain the contamination of insect pads by these crystals. Crystals with high and low aspect ratios are examined, and examples for both cases are discussed. The models are presented for crystals of a tubular shape, but the general approach is valid also for crystals of other shapes.

2 Mechanical properties of plant waxes

It is known that for moderate values of loads, i.e. if the applied stresses are equal to a specific stress, the proportional limit (σ_{pr}) of crystalline materials follow the linear Hook's

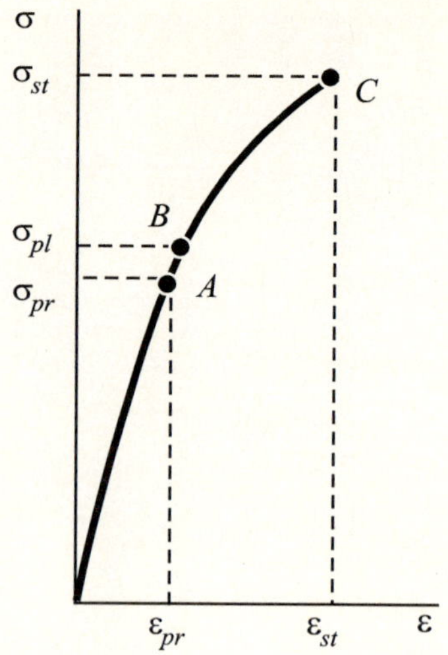

Fig. 3 Typical stress-strain diagram for a brittle crystalline material. Point A is the proportional limit (stress σ_{pr}), point B is the yield limit (stress σ_{pl}), and point C is the fracture stress (the material strength σ_{st})

law between stress (σ) and strain (ε), or between the tensile elongation or reduction (δ) and the force (P) applied to a tested specimen:

$$\delta = \frac{PL}{AE} \quad \text{or} \quad \sigma = E\varepsilon. \tag{1}$$

Here A is the area of the cross section of the specimen, E is a material characteristic (Young or elastic modulus of the material) and L is the length of the sample; $\varepsilon = \delta/L$ and $\sigma = P/A$.

If the stresses are between the proportional limit and the yield stress σ_{pl} ($\sigma_{pr} \le \sigma < \sigma_{pl}$), there is a small region of non-linear elastic behaviour, i.e. after unloading of the sample, it has the same shape as before the loading; however, the material does not follow the linear relation (1). When the value of the tensile (or compressive) stress is greater than the yield stress σ_{pl}, then after unloading of the sample, its shape differs from the shape it had before the loading, i.e. the sample got plastic deformations. Usually $\sigma_{pr} \cong \sigma_{pl}$ and hence, it can be assumed that they are the same. If materials are brittle, then they fail with only little (just a few percents) elongation or reduction after the yield stress (point B in Fig. 3).

In spite of the biological importance of plant waxes, there is surprisingly rather poor information on their mechanical properties. To the best of our knowledge, properties of plant waxes were studied only by Gorb et al. [14]. Other studies were related mainly to beeswax, dental waxes and some natural waxes used in dental applications [16–18].

It was found that all mechanical characteristics like E, σ_{pr} and the compressive strength σ_{st} of waxes in general are temperature dependent. As an example, one can consider beeswax, produced by honey bees for building honeycomb cells [19]. It is a mixture of wax esters, fatty acids and hydrocarbons. For this wax, $E = 75.83 \cdot 10^6$ Pa, $\sigma_{pr} = 551.5 \cdot 10^3$ Pa and $\sigma_{st} = 868.6 \cdot 10^3$ Pa were measured at room temperature ($T = 23°$C), while $E = 35.16 \cdot 10^6$ Pa, $\sigma_{pr} = 413.7 \cdot 10^3$ Pa and $\sigma_{st} = 606.7 \cdot 10^3$ Pa at $T = 30°$C [16]. The carnauba wax, which derives from the leaves of the carnauba wax palm (*Copernicia prunifera*), shows similar strong dependence on temperature. Its Poisson's ratio (ν) and density are respectively $\nu = 0.49$ and $\rho = 951$ kg/m^3. It has the highest values of mechanical properties compared to all other waxes tested by Craig et al. [16]. For this wax, $E = 1806.5 \cdot 10^6$ Pa, $\sigma_{pr} = 10.94 \cdot 10^6$ Pa and $\sigma_{st} = 18.775 \cdot 10^6$ Pa at room temperature ($T = 23°$C), while $E = 772.2 \cdot 10^6$ Pa, $\sigma_{pr} = 5.72 \cdot 10^6$ Pa and $\sigma_{st} = 9.03 \cdot 10^6$ Pa at $T = 37°$C.

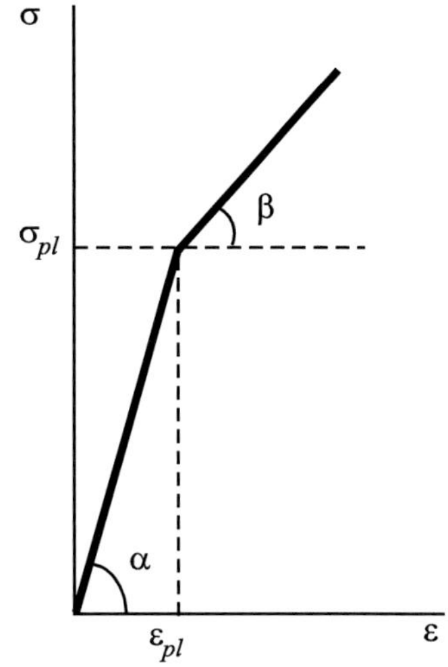

Fig. 4 An idealised bilinear stress-strain diagram. For stresses whose absolute values are less than the yield stress σ_{pl}, the material follows Hooke's law (1) with $E = \tan\alpha$. If the absolute value of stress is greater than the yield stress, the strain hardening is approximated by a line with another slope: $E_1 = \tan\beta$. If $\beta = 0$, then $E_1 = 0$. This means that the wax yields without further increasing of the external load

Craig et al. [16] found that for all waxes tested, the strain before fracture under compression ranged from 2.7 to 4.3%. The samples showed brittle fracture behaviour. This observation of brittle character of fracture of the crystalline waxes is in agreement with [14] experiments on the plant waxes. Shellhammer et al. [19] found that natural candelilla (derived from leaves of the candellilas *Euphorbia cerifera* and *E. antisyphilitica*) and carnauba waxes behaved similar to hard and elastic materials, following the 2% compressive strain input, while candelilla wax had more viscous behaviour than the carnauba wax. However, both waxes behaved more as elastic materials compared to the beeswax exhibiting considerably more viscous and less elastic behaviour.

There is a common claim that the carnauba wax is the hardest wax known. Indeed at corresponding temperature, it possessed the highest values of measured mechanical properties among all waxes tested by Craig et al. [16]. However, our previous experiments [14] showed that other plant waxes may have higher values of hardness and elastic modulus than the carnauba wax. By applying depth-sensing nanoindentation to the wax of the carnivorous plant *Nepenthes alata*, we found that at $T = 22°\mathrm{C}$, elastic modulus was about $E = 2.5$ GPa, while hardness was $H = 0.1$ GPa.

In mechanics of elastic-plastic contact, it is usually accepted that hardness is such an average stress under the indenter that the whole material under the indenter yields. Usually, this gives $H \approx (2.8 \div 3)\sigma_{pl}$, i.e.

$$\sigma_{pl} \approx 35 \text{ MPa}.$$

If one assumes that the waxes have a bilinear diagram (Fig. 4), whose reference points depend on the temperature, then in order to build this diagram, one needs to know its reference points. Using Hooke's law, one can calculate

$$\varepsilon_{pl} = \frac{\sigma_{pl}}{E} = \frac{35 \cdot 10^6}{2.5 \cdot 10^9} = 0.014.$$

Based on experiments of Craig et al. [16], one may assume that the strain at fracture is 2.7%, i.e. $\varepsilon_{st} = 0.027$. To draw the bilinear diagram, we need to know either the inclination of the linear hardening part (the tangent modulus E_1) or the stress at fracture σ_{st}. From the bilinear diagram, we obtain the following relation:

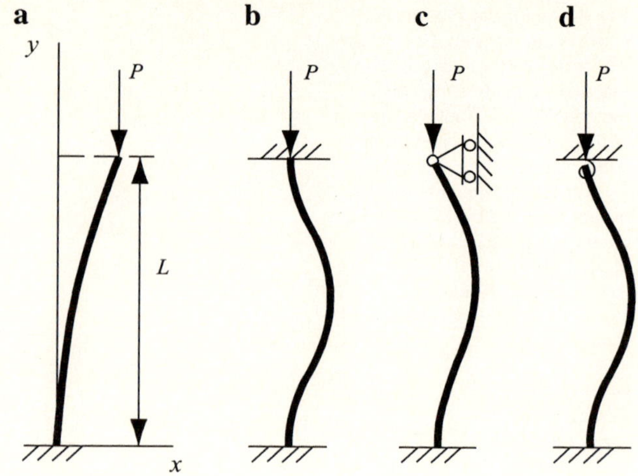

Fig. 5 Boundary conditions and corresponding effective lengths for compressed columns. A. A column fixed at the base and free at the top. B. A column with both ends fixed against rotation. C. A column fixed at the base and pinned at the top. D. A column fixed at the base and having an elastic torsion spring at the top. The torsion spring models adhesive interaction between an insect seta and the wax column. If spring stiffness is relatively high, then one may adopt $K \approx 0.6$

$$\varepsilon_{st} - \varepsilon_{pl} = \frac{\sigma_{st} - \sigma_{pl}}{E_1}.$$

We may assume that the ratio σ_{st}/σ_{pl} of the *N. alata* wax is the same as in carnauba wax, i.e. 1.575. Then, we have

$$\varepsilon_{st} - \varepsilon_{pl} = 0.013 = \frac{0.575\sigma_{pl}}{E_1},$$

and hence,

$$E_1 = 20.125 \cdot 10^6/0.013 = 1.548 \cdot 10^9 \text{ Pa}.$$

3 Buckling of elastic-plastic wax columns

In mechanics of materials, long and slender structures loaded axially in compression are called columns. Let us consider first the stability of tubular columns corresponding to the tubular wax crystals under compressive loading. If applied compressive load P increases, a slender elastic column will buckle at the critical load P_{cr}, which can be calculated according to the Euler formula:

$$P_{cr} = \frac{\pi^2 EI}{(KL)^2} = \frac{\pi^2 EI}{(L_{ef})^2}, \tag{2}$$

where K is an effective length factor, which depends on the support conditions, and the effective length is $L_{ef} = KL$. For a column with pinned ends, $K = 1$. If it is fixed at the base (built-in condition) and free at the upper end (Fig. 5A), $K = 2$. If it is fixed at the base and pinned at the top (Fig. 5C), then $K = 0.7$. If both ends are fixed (Fig. 5B), then $K = 0.5$.

It is reasonable to assume that one end of the wax crystal is fixed to the plant surface, i.e. it is in built-in condition, while the top end is in adhesive interaction with the insect attachment organ (adhesive pad). Insects always use capillary adhesion that is created between the pad and substrate due to secreting fluid (pad secretion) into a contact zone [20]. These conditions can be described as the column has an elastic torsion spring at this end (Fig. 5D). Hence, $L_{ef} = 0.6L$ may be taken as a good approximation for the effective length of the column.

The Euler formula (2) was derived under the assumption that the stress-strain relation is linearly elastic, and it is not valid, if the values of the compressive stresses in the column are over the yielding stress. Hence, one has to consider inelastic buckling. This theory was developed by Engesser and Jasinski (see [21]). According to it, a slender elastic-plastic column will buckle at critical load P_{cr}, which can be calculated according to the formula

$$P_{cr} = \frac{\pi^2 E_r I}{(KL)^2}, \tag{3}$$

where E_r is the reduced modulus of the column. For a rectangular cross section, this modulus can be calculated using the following formula:

$$E_r = \frac{4EE_1}{(\sqrt{E}+\sqrt{E_1})^2}.$$

It may be used as an approximation for a column with a circular cross section. For the above values of elastic moduli of the plant wax $E = 2.5$ GPa and $E_1 = 1.548$ GPa, one obtains

$$E_r = \frac{4 \cdot 2.5 \cdot 1.548}{(\sqrt{2.5}+\sqrt{1.548})^2} = 1.94 \text{ GPa}.$$

The above formulae are valid for slender columns. To characterise the behaviour of a column, one has to introduce the slenderness ratio $\lambda = L_{ef}/r$, where r is the radius of gyration of the cross section of the column in the plane of bending:

$$r = \sqrt{\frac{I}{A}}.$$

Here A is the area of the cross section and I is the moment of inertia of the cross-sectional area. For a circular cross section of diameter D, one has $A = \pi D^2/4$ and $I = \pi D^4/64$, and hence

$$r = \sqrt{\frac{I}{A}} = \sqrt{\frac{4D^4}{64D^2}} = \frac{D}{4}. \tag{4}$$

One has to realise the difference between the aspect ratio and the slenderness of the column. The former is a pure geometric characteristic that is equal to the ratio between the largest and smallest dimensions of the column, while the latter characteristic depends not only on the geometry represented by the length L and the radius of gyration of the cross section of the column r, but also on the conditions of the loading represented by the effective length factor K. For example, all cases A–D in Fig. 5 have the same aspect ratio *L/D*, while the values of the slenderness vary considerably from one case to another.

Since the Euler formula (2) is valid only for pure linear elastic materials, the critical stress, when one has to start using formula (3) instead of the Euler formula (2), is (see Fig. 6)

$$\sigma_{cr} = \frac{P_{cr}}{A} = \frac{\pi^2 EI}{AL_{ef}^2} = \sigma_{pl}.$$

And hence, the critical value of the slenderness ratio is

$$\lambda_c = \left(\frac{L_{ef}}{r}\right)_c = \sqrt{\frac{\pi^2 E}{\sigma_{pl}}}.$$

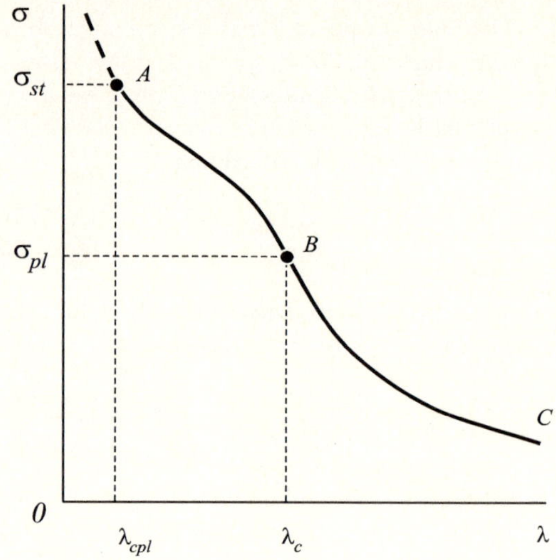

Fig. 6 Diagram of average compressive stress σ versus slenderness ratio λ for plant waxes. For the region BC ($\lambda \geq \lambda_c$), the critical stress is calculated according to Euler's hyperbola $\sigma_{cr} = \frac{\pi^2 E}{\lambda^2}$. For the region AB ($\lambda_c \geq \lambda \geq \lambda_{cpl}$), the critical stress is calculated according to the theory of buckling of elastic-plastic columns developed by Engesser and Jasinski $\sigma_{cr} = \frac{\pi^2 E_r}{\lambda^2}$. If $\lambda_{cpl} \geq \lambda$, then the compressed column fails by crushing of the material without buckling, because the theoretical buckling stress would exceed the strength limit σ_{st}

Substituting the above values for plant wax, we obtain

$$\lambda_c = \sqrt{\frac{9.87 \cdot 2.5 \cdot 10^9}{35 \cdot 10^6}} = 26.5.$$

Using the same arguments as above, one can calculate the values of the slenderness ratio for columns with $K = 0.6$ and the radius of gyration $r = D/4$ as

$$\lambda = 0.6L/r = 0.6L/(D/4) = 2.4L/D.$$

According to Gorb and Gorb [13], some plants, e.g. the European columbine *Aquilegia vulgaris*, common barberry *Berberis vulgaris*, white goosefoot *Chelidonium majus* and European plum *Prunus domestica* have tubular shape of epicuticular wax crystals. Crystals of the white-veined dutchman's pipe *Aristolochia fimbriata* have the same shape (Fig. 7). For these five plants, the average lengths of the crystal columns were 580, 730, 830, 580 and 6310 nm, respectively, and the diameters were 170, 160, 180, 260 and 92 nm, respectively. This gives the following values of the slenderness ratio $\lambda = L_{ef}/r$: 8.18, 10.94, 11.06, 5.35 and 164.6, respectively. Evidently, from all considered plant wax crystals, having tubular shapes, only those of *A. fimbriata* can buckle elastically, because the calculated values of the slenderness ratio for other plants are less than the critical value (26.5). Thus, crystals of these four plants are not long enough for Euler formula approximation.

Since formula (2) is valid only for stresses that are below the ultimate compressive stress of the material σ_{st}, the critical value of the slenderness ratio λ_{cpl}, when one has to stop using formula (2) of elastic-plastic buckling, is

$$\lambda_{cpl} = \left(\frac{L_{ef}}{r}\right)_{cpl} = \sqrt{\frac{\pi^2 E_r}{\sigma_{st}}}.$$

Substituting the above values for plant waxes, we obtain

$$\lambda_{cpl} = \sqrt{\frac{9.87 \cdot 1.94 \cdot 10^9}{55 \cdot 10^6}} = 18.66.$$

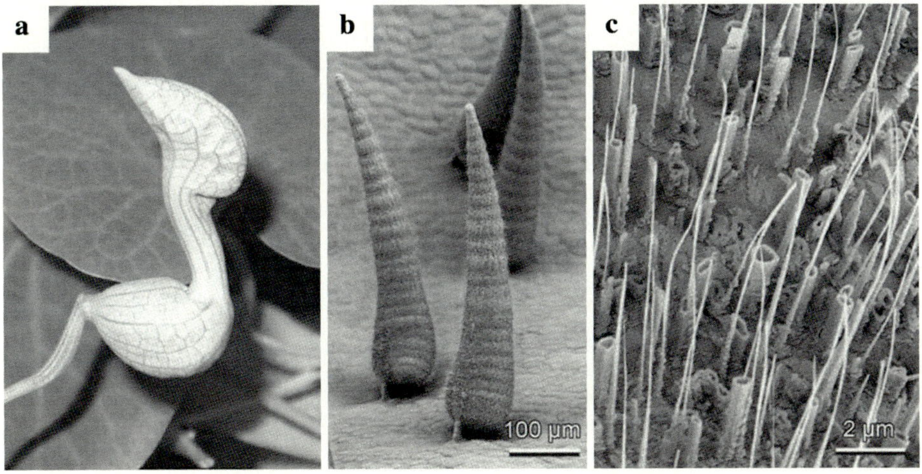

Fig. 7 **a** White-veined dutchman's pipe (*Aristolochia fimbriata*) flower. **b** SEM micrographs of trichomes on the inner surface of the flower trap. **c** Wax crystals covering the trichome surface

If one draws a diagram of average stress versus the slenderness ratio (Fig. 6) using the above formulae, then one may see that excluding *A. fimbriata* plant crystals that buckle in elastic regime, all other plants with tubular wax crystal columns with the above values of the slenderness ratio buckle neither in the elastic, nor in the elastic-plastic regime. In fact, such short columns cannot be broken by pure buckling mechanism. They could be broken only when the compressive stress is equal to the strength limit of the wax. Hence, one has to consider bending of the wax crystals loaded simultaneously by axial and orthogonal loads.

4 External loads applied to wax crystals by insects

4.1 Weights of flies

Let us estimate the loads that are applied to the wax crystals by insect feet. The mass of an insect (m) may be estimated as $4.8 \cdot 10^{-6} \leq m \leq 160 \cdot 10^{-6}$ kg, since the masses of syrphid flies range from $m_1 = 4.8 \cdot 10^{-6}$ kg in the hoverfly *Platycheirus angustatus* to $m_2 = 160 \cdot 10^{-6}$ kg in the pellucid hoverfly *Volucella pellucens* [22]. Because during walking, the weight is passed only through a half of the legs, the load on one foot is

$$\frac{4.8 \cdot 10^{-6} \cdot 9.81}{3} \leq G_1 \leq \frac{160 \cdot 10^{-6} \cdot 9.81}{3} \text{ N.}$$

Thus, the range for values of the load applied to a leg of a fly may be estimated as

$$15.7 \cdot 10^{-6} \leq G_1 \leq 523.2 \cdot 10^{-6} \text{ N.}$$

4.2 Structure of the fly foot

Each foot of a fly consists of two pulvilli with N_{set} setae on each of them. Pulvilli are the hairy adhesive pads situated at the end of legs (Fig. 8a). Pulvilli are covered by adhesive setae (Fig. 8b), sometimes termed tenent hairs, which serve to increase the actual area of

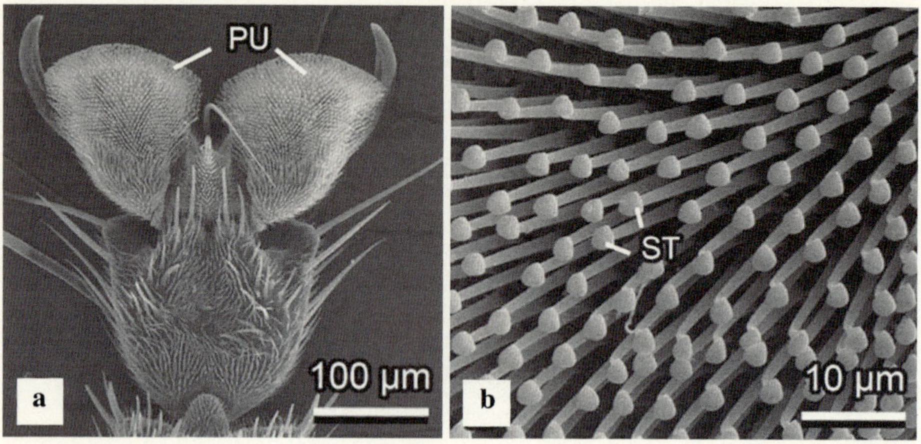

Fig. 8 SEM micrographs of the attachment devices in the hoverfly *Eristalis pertinax*. PU, pulvilli; ST, tenent setae

attachment to the surface. Nominal area of a puvillus for the above described representative flies is $5 \cdot 10^3\ \mu m^2$ and $30 \cdot 10^3\ \mu m^2$, while the area of a setal tip A_{set} is $4.5\ \mu m^2$ and $6.5\ \mu m^2$, respectively. The density of setae is respectively 55 and 50 per $1000\ \mu m^2$, hence one obtains respectively $N_{set} = 55 \cdot 5 = 275$ and $N_{set} = 50 \cdot 30 = 1500$ for a singe pulvillus.

4.3 The force applied to a plant wax crystal

The force applied to a seta can be calculated as

$$G_{seta} = \frac{G_1}{2 \cdot N_{set}}.$$

This gives $G_{seta} = \frac{15.7 \cdot 10^{-6}}{2 \cdot 275} = 28.5 \cdot 10^{-9}$ N and $G_{seta} = \frac{523.2 \cdot 10^{-6}}{2 \cdot 1500} = 174 \cdot 10^{-9}$ N, respectively.

The number of plant wax crystals under one seta is $N_{cryst} = A_{set} \cdot \rho_{cryst}$, where ρ_{cryst} is the density of crystals per unit area. For the plants under consideration, $0.17 < \rho_{cryst} < 29.1$ per μm^2.

Now one can estimate the force applied to a wax crystal:

$$G_{cryst} = \frac{G_{seta}}{N_{cryst}} = \frac{G_{seta}}{A_{set} \cdot \rho_{cryst}}.$$

Substituting the above values of density of crystals, the area of setal tip, and the force applied to setal tip, one obtains

$$\frac{28.5 \cdot 10^{-9}}{4.5 \cdot 29.11} \leq G_{cryst} \leq \frac{174 \cdot 10^{-9}}{6.5 \cdot 0.17}\ \mathrm{N}$$

or

$$2.2 \cdot 10^{-9} \leq G_{cryst} \leq 157.5 \cdot 10^{-9}\ \mathrm{N}.$$

One has to take into account that the attachment of an insect to a crystal is in the presence of some fluid that enhances capillary adhesion to surfaces. Normally, the force is applied under some angle to the crystal and hence, the crystals are loaded not only axially in compression, but also by forces acting transversely to the longitudinal axis.

5 Bending of tubular wax crystals

Let us consider a tubular wax crystal as a cantilever beam loaded at its free end by a concentrated load P. The bending moment M at the built-in end is PL. It is assumed that material of the beam follows Hooke's law up to the yield stress σ_{pl} and then yields plastically. The moment acting on the beam, when the stress in the beam at its farthest fibres reaches the yield stress, is called the yield moment M_{pl}. For a beam of square cross section of size D, one has

$$M_{pl} = \frac{\sigma_{pl} D^3}{6}.$$

For the wax crystals of the above considered five plants, one has $D = \{170, 160, 180, 260, 92\}$ nm respectively, and the corresponding yield moment has the following values:

$$M_{pl} = \{28.7, 23.9, 34.0, 102.5, 4.5\} \cdot 10^{-15}\ \mathrm{N\,m}.$$

If the transverse component of the applied force is F_H, i.e. this is the force applied transversely to the crystal axis, then the bending moment is $M(L) = F_H L$. If the plant surface has an angle θ with the horizontal plane, then $F_H = G_{\mathrm{cryst}} \sin\theta$. If $\theta = 90°$, then for $G_{\mathrm{cryst}} = 157.5$ nN and $L = \{580, 730, 830, 580, 6310\}$ nm, one obtains $M(L) = \{91, 115, 131, 91, 994\} \cdot 10^{-15}$ N m, and for $G_{\mathrm{cryst}} = 2.2$ nN, $M(L) = \{1.3, 1.6, 1.8, 1.3, 138.8\} \cdot 10^{-15}$ N m.

The simplest case of inelastic bending is ideal plastic bending, when it is assumed that material yields plastically under constant stress. According to this model, the plastic yield starts at the farthest from neutral axis fibres. As the bending moment increases, the plastic region extends farther inward. The beam reaches its ultimate moment-resisting capacity, when the whole cross section is in the plastic regime. The corresponding moment is called the plastic moment M_p. The ratio of the plastic moment for a beam to its yield moment is solely a function of the shape of the cross section and is called the shape factor f. For circular cross section $f = 1.7$, while for square cross section it is 1.5. In the latter case, one obtains for the above crystals

$$M_p = \{43, 39, 51, 154, 6.8\} \cdot 10^{-15}\ \mathrm{N\,m}.$$

Now we can compare these values of the plastic moment with the above values of the bending moment that could be created by the fly *Platycheirus angustatus* ($m_1 = 4.8 \cdot 10^{-6}$ kg). We see that this fly could not cause bending moment, which could break the crystals except very flexible crystals of *Aristolochia fimbriata*.

For the much heavier fly *Volucella pellucens* ($m_2 = 160 \cdot 10^{-6}$ kg), it is likely that the fly may break crystals, if the load is applied at an appropriate angle. For example, if the force is applied at an angle of 45° to the free end of the crystal, then $F_H = 0.707 G_{\mathrm{cryst}}$ and corresponding values of the bending moment are

$$M(L) = \{64.3, 81.3, 92.6, 64.3, 702.7\} \cdot 10^{-15}\ \mathrm{N\,m}.$$

These values are higher than the corresponding values of the ultimate plastic moment for all considered plant crystals except for crystals of *Prunus domestica* fruits.

6 Discussion

The possible mechanisms of the plant wax crystals fracture during adhesive contact formation between insect adhesive organs and plant surface have been examined, in order to explain the contamination of insect pads by wax crystals. Using the methods of mechanics of materials, one can obtain some quantitative conclusions. It is shown here that mechanisms of the wax crystal fracture may be rather different, depending on the slenderness ratio of the crystals. If the values of the slenderness ratio are higher than the critical values, one has to apply the formulae of either elastic or elastic-plastic buckling. For five plants species under consideration (*Aquilegia vulgaris*, *Berberis vulgaris*, *Chelidonium majus*, *Prunus domestica*, and *Aristolochia fimbriata*), the critical value of the slenderness ratio for elastic buckling is 26.5, while for elastic-plastic buckling it is 18.7. If the values of the slenderness ratio are lower than the critical values, one has to consider bending of the crystals, because rather short crystals may be bent under the weight of insects as elastic-plastic beams. The bending moment in the crystal at the fixed end is proportional to the crystal length. These conclusions are not only in agreement with the observations by Gorb and Gorb [15] that the contamination ability of insects is related to both the largest dimension and the largest aspect ratio of crystals, but they give also more quantitative insight into these observations. The above calculations show that from all considered plant wax crystals having tubular shapes, only crystals of *A. fimbriata* could buckle elastically. In this plant species, wax crystals on specialised weir trichomes in flowers contribute to trapping and retention of insects and consequently, to cross-pollination [23]. Failure behaviour of wax crystals of *A. fimbriata* (Appendix) is important, in order to capture extremely small brachyceran flies, associated with this plant.

However, since the calculated values of the slenderness ratio for most tubular plant crystals are lower than the critical value and, therefore, formulae of the elastic-plastic buckling cannot be applied, these crystals should be examined for their resistance to elastic-plastic bending. The above examination of possible fracture mechanisms of plant waxes showed that it is very unlikely that plant crystals could buckle, while it is likely that an insect may break crystals by elastic-plastic bending and contaminate attachment pads. Although the model of ideal plastic bending has been employed, the results will not change drastically if more sophisticated theories (e.g. bending of beams, whose material has the bilinear stress-strain diagram) are employed.

The above calculation showed that neither buckling, nor elastic-plastic bending mechanisms are applicable for description of the fracture in the wax crystals of *P. domestica* fruits. One possible explanation for this result is that our assumption that a crystal is connected to the surface of the fruit at all points of its cross section is not correct. Perhaps the connection is fulfilled through several slim and breakable contact points that are not strong enough to resist bending moment created by an insect. To answer these questions, additional experimental studies should be performed.

The present study has considered crystals of tubular shapes. In this case, one can employ equations described in classical books on mechanics of materials (see e.g. [21]). When wax crystals are not tubular, one has to employ more sophisticated formulae. However, the general approach will be valid also for crystals of various shapes.

Acknowledgements This work was fulfilled in the framework of the ADHESINT International Network. The authors are grateful to the Leverhulme Trust for financial support of the Network. This work was supported by the SPP 1420 priority program of German Science Foundation (DFG) "Biomimetic Materials Research: Functionality by Hierarchical Structuring of Materials" (project GO 995/9-1) to SG.

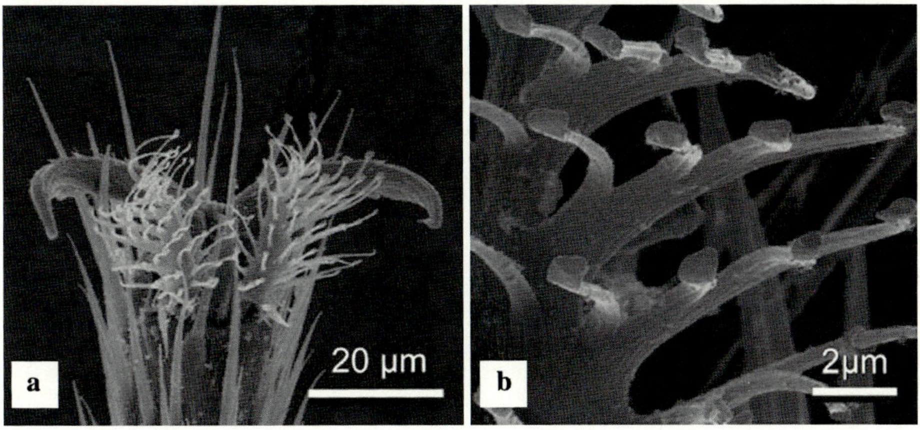

Fig. 9 SEM micrographs of the attachment devices in the brachyceran fly *Megaselia rufipes* (after Schuppert [25])

Appendix

For wax crystals of *Aristolochia fimbriata*, we have Young modulus $E = 2.5$ GPa, the length $L = 6310$ nm, and the diameter $D = 92$ nm. Substituting these values into the Euler formula (2), we obtain

$$P_{cr} = \frac{\pi^2 EI}{(KL)^2} = \frac{3.14^2 \cdot 2.5 \cdot 10^9 \cdot (3.14 \cdot 92^4 \cdot 10^{-36})/64}{(0.6 \cdot 6310 \cdot 10^{-9})^2}$$
$$= \frac{8.68 \cdot 10^{-20}}{14.33 \cdot 10^{-12}} = 6 \text{ nN}.$$

Hence, the wax crystals of this plant species can buckle elastically under external load.

However, the insects that interact with *A. fimbriata* are small brachyceran flies, e.g. those of the genus *Megaselia* (Phoridae) [24], whose mass $m = 0.3$ mg [25] is lower than that of the syrphid fly *Plactycheirus angustatus* considered in the above calculations. Since during walking, the weight is passed only through a half of the legs, the load on one foot of the *Megaselia* fly is

$$G_1 = \frac{0.3 \cdot 10^{-6} \cdot 9.81}{3} = \frac{2.94 \cdot 10^{-6}}{3} = 0.981 \cdot 10^{-6} \text{ N}.$$

For this fly, we have the number of setae per one pulvillus $N_{\text{set}} = 38$ (Fig. 9). The force applied to a setal tip is

$$G_{\text{seta}} = \frac{G_1}{2 \cdot N_{\text{set}}} = \frac{0.981 \cdot 10^{-6}}{2 \cdot 38} = 12.9 \cdot 10^{-9} \text{ N}.$$

Multiplying the area of one setal tip $A_{\text{set}} = 0.51\ \mu\text{m}^2$ on the density of crystals $\rho_{\text{cryst}} = 0.17\ \mu\text{m}^{-2}$, we obtain the number of plant wax crystals under one setal tip: $0.51 \cdot 0.17 = 0.087$. Hence, the total load acting on a crystal is

$$G_{\text{cryst}} = 12.9 \cdot 10^{-9}/0.087 = 148.4 \cdot 10^{-9} \text{ N}.$$

Since it is much higher than P_{cr}, the crystal failure in this plant species, caused by the associated *Megaselia* fly, is due to buckling.

Let us check if the crystal fracture in *A. fimbriata* also happens due to bending, as in the case in other considered plants. The maximum bending moment of a cantilever beam is at the built-in end, and under action of a concentrated load P it equals

$$M_{\max} = PL.$$

Assuming that the external force from pulvilli is applied at an angle of 45° to the free end of the crystal, then $P = 0.707G_{\text{cryst}} = 111.313$ nN, hence the maximum bending moment is

$$M_{\max} = PL = 111.313 \cdot 10^{-9} \cdot 6310 \cdot 10^{-9} = 702.388 \cdot 10^{-15}\ \text{N m}.$$

This value is much higher than the plastic moment M_p, which is equal to $6.8 \cdot 10^{-15}$ N m. Therefore, the wax crystals of *A. fimbriata* can be easily broken by *Megaselia* flies.

The above calculations allow us to conclude that the failure mechanism for wax crystals in *A. fimbriata* flowers is elastic-plastic bending, while these long and slender crystals first buckle elastically, and then they can be broken under the weight of insects.

References

1. S. Gorb, *Attachment Devices of Insect Cuticle* (Kluwer Academic, Dordrecht, 2001)
2. R. Beutel, S.N. Gorb, J. Zool. Syst. Evol. Res. **39**, 177 (2001)
3. S.N. Gorb, R.G. Beutel, Naturwissenschaften **88**, 530 (2001)
4. C.E. Jeffree, in Biology of the Plant Cuticle, ed. by M. Riederer, C. Müller (Blackwell, Oxford, 2006), pp. 11–125
5. R. Jetter, L. Kunst, L. Samuels, in Biology of the Plant Cuticle, ed. by M. Riederer, C. Müller (Blackwell, Oxford, 2006), pp. 148–181
6. G. Bianchi, in Waxes: Chemistry, Molecular Biology and Functions, ed. by R.J. Hamilton (The Oily Press, Bridgwater, 1995), pp. 177–222
7. W. Barthlott, N. Ehler, Trop. Subtrop. Pflanzenw. **19**, 367 (1977)
8. W. Barthlott, C. Neinhuis, D. Cutler, F. Ditsch, I. Meusel, I. Theisen, H. Wilhelmi, Bot. J. Linn. Soc. **126**, 237 (1998)
9. W. Barthlott, E. Wollenweber, Trop. Subtrop. Pflanzenw. **32**, 7 (1981)
10. W. Barthlott, in Scanning Electron Microscopy in Taxonomy and Functional Morphology, ed. by D. Claugher (Clarendon, Oxford, 1990), pp. 69–94
11. H. Bargel, K. Koch, Z. Cerman, C. Neinhuis, Funct. Plant Biol. **33**, 893 (2006)
12. E.V. Gorb, S.N. Gorb, in Scaling in Solid Mechanics, ed. by F.M. Borodich (Springer, Dordrecht 2009), pp. 243–252
13. E. Gorb, S. Gorb, Entomol. Exp. Appl. **105**, 13 (2002)
14. E. Gorb, K. Haas, A. Henrich, S. Enders, N. Barbakadze, S. Gorb, J. Exp. Biol. **208**, 4651 (2005)
15. E.V. Gorb, S.N. Gorb, in Ecology and Biomechanics—A Mechanical Approach to the Ecology of Animals and Plants, ed. by A. Herrel, T. Speck, N.P. Rowe (CRC Press, Boca Raton, 2006), pp. 147–162
16. R.G. Craig, J.D. Eick, F.A. Peyton, J. Dent. Res. **46**, 300 (1967)
17. M. Ito, T. Yamagishi, Y. Oshida, C.A. Munoz, J. Prosthet. Dent. **75**, 211 (1996)
18. E. Kotsiomiti, J.F. McCabe, J. Oral Rehabil. **23**, 114 (1996)
19. T.H. Shellhammer, T.R. Rumsey, J.M. Krochta, J. Food Eng. **33**, 305 (1997)
20. S.N. Gorb, Proc. R. Soc. Lond. B **265**, 747 (1998)
21. S.P. Timoshenko, J.M. Gere, Mechanics of Materials (Van Nostrand-Reinhold, New York, 1972)
22. S. Gorb, E. Gorb, V. Kastner, J. Exp. Biol. **204**, 1421 (2001)
23. B. Oelschlaegel, S.N. Gorb, S. Wanke, C. Neinhuis, New Phytol. doi: 10.1111/j.1469-8137.2009.03013.x (2009)
24. B. Rulik, S. Wanke, M. Nuss, C. Neinhuis, Flora **203**, 175 (2008)
25. J. Schuppert, Vergleichende Strukturuntersuchungen und Kontaktbildung des Haftsystems bei Zweifluegle rn (Diptera) (Saarbruecken, 2003)

2. Friction

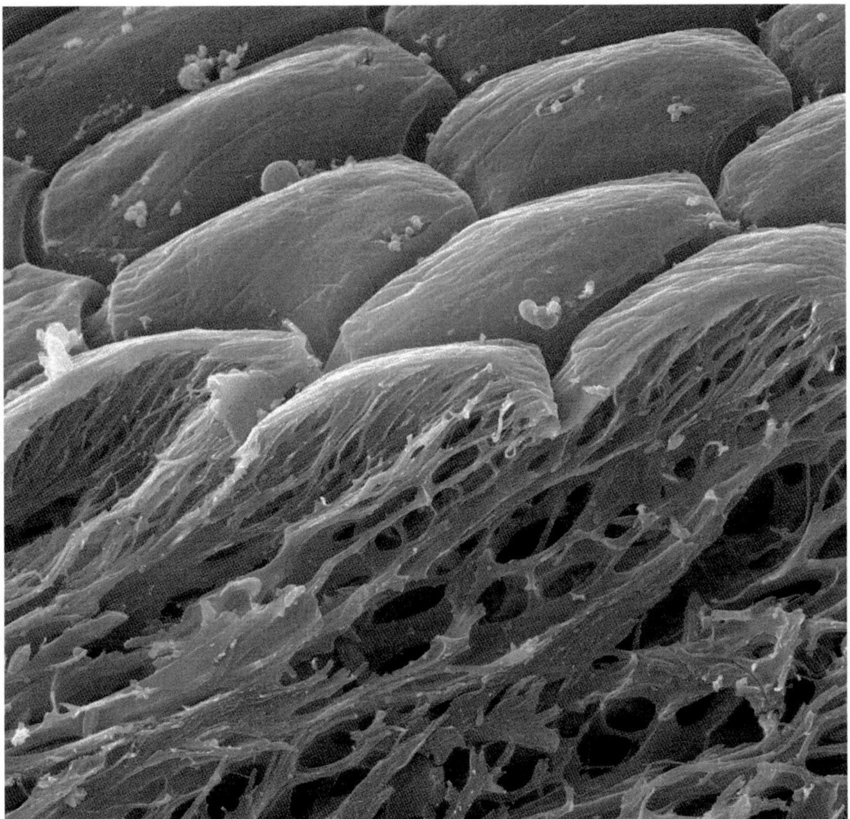

Tribology Letters 8 (2000) 1–7
DOI 10.1023/A:1019158428645

Microtribology of biological materials

M. Scherge · S. N. Gorb

Received: 11 November 1999 / Accepted: 7 January 2000

Abstract This paper presents a study on friction and indentation of attachment pads of the great green bush cricket, using a system involving a biological material and a flat surface. The pads are evolutionarily optimized frictional devices that are adapted to attach the insect to a variety of natural surfaces. The attachment pads of both living and dead insects were tested and evaluated using several analysis techniques. These techniques included friction analysis by microtribometry, microindentation, scanning electron microscopy (SEM), histologic sectioning, and shock freezing. Microtribological analysis showed that the friction between pad and surface is anisotropic. The anisotropy increases with increasing normal force. Microindentation revealed that the pads respond very flexibly to applied loads leading to an optimized morphology with respect to contact area. All results make clear that the microtribological properties are a perfect combination of surface structure, lubrication and active interface control.

Keywords Biotribology · Attachment pads · Insects · *Tettigonia Viridissima L*

Introduction

Biological motion systems are highly adapted micromechanical units. Due to the long period of evolution these systems exhibit an optimum combination of mechanical (elasticity) and tribological (friction, adhesion, wear) properties. Therefore, biological systems are excellent candidates as model systems for the design of new materials and for micromechanical applications. A rich source of design principles are to be found among insects [1]. The main advantage of biological micro systems is their strong ability to adapt to different substrate profiles as well as their excellent attachment/detachment performance [2, 3]. However, natural frictional systems have been poorly studied with respect to their mechanical as well as their structural properties. In contrast, existing industrial macro scale

M. Scherge (✉)
Institut für Physik, TU Ilmenau, PF 100565, D-98684 Ilmenau, Germany

S. N. Gorb
Max-Planck-Institut für Entwicklungsbiologie, PO Box 21 09, D-72011 Tübingen, Germany

applications for materials with particular frictional properties ranging from velcro zippers to tire material have been extensively studied. Materials scientists still continue to look for new types of materials with controlled friction and adhesion [4, 5]. This is of special interest for the introduction of new types of adhesives. Research on the microtribological properties of biological systems also needs to be conducted. Knowledge gained from these examinations will not only extend our insights concerning friction, adhesion, lubrication and wear, but these results may also prove indispensable in real-world applications, for instance in micro electro mechanical systems (MEMS).

In this paper, the tribological properties of attachment pads (*euplantulae*) of the great green bush cricket *Tettigonia viridissima L.* (*Orthoptera: Tettigoniidae*) were examined. The functional units have dimensions similar to real micromechanical applications. However, the physiological activity makes the measurement of the mechanical properties a challenging process. Dead substrates deteriorate in just a few minutes due to the loss of water, while testing living substrates requires sophisticated preparation techniques.

The paper is organized as follows. In sections 2.1 and 2.2 the biological tribosystem and the tester are described. Section 3 focuses on indentation as well as friction as a function of normal force and sliding direction. The following discussion tries to interpret the data by means of biological structure information obtained by shock-freezing and fracturing followed by scanning electron microscopy (SEM).

Testing conditions

The biological system

Nature offers an overwhelming variety of different surfaces, ranging from rough materials like bark to very smooth materials, such as the leaves of many plant species. The attachment system is very well adapted to all of these surfaces [6, 7]. To hold onto a rough surface, the insect uses small claws at the end of each leg (see figure 1(a)). To adhere to a smooth surface, however, the insect uses attachment pads. These pads have a cross-sectional shape of a hemisphere [8, 9]. The micro structure of the attachment system is composed of hexagonal outgrowths of highly flexible material, which is designed to adapt quickly to the surface profile, figure 1(b). If no pressure is exerted, then the surface is patterned by regular hexagons with a diameter of about 5 μm. Each hexagon is separated from its neighbor by a trench with a width of about 1 μm resulting in a ratio of area covered by hexagons to the area of hexagon plus trenches of 7/8. The inner construction is characterized by a dendrite-like structure branching in the proximity of the surface, figure 1(c). The main rods of the dendrites contact the surface at an angle α of about 70°. The remaining volume is filled up with a liquid (haemolymph). The insects have an average weight of 1 g. Since each of the six legs has two major pads the normal force can be calculated as 800 μN for each attachment pad.

The tester

In friction experiments, the interaction is measured between the attachment pad and a flat silicon surface. The leg containing the pad (tarsus) was attached to a sample holder that is oscillated in tangential direction over a length of 30 μm by a piezo transducer (x-piezo) at constant velocity (30 μm/s). The silicon sample – a 5 × 5 mm piece cut from a silicon wafer (Ra = 0.3 nm) – was attached to a double leaf spring as shown in figure 2. Friction

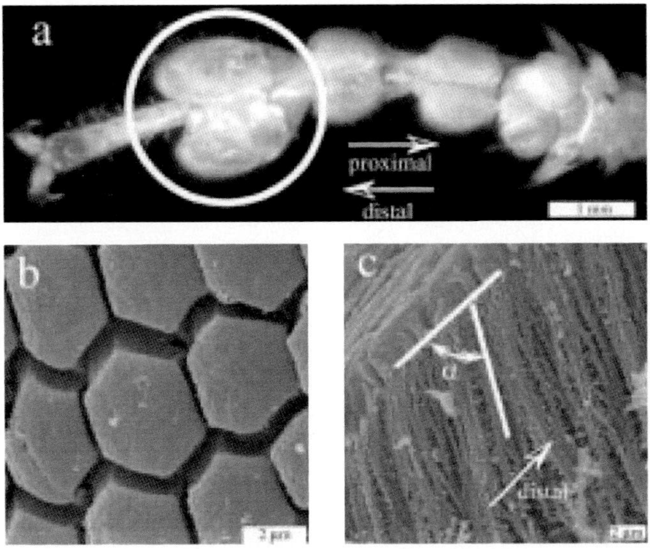

Fig. 1 *Tettigonia viridissima*. (a) Last segment of the leg. The circle denotes the attachment pad studied here. The descriptions distal and prox- imal describe the direction towards the body of the insect. (b) Top view of the surface micro structure of an attachment pad. (c) Cross section of a fractured attachment pad (*euplantula*). In the upper left corner the surface micro structure can be seen

causes the spring to deflect and is measured by a single-beam laser interferometer. The measured length is then multiplied by the spring constant to obtain the tangential force. The laser interferometer has a resolution of 1 nm. Therefore, with a spring constant of 50 N/m, a force range from 50 nN to 1.5 mN can be covered. The normal force, adjustable in a range between 500 nN and 1 mN, is set by a second piezo transducer (z-piezo) acting in the normal direction. This deflection was calibrated by means of a fiber optical length sensor placed above the spring. During the tests the normal force was set by expanding the z-piezo. The spring constant in the normal direction was 100 N/m. By means of mechanical micro positioners, both samples are first approached manually. The final engagement is achieved by expanding the z-piezo (see [10, 11]). When the first tangential deflection of the spring is detected, contact is established with an average error of 50 nN (due to mechanical noise). The normal force can be increased by increasing the control voltage of the piezo transducer.

To immobilize the insect, its body was attached to the holder connected to the x-piezo using adhesive tape. The body was positioned upside down (back was attached), and oriented in a horizontal direction along the x-axis. The leg tarsus was firmly connected to the sample holder using wax. The silicon surface was positioned parallel to the lower sample holder in a way that limited the contact of the leg to the attachment pads as encircled in figure 1(a). In this condition, the whole insect was moved together along with its leg section by periodic expansion and retraction of the x-piezo. The force acting on the leg was set appropriate to the natural level of mechanical interaction. It is therefore assumed that the pads were not damaged during the tests.

The complete tester is housed in an environmental chamber where both temperature and relative humidity can be kept constant. All experiments were carried out in air at the

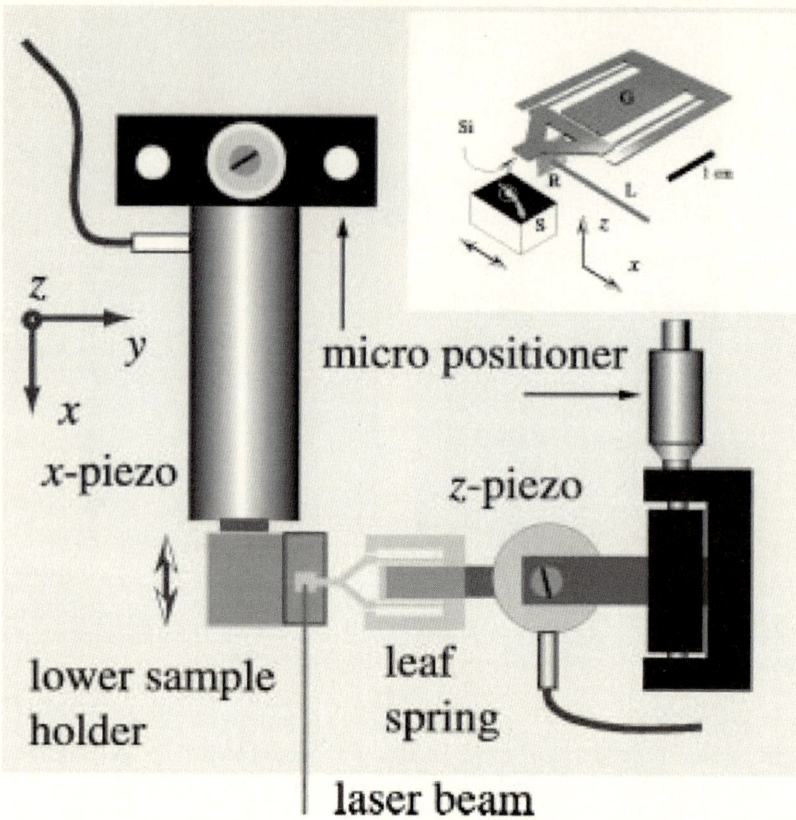

Fig. 2 Schematic of the tester. S: sample, i.e., the last segment of the leg (the body of the insect is not shown.). G: glass spring consisting of the glass body and two 100 μm wide beams serving to detect the deflection. The Si sample is located at the lower side of the triangular section of the spring. R: reflector for the laser beam. L: laser beam. The laser beam is reflected back to the interferometer head where it interferes with itself

prevailing temperature of 25 °C and a relative humidity of 58%. To ensure minimum mechanical interference, the tester and the chamber are positioned on a concrete block with a mass of 3 tons.

Experimental

Sample ageing

One of the most important issues for testing biological systems is sample ageing. When tests are performed with parts cut off from the insect body, the samples undergo rapid structural changes caused by the loss of water. To investigate this effect the mass of the detached parts was measured during ageing. Pieces of the leg – with and without pads – were severed. Judging from the morphology it was expected that parts containing the attachment pads will evaporate a higher amount of water than other parts of the leg. Therefore, the evaporation was measured for both types of material. The initial mass was

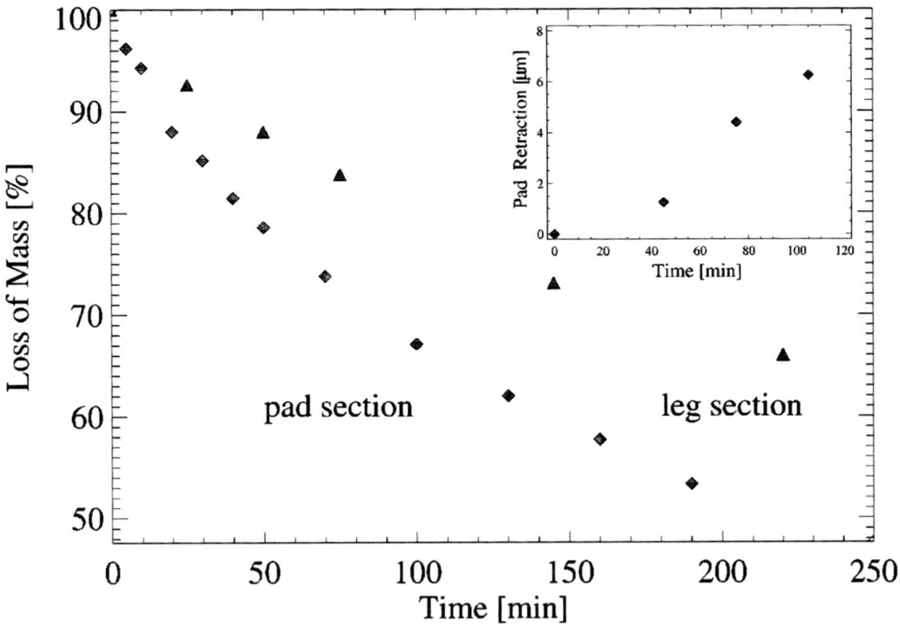

Fig. 3 Loss of mass versus time for a section of the leg that contains the attachment pad and one with no pads. The inset depicts the retraction of the upper surface of the pad due to the loss of water

taken as 100%. Figure 3 shows the results for the parts with and without pads. It can be seen that after 1 h about 25% of the weight is lost. This must have a drastic impact upon the micro structure of the pad, and the hexagonal structure must change significantly. The evidence was found in friction tests performed with cut legs. The samples were approached during oscillation. Contact was achieved when the spring showed tangential deflection. As a measure of the structural change the point of engagement was taken. Thus, the change in piezo expansion corresponds to the distance of pad retraction. In other words, when at the beginning of the experiment the engagement occurred at an expansion z_0 and at a later time the engagement was possible at z_1, with $z_1 > z_0$, then $\Delta z = z_1 - z_0$ is equal to the retraction of the pad. As shown in the inset of figure 3 the pad retracts due to the loss of water. A retraction by 6 μm results in a change in the normal force of 600 μN. This change in the normal force with time leads to different results and makes comparison and analysis difficult. As a consequence, all following tests were carried out with living insects. However, it has to be pointed out that freshly cut legs showed the same tribological behavior as the legs of living insects. This shows that the mechanics in the pad is not controlled by the insect. Therefore, intrinsic tribological properties of the material can be distinguished from the action of the insect.

Indentations

The initiation of friction is always accompanied by the indentation of the soft pad material. From our experience resulting from the sample handling as well as from observations of the insect it was deduced that the pad material must be also very flexible in its response to mechanical interaction. This assumption was strongly supported by shock-freezing

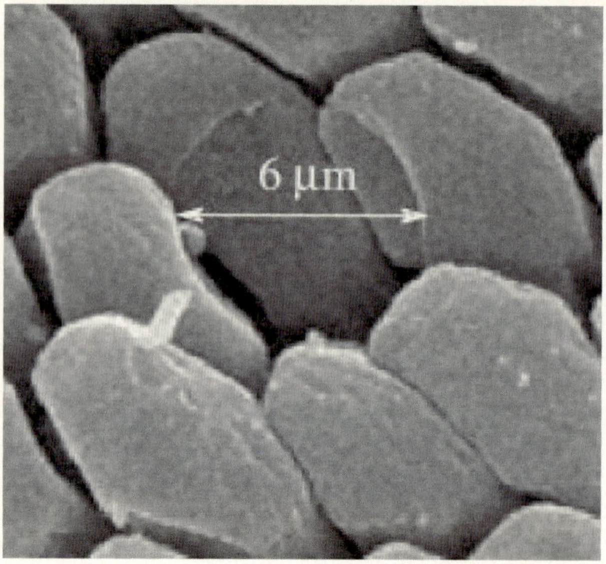

Fig. 4 Indentation of a single bump. The pad completely recovers after the retraction of the bump

experiments. The pad was pressed against a matrix of circular bumps with a diameter of 6 μm and a height of 5 μm. In this state the pad was frozen, sputtered with gold and imaged by SEM. Figure 4 clearly shows the indent. The test was repeated several times without freezing. After multiple ondentations the pad was again frozen and imaged. The SEM images showed a completely unaffected hexagonal pad surface proving that the pad material is extremely flexible and capable of replicating a given roughness profile. To measure the indentation of the whole pad, the deflection of the spring in z-direction was taken, see inset of figure 5. Pad and silicon sample were approached without oscillation. The position of the silicon sample before indentation is z_0, this means that the spring is not deflected. With established contact the silicon sample is moved to z_1, thus $\Delta z = |z_1 - z_0|$ is equal to the deflection of the spring in z-direction. As a reference the experiment was repeated with a silicon ball (similar dimension as the pad) where the indentation can be neglected due to hardness. The difference between this curve – a straight line according to Hooke's Law – and the curve of the soft material is equal to the indentation. It can be followed that with larger indentation the contact area increases also. Further tests revealed that the contact area changed with elapsing time. When the indentation is stopped, an exponentially decreasing relaxation of Δz was measured. This decrease is stronger at higher applied loads. As a consequence the area of contact changes. The pad material shows a higher stiffness for quick compression. During slow compression the material starts to creep leading to the increase of contact area. These effects obviously limit the use of the elastic contact models since the visco-elastic material properties cannot be neglected.

Load dependence and friction hysteresis

As described in section 2.2, each friction test was initiated by the approach of the Si sample while oscillating the pad. If dry samples are approached, then the friction force increases gradually. For the pad samples, however, a distinct jump to contact was observed. In order to have a reference experiment the pad was again replaced by the silicon

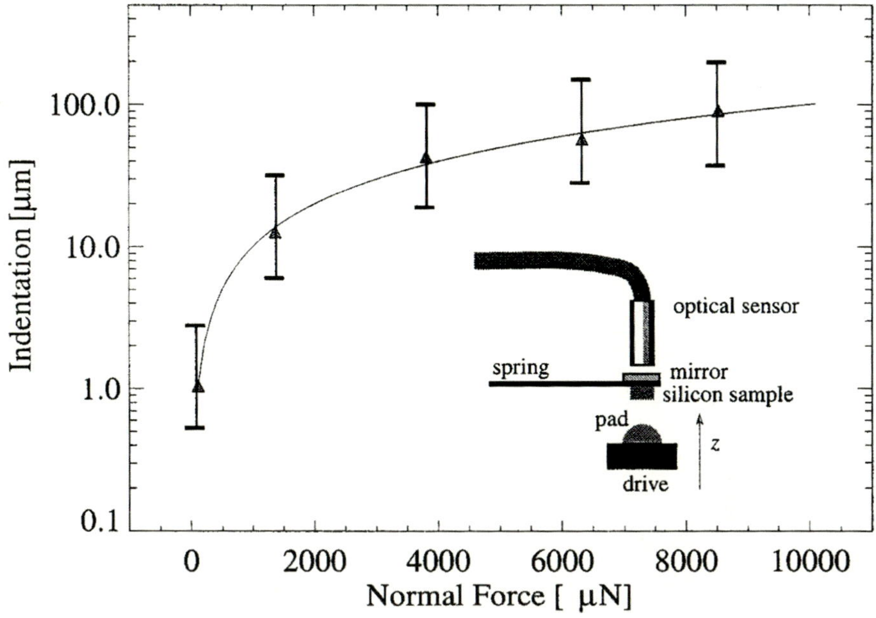

Fig. 5 Indentation into the soft pad surface. Inset: experimental setup. The distance between the sensor head and the mirror is measured optically. A deflection can only be measured when the samples are in contact

ball. Silicon was the subject of numerous experiments [12]. It is known that hydrophilic silicon (SiO_2) has a water coverage between 0.3 and 0.6 nm [13]. For those samples also a distinct jump into contact was measured and the friction force was approximately of the same magnitude. This indicates that liquid bridges between pad and surface induce capillary forces. Studies investigating the chemical composition of insect footprints and cuticula lipids [14] indicate that the liquid that covers the pads is a hydrocarbon with hydrophobic surface properties. Therefore, capillary action due to water can be ruled out as origin of the jump into contact, since the hydrocarbon surface does not wet in order to form a capillary neck. However, since the secretion is also a liquid capillary action based on hydrocarbons is conceivable. This can be seen in figure 6. Besides the jump into contact also the negative normal force (pulling force) underlines the concept of capillarity. The capillary neck breaks suddenly at a normal force of $-450\ \mu$N.

Due to the action of capillarity the friction force versus load curves exhibit a clear hysteresis. Figure 6 demonstrates that the hysteresis has larger values for proximal than for distal motion. Proximal means that the pad is pulled towards the center of the insect body, whereas distal refers to the opposite direction. The inset of figure 6 shows the force curve as a function of time. F_d and F_p denote the static friction force in the distal and proximal directions. Clearly visible is the increase of dynamic friction force in the distal direction, whereas in the proximal direction the force remains constant. For low normal forces (up to 600 μN) the indentation experiments indicate a small contact area leading to low friction forces in both distal and proximal direction. For loads higher than 600 μN the samples snap into permanent contact. This means that the pad and the silicon sample move synchronously without sliding. Accordingly, friction cannot be measured for the given range of

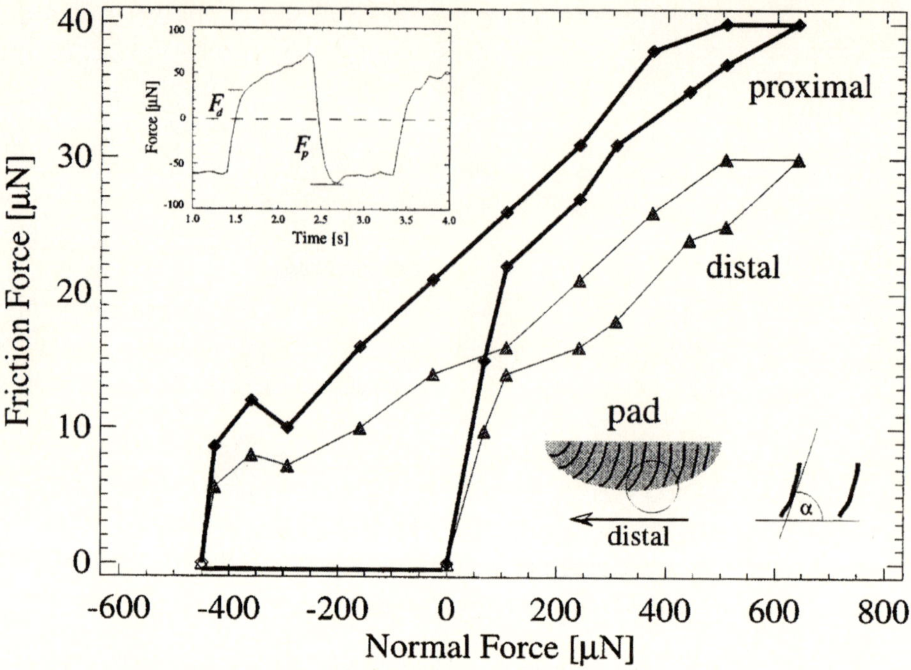

Fig. 6 Friction force versus normal force. Inset upper left corner: Tangential force versus time indicating proximal (p) and distal (d) direction. Inset lower right corner: Pad with dendrite system. These measurements were reproducible with an error of about 5 μN

tangential force. The more area involved in the interaction, the stronger is the capillary force. Capillarity contributes to the normal force and increases the friction force.

The formation of the permanent contact always occurred instantaneously for the case of a smooth Si sample. To slow down the contact formation in order to visualize the process the smooth surface was replaced by the patterned Si surface, as introduced above. The pattern consists of equally spaced circular bumps with a surface finish like regular wafer silicon. After exceeding 600 μN in normal force the propagation of the contact front was observed. This is shown in figure 7. The normal force was set slightly above 600 μN. As a result the contact was formed gradually indicated by increasing friction. For higher normal forces the time to establish the contact decreased significantly.

Discussion of the results with respect to biological structure

The indentation experiments proved that the pad material has a very soft constituency. This always allows the insect to flatten out the pads, in order to achieve maximum contact, even for the upside down position. The shock-freezing experiments provided useful information concerning material properties. It was shown that the inner structure of the pad is composed of flexible rods that split into even more flexible branches close to the inner surface. We therefore assume that this construction is capable of adjusting the pad perfectly to micro roughnesses at different length scales.

The anisotropy of the friction force can also be explained by the inner structure of the pad. If the pad does not contact the substrate, the main rods of the dendrite system contact

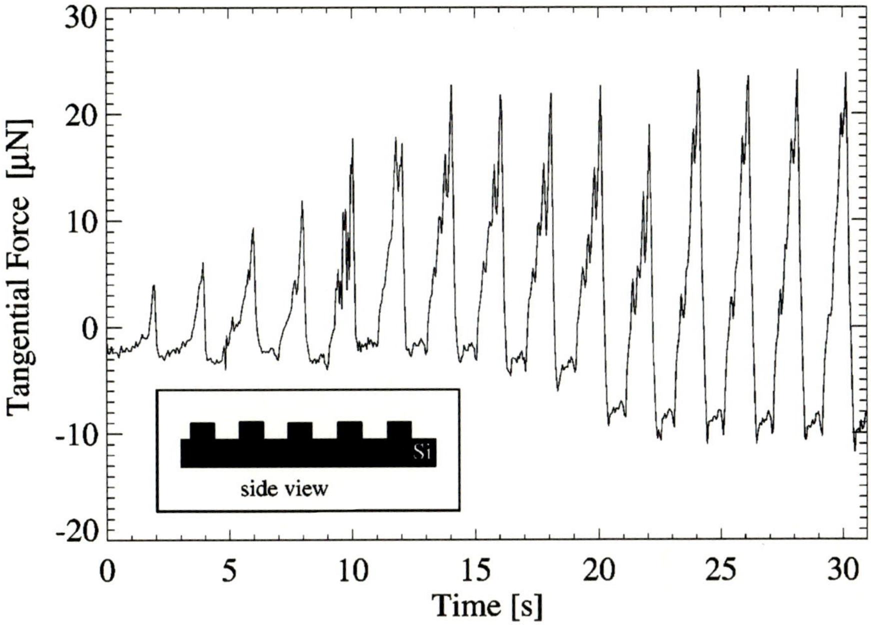

Fig. 7 Increasing friction force due to the time-dependent formation of the contact between the pad and a structured Si sample

the inner surface at an angle of 45°–70°, see inset of figure 6. When the pad is pressed against a smooth substrate, α decreases to 50°–100°. Since the inner structure of the pad has an initial predetermined slope in one preferred direction, the applied normal force will cause an even stronger deformation of the rods in the same direction (decreasing α). The leg movement in the proximal direction increases the rod deformation further. We assume that the brush-like structure of the rods is designed to maintain a constant contact area between pad and substrate. If the pad is moved distally, a critical force is needed to overcome the rod stiffness and to tilt the rods in the other direction. This particular energy-loss mechanism could be the reason for the increasing friction. In addition to mechanical interlocking during sliding, energy is consumed by the bending of the rods. Since the friction force is smaller in the distal than in the proximal direction, we assume that the bending of the rods decreases the contact area by deformation of the hexagonal outgrowth.

The experiments involving pad and patterned Si surface seem to suggest that after exerting a sufficiently large normal force the pad tries to achieve optimum contact resulting in the firm attachment of the insect. Permanent contact was established after exceeding $F_n > 600\ \mu$N. This is approximately equal to the force that is necessary to balance gravity in the upside-down position.

The pads have the function comparable to that of a clutch. Friction under low normal forces leads to the tilting of the flexible structures inside the pad to achieve an optimized contact area. The tilting mechanism is supported by the action of the normal force and the proximal leg motion which is always present as a component of gravity. After exceeding the critical load of about 600 μN permanent contact is established and sliding vanishes. The attraction that is necessary to balance gravity results from a combination of adhesion provided by capillarity due to a liquid secretion and the formation of a self-mating mechanical contact between the pad and any irregular surface.

Conclusions

The experimental work with biological samples has shown that the tests have to be performed with living substrates. Due to evaporation, water is lost, rendering dead samples unsuitable. Friction and indentation experiments made clear that the interpretation of the data must result from the combination of physical, chemical and biological approaches.

It was shown that living organisms – in our case the great green bush cricket – are perfectly optimized frictional systems. The inner architecture of the pads provides stability and simultaneously extreme flexibility. This ensures the adaptation to a variety of surfaces, which are unpredictable for mobile insects. Due to the particular orientation of stiff components in the bio-composite, the material is optimized to produce maximum friction.

The key issues of the perfect mechanical functionality of bio-systems is the combination of surface structuring and tribology. It was shown that the surface is covered by sticky hydrocarbon layer. It was also shown that the insect varies the contact area between pad and surface to optimize the contact force. The release mechanism, however, is still unknown and needs to be addressed in further studies.

One of the main differences to technical systems is the permanent self-maintenance of the tribological contact in bio-systems. This study was thought as a first approach to discover the micro-mechanical properties of biological systems. The outcome of further studies could be used to foster new design philosophies for mechanical microsystems (MEMS).

Acknowledgement This work was supported by the German Ministry of Education and Research (BMBF, BioFuture 0311851).

References

1. S.N. Gorb, *Technische Biologie und Bionik*, 4. Bionik Kongress, Mü nchen, 1998 (Gustav Fischer Verlag) pp. 185–189
2. G. Walker, A.B. Yule and J. Ratcliffe, J. Zool. 205 (1985) 297
3. N.E. Stork, J. Exp. Biol. 88 (1980) 91
4. D.G. Crevoisier, P. Fabre, J.-M. Corpart and L. Leibler, Science 285 (1999) 1246
5. T.P. Russell and H.C. Kim, Science 285 (1999) 1219
6. L.M. Roth and E.R. Willis, J. Exp. Biol. 119 (1952) 483
7. J.W. Arnold, Int. J. Insect Morphol. Embryol. 3 (1974) 317
8. W. Hennig, *Handbuch der Zoologie* (Berlin, 1973) pp. 1–335
9. E.H. Slifer, Annals of the Entomological Society of America 43 (1950) 173
10. M. Scherge and J.A. Schaefer, *Tribology Issues and Opportunities in MEMS* (Kluwer Academic, 1998) pp. 529–538
11. M. Scherge and J.A. Schaefer, Tribol. Lett. 4 (1998) 1, 37
12. M. Scherge, X. Li and J.A. Schaefer, Tribol. Lett. 6 (1999) 3, 215
13. M. Heim, R. Steigerwald and R. Guckenberger, J. Struct. Biol. 119 (1997) 212
14. A. Kosaki and R. Yamaoka, Jpn. J. Appl. Entomol. Zool. 40 (1996) 47

J Comp Physiol A
DOI 10.1007/s00359-006-0157-y

ORIGINAL PAPER

Friction and adhesion in the tarsal and metatarsal scopulae of spiders

Senta Niederegger · Stanislav N. Gorb

Received: 10 November 2005 / Revised: 25 January 2006 / Accepted: 15 May 2006

Abstract Friction and adhesion forces of the ventral surface of tarsi and metatarsi were measured in the bird spider *Aphonopelma seemanni* (Theraphosidae) and the hunting spider *Cupiennius salei* (Ctenidae). Adhesion measurements revealed no detectable attractive forces when the ventral surfaces of the leg segments were loaded and unloaded against the flat smooth glass surface. Strong friction anisotropy was observed: friction was considerably higher during sliding in the distal direction. Such anisotropy is explained by an anisotropic arrangement of microtrichia on setae: only the setal surface facing in the distal direction of the leg is covered by the microtrichia with spatula-like tips. When the leg is pushed, the spatula-shaped tips of microtrichia contact the substrate, whereas, when the leg is pulled over a surface, setae bend in the opposite direction and contact the substrate with their spatulae-lacking sides. In an additional series of experiments, it was shown that desiccation has an effect on the friction force. Presumably, drying of the legs results in reduction of the flexibility of the setae, microtrichia, spatulae, and underlying cuticle; this diminishes the ability to establish proper contact with the substrate and thus reduces the contact forces.

Keywords Legs · Locomotion · Friction · Adhesion · Attachment

Introduction

Many walking spiders, such as those from the group Mygalomophae, as well as representatives of Thomisidae, Salticidae, Lycosidae, and Ctenidae are able to climb smooth, steep, vertical surfaces or even to walk on a smooth ceiling by using attachment organs

S. Niederegger · S. N. Gorb (✉)
Evolutionary Biomaterials Group, Department Arzt, Max Planck Institute for Metals Research, Heisenbergstr. 3, 70569 Stuttgart, Germany
e-mail: s.gorb@mf.mpg.de

S. Niederegger
Institute of Forensic Medicine, Friedrich Schiller University of Jena, Fürstengraben 23, 07743 Jena, Germany

called scopulae, located on the ventral side of the pretarsus near the claws, tarsus, and metatarsus (Homann 1957; Rovner 1978; Foelix 1996; Gorb 2001). The scopulae are composed of thousands of hairs called setae, each covered with numerous fine cuticular extensions (microtrichia, setulae) terminating in spatula-shaped tips (Roscoe and Walker 1991). The functional role of the tarsal and metatarsal scopulae is not fully explained. It has been previously reported that they are used for prey capture (Rovner 1978; Foelix et al. 1984). Some spiders also use them for attachment while resting on smooth vertical surfaces in the downward position (personal observation). Measurements of the frictional and adhesive forces of tarsal and pretarsal scopulae, performed in the present study, might aid in explaining their function.

Many insects and lizards have developed similar mechanisms to attach themselves to a variety of substrata by using hairy spatula-bearing structures (Gorb 2001; Gorb and Beutel 2001). For representatives of many insect groups, the presence of secretion in the contact area between spatulae and substrate has been previously reported (Ishii 1987; Kosaki and Yamaoka 1996; Eisner and Aneshansley 2000; Gorb 2001; Stadler et al. 2001). The fluid secretion results in the enhancement of attractive capillary forces (Langer et al. 2004). Hairy attachment devices of spiders and geckos seem, however, not to be supplemented by secretory substances. Therefore, they were often reported as dry adhesive devices relying on the van der Waals interactions between spatulae and substrate (Autumn et al. 2000; Kesel et al. 2003). Recent measurements of adhesive forces of single gecko spatulae using atomic force microscopy (AFM) demonstrated a strong dependence of adhesion on the humidity of the environment (Huber et al. 2005b). Similar effects have been previously predicted for the spider scopulae (Homann 1957). It is generally accepted that in the apparently "dry" adhesive devices of geckos and spiders, attachment forces are similar for the fresh and dried samples. However, no rigorous comparison between fresh and dried samples has been previously done. In this study, we compare contact forces measured on spider tarsi and metatarsi in both fresh and dry conditions.

Recent experimental work using AFM has demonstrated that the attachment system of the spider *Evarcha arcuata* has a safety factor (relationship between the maximum adhesive force and the body mass) of 160 (Kesel et al. 2003, 2004). However, this estimation is only true when all spatulae (which are responsible for the contact formation) form the contact with the substrate. Previous estimates of the number of contacting spatulae in the gecko toe provided numbers in the range of 0.1–0.3% of the total number of spatulae (Autumn et al. 2000; Peressadko and Gorb 2004a). For the spider system, such an estimate is only possible by comparing the forces at the local scale (spatulae) with those at the global scale (entire leg). In the present study, the measurements at the global scale were carried out.

Previous authors showed that a slight proximal pull of the individual setae is necessary to generate strong adhesion in the gecko (Autumn et al. 2000). A proximal pull is also an important functional element for proper contact formation in the attachment devices of flies (Niederegger and Gorb 2003). In our previous studies, we have shown that there is a friction asymmetry in the smooth attachment devices of the grasshopper *Tettigonia viridissima*, because of the preferred orientation of fibres in the pad material (Gorb and Scherge 2000). Since spider setae and their subdivisions also have a preferred orientation, strong frictional anisotropy may be predicted for this attachment system as well.

In this study, the adhesive and frictional properties of tarsal and metatarsal scopulae were estimated using an experimental set-up, which allowed 3D fine manipulation of samples and 2D force measurement with a resolution of 10 μN. The distribution, orientation, and structure of the adhesive setae were observed by using scanning electron microscopy (SEM). Contact formation between the spatulae and a glass substrate was visualised by using cryo-SEM and reflection light microscopy. The following five questions were asked. (1) What is the relationship between friction and adhesion in the attachment system of spiders? (2) Is there an asymmetry in the friction force in regard to sliding of the leg in the proximal and distal direction? (3) Are there differences in friction force between fresh and dried samples? (4) How do the morphology and orientation of setae relate to attachment forces? (5) How is contact established between seta and substrate?

Materials and methods

Animals

Aphonopelma seemanni (Cambridge 1897) (Araneae, Theraphosidae) lives in Costa-Rica, where it inhabits long burrows in stone-sand soil of the tropical rainforest. Adult females were obtained from pet shops (Stuttgart, Germany). The animals were kept in plastic terraria and fed with insects and small mammals.

Cupiennius salei (Keyserling 1877) (Araneae, Ctenidae) lives in Central America, where it dwells on plant surfaces in the tropical rainforest. This species belongs to the category of wandering or hunting spiders. *C. salei* does not build webs in order to catch prey. It typically spends the day hidden in retreats on the plants on which by night it hunts, courts a partner or moults (Barth 2002). The animals were obtained from the laboratory stock of the Department of Neurobiology and Behavioural Sciences of the University of Vienna (Austria), kept in plastic terraria and fed with insects.

The individual spiders were weighted using a precision balances (Mettler Toledo AG204 DeltaRange).

Photo- and video-recordings

To obtain information about spider locomotion on the smooth substrate, the spiders were placed in a large glass Petri dish (diameter = 190 mm) and photographed with a digital camera (Nikon Coolpix 995) (Fig. 1a, c). The clinging abilities of spiders were observed while the Petri dish was continuously tilted from 0° (horizontal plane) to 90° (vertical plane) in <60s. Spider behaviour during substrate tilting was video-recorded with a camcorder (Sony DCR-PC4E).

Morphology and ultrastructure

Morphological studies on freshly ablated legs were performed with a dissecting microscope (Leica MZ 12.5). For SEM observations, air-dried tarsi and metatarsi were mounted on metal stubs, sputter-coated with an 8-nm thick layer of gold–palladium, and viewed in a SEM (Hitachi S-800) at 20 kV. The terminology used in this work for the description of cuticle outgrowths is based on Richards and Richards (1979).

To evaluate the thickness of the cuticle in the dorsal and ventral areas of the tarsi, desired pieces were severed from fresh legs, embedded in water-soluble tissue mounting medium (Tissue-Tek®) and frozen at −20°C in a cryomicrotome (Leica). The 8–10-μm thick cross sections were prepared using a steel knife, transferred on Superfrost® glass slides, dried on a heating plate at 30–40°C, mounted in the water-soluble mounting medium (Moviol), and observed in a phase contrast light microscope (Zeiss Axioplan).

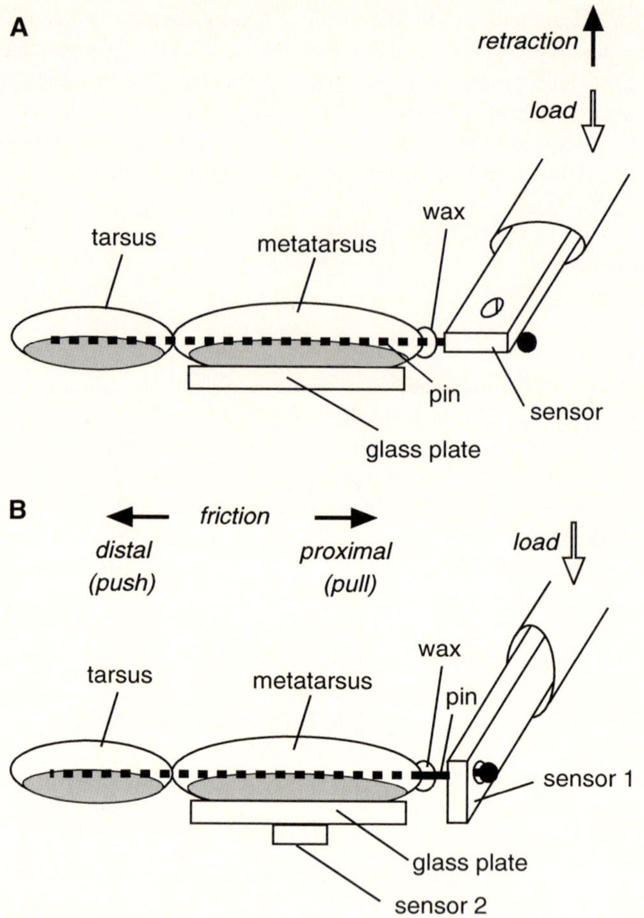

Fig. 1 Experimental set-up for force measurements. **a** Adhesion test: the force transducer (*sensor*) registered both load and adhesive forces while pressing onto (*white arrow*) and retracting the segments from (*black arrow*) a glass surface without sliding. **b** Friction test: two force transducers simultaneously registered normal (*sensor 2*) and friction (*sensor 1*) forces generated after applying a load (*white arrow*) and sliding the segments over a glass plate in both distal and proximal directions (*black arrows*)

Observation of the contact formation between setae and smooth glass

Light microscopy

Tarsi and metatarsi were cut off legs autotomised by a living animal, and attached with their dorsal parts by melted wax to a glass holder mounted on a mobile stage of a micromanipulator. The ventral surface of the tarsi was brought in contact with a glass surface, and the interface of the contact between setae and glass was observed using a light microscope (Olympus BH2-UMA). Tarsi and metatarsi were moved over the glass surface in both proximal and distal directions. Digital photographs of the leg ventral surface during sliding over the glass surface were made by the video camera built into the interference microscope. In the microscope, the dark areas corresponded to the spatulae making true contact with the glass surface. The total size of these areas was measured from digital images using Sigma Scan Pro 5.0 software for Windows.

Cryo-SEM

Setae attached to the glass substrate were analysed using a Hitachi S-5200 in-lens field emission SEM with a Gatan 525 cryo stage. After sliding small pieces of the tarsal scopula over a glass surface, the glass with the hairs attached to it was immediately transferred to liquid nitrogen. The frozen sample was mechanically attached to a brass holder, rotary sputter-coated with platinum-carbon (5 nm) at a temperature of −100°C in a Balzers 300 freeze etching device (Bal-Tec, Liechtenstein) and viewed in a frozen condition at 4.0 kV in the cryo-SEM.

Force measurements

For the adhesion measurements, severed tarsal and metatarsal segments were pronged on metallic pins (diameter = 1 mm), fixed with a melted mixture of bees wax and rosin gum. The pin was then firmly connected with a metal screw to a force sensor (load cell force transducer, 10 g force range, Biopac Systems Ltd., Santa Barbara, CA, USA) (Fig. 1a). Setae of the pretarsal scopula were shaved off to prevent their influence on the results obtained for the tarsal segment. The force transducer was connected to a stepping motor (DC3001R with controller MS314, World Precision Instruments, Sarasota, FL, USA) to control the movements of the leg segments. The leg segments were pressed with their ventral surfaces to a glass cover slip surface and then retracted to the original position. Both the load force and adhesion force were simultaneously recorded.

For the friction measurements, severed leg segments were fixed to the first force transducer as shown in Fig. 1b. To measure the normal force, generated by the segments pressed on a glass surface, a cover slip was connected to a second force sensor. The experiment was run as follows. (1) The ventral surface of the leg segment (metatarsus or tarsus) was pressed against the glass surface. While placing the segment onto the glass plate, not only did the normal force increase but also the friction force sensor monitored a slight signal change, due to the leverage caused by the pin (this force was set to 0 for further data processing). (2) Then the leg segment was pulled in the proximal direction of the leg for 3s corresponding to a distance of 600 μm. (3) After pulling, the leg segment was pushed in the opposite direction for about 3s (600 μm). (4) Finally, the leg segment was retracted from the glass surface.

The severed tarsi and metatarsi were measured in the fresh condition, immediately after amputation, and then in a dry condition, up to 10 days after the first experiment. For statistical analyses, Sigma Stat 2.03 software (SYSTAT Software Inc., Chicago, IL, USA) was used. For data comparison, a *t*-test was employed if the data were normally distributed; otherwise a Mann–Whitney rank sum test was applied.

Results

Morphology and climbing ability

Both, *A. seemanni* and *C. salei* are able to climb steep smooth glass surfaces. While *C. salei* (Fig. 2a) almost always rested on vertical surfaces with the prosoma pointing downward, *A. seemanni* (Fig. 2e) prefers resting on horizontal substrata and climbs a wall only in case of emergency. *A. seemanni* was not able to adhere to smooth glass surfaces at angles above 90°, whereas *C. salei* could walk on the ceiling.

Tarsi and metatarsi of both species bear scopulae on their ventral sides. Although both species belong to different lineages, and *A. seemanni* is considerably heavier (17.9 ± 3.0 g, $N = 3$) than *C. salei* (3.3 ± 1.0 g, $N = 5$), the scopulae of both species have similar morphology (Fig. 2b, f). They are composed of setae (Fig. 2c, g) bent distally at their tips (Fig. 3d). The concave part of the bent seta is covered with fine cuticular outgrowths called setulae (Kesel et al. 2003, 2004) or microtrichia (Richards and Richards 1979) (Fig. 2d, h) so that each seta resembles a broom or a toothbrush.

The entire seta with its microtrichia is able to adapt to a surface profile, because of its hierarchical organisation. The most terminal level of this hierarchy comprises the spatulae-like ends (terminal elements) of the microtrichia, which have a degree of flexibility, as can be seen by examining shock-frozen attached spatulae (Fig. 3b, d). During contact formation single spatulae are pulled onto the substrate (Fig. 3b). The next basal hierarchical level is the microtrichia, small cuticular outgrowths 2–4 μm long and 0.1–0.3 μm wide (Fig. 3b–d). The microtrichia of *A. seemanni* and *C. salei* have basically same structure, but the spatulae of the much larger spider *A. seemanni* are about half as big (0.13 ± 0.02 μm^2) as the spatulae of *C. salei* (0.30 ± 0.05 μm^2). Setae, which are several hundreds of micrometres long, belong to the next hierarchical level (Fig. 3a, c). The deflection of setae in the proximal direction is limited by the asymmetric socket at its base (Fig. 3f). Lateral deflection is also limited in some setae due to lateral plate-shaped structures forming a guide groove (Gorb 2001). The fourth, most basal level of the

Fig. 2 Spider species studied: **a–d** *C. salei*; **e–h** *A. seemanni*. **a, e** Spiders resting on a tilted glass plate. **b, f** Claws, pretarsal scopula (claw scopula), and tarsal scopula in contact with the glass surface viewed in a binocular microscope. **c, g** Tip of a single seta, SEM. **d, h** Microtrichia with spatulae at their tips, SEM. *CL* claws, *CS* claw scopula, *L1–L4* left legs from the front to back, *PP* pedipalps, *R1–R4* right legs from the front to back, *SE* seta, *SP* spatulae, *TS* tarsal scopula

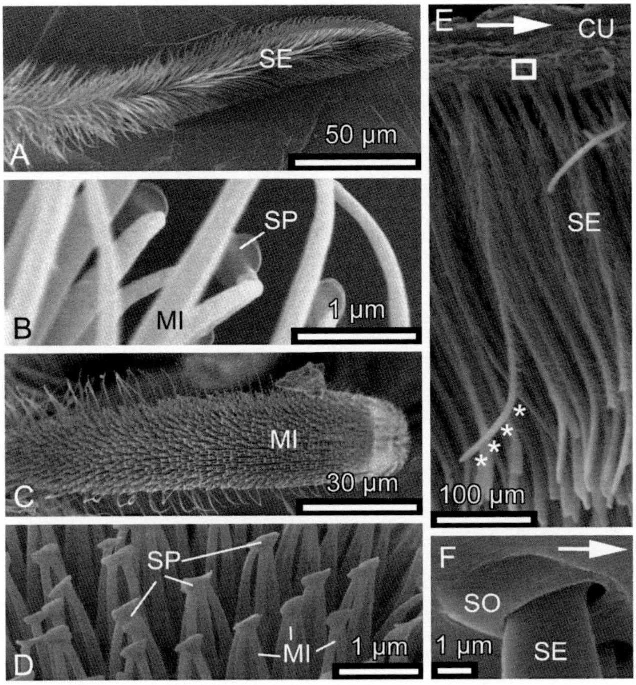

Fig. 3 Setae of *A. seemanni*. **a** Seta attached to the glass surface by its distal side, cryo-SEM. **b** Spatulae in contact with the glass surface, cryo-SEM. **c**, **d** Ventral view of the seta after being frozen in contact with the glass cover slip (some spatulae are stuck together), cryo-SEM. **e** Orientation of the setae within the tarsal scopula, SEM. **f** Socket of a seta, SEM. *Arrows* indicate distal direction; *asterisks* indicate distal surface of the seta covered with spatula-bearing microtrichia. *CU* cuticula, *MI* microtrichia, *SE* seta, *SO* socket, *SP* spatulae; *white rectangle* in (**e**) indicates area enlarged in (**f**)

hierarchy is the underlying cuticle. The setae are located on considerably thinner cuticle (*A. seemanni*—thickness = 48.3 µm, SD = 2.2, $n = 10$, ventral wall of the leg segment) than the regular cuticle of the dorsal side of the segment (*A. seemanni*—84.2 µm, SD = 5.3 µm, $n = 10$). In *A. seemanni*, the cuticle underlying the tarsal scopula is thinner and thus more flexible than that under the metatarsal scopula.

Adhesion and friction

When the leg segments were loaded perpendicularly to the glass surface and retracted back without sliding along the surface, almost no adhesive force was detected (Fig. 4a). In the friction experiments, the segments slid over a glass surface, and friction force was recorded together with the normal force (Fig. 4b). After orientation of the leg segment and glass surface parallel to one another, an initial loading was achieved by pressing the leg to the glass plate. Then the leg was oscillated proximally (pulling)—distally (pushing). During pulling, the friction force remained unchanged or dropped slightly below the initial loading value. In pushing, the frictional signal always strongly rose above the initial loading value. When the segment retracted from the glass surface, both friction and normal forces returned to zero (Fig. 4b). Friction force plotted versus normal force demonstrated a correlation between both forces on load (Fig. 4d); however, friction in the pushing mode was considerably higher than in the pulling mode at the same loads (Fig. 4c, d). The friction coefficient, which is the friction force divided by normal force ($\mu = F_f/F_n$), decreased slightly with increasing load. It was considerably higher in pushing than in pulling mode for the same loads (Fig. 4e).

Average friction coefficients were compared for *A. seemanni* and *C. salei*. For *A. seemanni*, the average friction coefficient was 0.12 in pulling, and 0.66 in pushing. For *C. salei*, these values were 0.20 and 0.62, respectively (Fig. 4c). In both species, the friction coefficient decreased when the legs became dry (Fig. 4f; Table 1) (push—*A. seemanni*, $P < 0.001$, $t = 10.2$, $df = 116$, *t*-test; *C. salei*, $P = 0.01$, $t = 2.633$, $df = 99$). In *A. seemanni*, the friction coefficient decreased from 0.15 to 0.08 in pulling and from 0.74 to 0.58 in pushing. In *C. salei*, friction coefficients decreased after drying out in pulling from 0.27 to

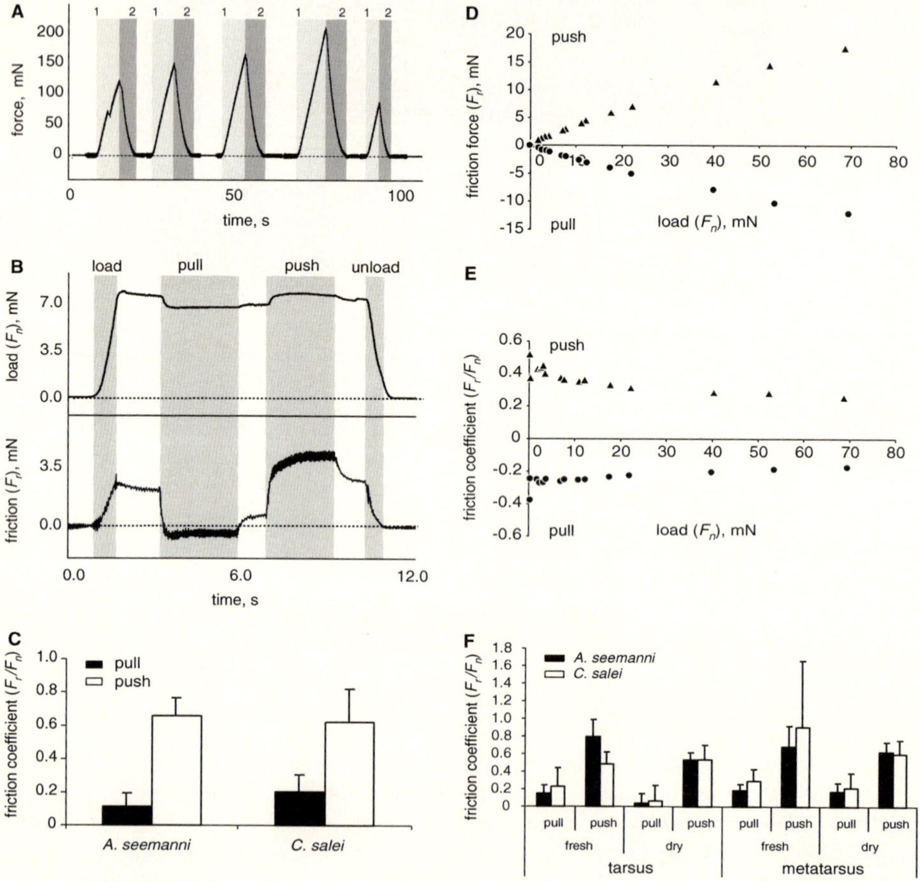

Fig. 4 Force measurements. **a** *C. salei*, fresh tarsus, typical recording of adhesive force. The leg was pressed against the glass surface (loading process, *1—light grey area*) and retracted (unloading process, *2—dark grey area*). **b** *C. salei*, fresh tarsus, typical friction force recording. The leg was loaded against the glass surface (*load*), pulled (*pull*), pushed (*push*), and retracted from the surface (*unload*). **c** Comparison of the friction coefficient during pushing and pulling in both species. **d** *C. salei*, fresh tarsus, friction force versus normal load during leg pushing and pulling. **e** *C. salei*, fresh tarsus, friction coefficient during leg pushing and pulling. **f** Comparison of friction coefficients at pulling and pushing in fresh and dry tarsi and metatarsi in both species

Table 1 Comparison of friction coefficients for tarsi and metatarsi of *A. seemanni* and *C. salei* for dry and fresh conditions and for pulling and pushing movements

	Tarsus				Metatarsus			
	Fresh		Dry		Fresh		Dry	
	Pull	Push	Pull	Push	Pull	Push	Pull	Push
A. seemanni	0.136	0.803	0.001	0.545	0.166	0.666	0.157	0.622
C. salei	0.239	0.457	0.052	0.542	0.302	0.906	0.212	0.590

	Tarsus + metatarsus				Tarsus + metatarsus		
	Fresh		Dry		Fresh + dry		Total
	Pull	Push	Pull	Push	Pull	Push	
A. seemanni	0.151	0.735	0.079	0.584	0.115	0.659	0.387
C. salei	0.271	0.682	0.132	0.566	0.201	0.624	0.413

0.13 and in pushing from 0.68 to 0.57 (Table 1). Although friction at pulling was higher in *C. salei*, the differences were significant only in the fresh segments (fresh—$P < 0.001$, $T = 13.622$, $n = 126/96$; dry—$P = 0.052$, $T = 10.501$, $n = 111/105$). Friction at pushing was higher in the tarsi of *A. seemanni* and in the metatarsi of *C. salei*. However, the difference between species was significant only for legs in the fresh condition (fresh—$P < 0.001$, $T = 8.590$, $n = 126/96$; dry—$P = 0.058$, $T = 12.264$, $n = 111/105$).

In *A. seemanni*, the forces measured for tarsi were higher than for metatarsi. In *C. salei*, however, the forces were higher for metatarsi than for tarsi (Fig. 4f). When all data for tarsus and metatarsus were pooled together, there was no significant difference between the friction forces measured in the two spiders (*A. seemanni* 0.39, *C. salei* 0.41; $P = 0.586$, $T = 58{,}701$, $n = 236/268$).

Experiments on the estimation of the contact area between setae and glass substrate in *C. salei* demonstrated that the proportion of spatulae in contact is considerably higher in pushing than in pulling (Fig. 5a, b, d, e). The same part of the leg generated about 28,000 μm^2 of contact area in pushing (Fig. 5c) and only 5,700 μm^2 in pulling (Fig. 5f). Without sliding the contact area between setae and substrate was almost zero.

Discussion

Functional principles of hairy attachment devices

Adhesive hairs on legs were reported for insects, such as flies (Bauchhenß 1979; Gorb 1998; Niederegger et al. 2002), bugs (Gillett and Wigglesworth 1932), beetles (Stork and Evans 1976), as well as for lizards, such as geckos (Ruibal and Ernst 1965), anoles (Irschick et al. 1996), and skinks (Williams and Peterson 1982). Whereas in most insects there is only one hierarchical level of outgrowths on the attachment organ, in spiders a second level of outgrowths is present (Fig. 2c, d, g, h). For geckos, even more levels have been reported (Ruibal and Ernst 1965; Hiller 1968, 1975). One of the morpho-functional principles of hairy attachment devices is the division of the contact area into many single areas (Scherge and Gorb 2001). According to the Johnson–Kendall–Roberts contact

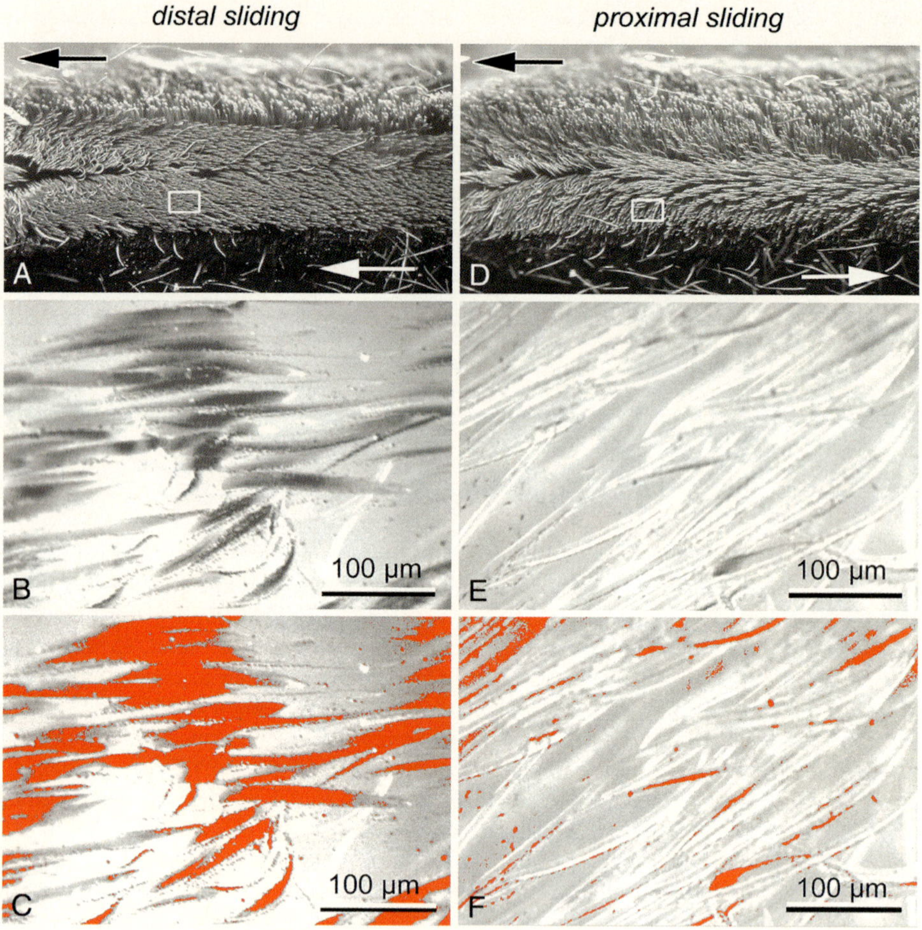

Fig. 5 Tarsal setae of freshly ablated legs of the spider *C. salei* during sliding in contact with a smooth glass surface in the distal (**a–c**) and proximal (**d–f**) directions. **a**, **d** View in the binocular microscope. **b**, **e** Selected part of the tarsus (*white rectangles* in **a** and **d**) in the reflection light microscope during distal (**b**) and proximal (**e**) sliding. *Black areas* correspond to the sites of contact between spatulae and glass surface. **c**, **f** Contact areas (*in red*) measured for distal (**c**) and proximal (**f**) sliding. *Black arrow*, distal direction; *white arrow*, sliding direction

theory, such multiple contacts generate stronger adhesion than one big contact (Johnson et al. 1971; Arzt et al. 2003).

A second functional principle of hairy attachment devices is the prevention of crack propagation. In a multicontact system, a crack will be stopped at the level of a single contact element and a new crack has to be formed at each subcontact during detachment. Formation of new cracks results in energy dissipation, which prevents contact breakage. Indeed, the behaviour of crack propagation on the patterned adhesive films leads to a significant enhancement of fracture energy (Chung and Chaudhury 2005).

A third principle is the very small effective elastic modulus of the fibre array, which is of fundamental importance for adhesion on smooth and rough substrates. During pull-off, the fibres may elongate many micrometres before the force in the fibre is high enough to break the bond to the substrate. Since the spring constant associated with a long (curved) fibre is very small, the displacement may be very large, leading to a very large effective surface energy. Therefore, it has been suggested that strong adhesion may result from "long bonds" rather than from "strong bonds" (Persson 2003).

Spatula-shaped tips and contact formation

A fourth functional principle is related to the spatula-shaped tips of the setae, which are responsible for proper contact formation with the substrate due to the low bending stiffness of the plates without or with a minimum of a normal load (Persson and Gorb 2003). While sliding over the surface, thin plate-like spatulae may easily make contact with the surface by adapting to the surface profile and replicating surface irregularities of certain length scales (Persson and Gorb 2003). Spatula-shaped contact elements of spider microtrichia strongly resemble those of gecko and insect tenent hairs (Foelix et al. 1984; Foelix 1996; Gorb 2001; Kesel et al. 2003, 2004; this study).

It has been previously shown for the tokey gecko *Gekko gecko* (Autumn et al. 2000; Gao et al. 2005; Huber et al. 2005a) and for the fly *Calliphora vicina* (Niederegger et al. 2002) that a slight proximal sliding motion is necessary for preloading setae with spatula-like tips. Adaptability effects have been previously described for the plate-like tips of insect hairs (Stork 1983; Niederegger et al. 2002; Peressadko and Gorb 2004b). However, in spiders, the terminal plates are considerably thinner than in insects, and therefore even more adaptable to an irregular surface profile.

In the present study, no adhesive forces were recorded in the tarsal and metatarsal scopulae of spiders when setae were just loaded onto a glass surface and retracted back without sliding (Fig. 4a). This indicates that the hairy attachment system of spiders also requires sliding to establish an intimate contact between the flat spatula-shaped tips and the surface. Our experiments demonstrate that distal sliding increases the contact area stronger than proximal sliding does, and results in a corresponding enhancement of contact forces. We may assume that adhesion measured after distal sliding of tarsal and metatarsal segments will be not of zero value. However, these experiments were not performed in this study.

Structural details of scopulae and behaviour

Our previous observations show that *C. vicina* flies usually move or sit on vertical surfaces with the head pointing upward or sideward in regard to the gravitational force, whereas *C. salei* spiders usually prefer sitting with the prosoma pointing downward. Such a posture results in pushing the front and partly the second legs against the substrate due to gravity. Spider setae are slightly bent in the distal direction (Fig. 3e). Also, the shape of the setal socket prevents an excessive bending of the seta in the distal direction (Fig. 3f). The part of the seta facing distally is covered by microtrichia with spatula-shaped tips, which contact the substrate, when the tarsal or metatarsal scopula is pushed (Fig. 6b). When the scopulae are pulled over the surface, the setae are strongly bent in the opposite direction (no stopper is present at the proximal side of the setal socket) and contact the substrate with cylindrically shaped surfaces of their spatulae-lacking microtrichia (Fig. 6a). This action results in the contact reduction between the setae and substrate.

The exoskeletal cuticle underlying the setae is thinner and therefore more flexible compared to that of the dorsal wall of tarsi and metatarsi. This may additionally contribute to the adaptability of setae to various substrata. The importance of the flexibility of the

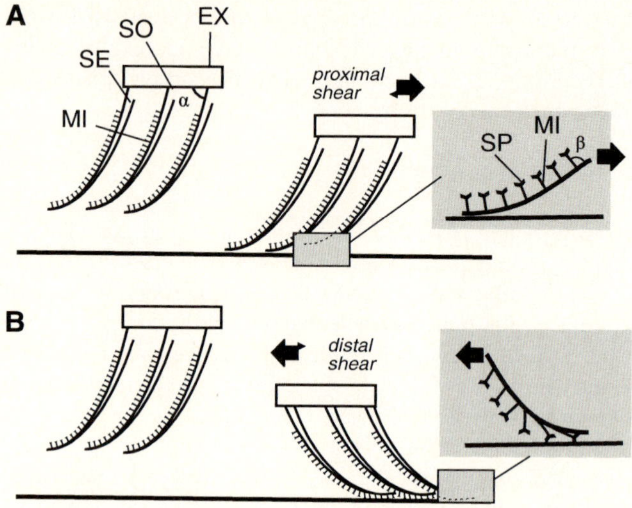

Fig. 6 Effect of the leg sliding direction on the orientation of setae and spatulae in tarsal and metatarsal scopulae. **a** Proximal pulls result in the alignment of setae with their proximal sides to the substrate. This side does not bear microtrichia with spatula-shaped tips. In this case the contact area is reduced, because of the cylindrical shape of the setal shaft contacting smooth glass. **b** Distal pushes result in the orientation of setae with their distal sides to the substrate. This side bears microtrichia with spatula-shaped tips. Thin flattened spatulae provide contact formation between setae and substrate. *EX* exoskeleton, *MI* microtrichia, *SE* seta, *SO* socket, *SP* spatula; α indicates slope of setae relative to the exoskeleton surface, β indicates slope of microtrichia relative to the seta

material that supports setae has been previously demonstrated on an artificial hairy attachment system (Geim et al. 2003). In *A. seemanni*, tarsi usually generated higher friction forces than metatarsi (Fig. 4f). This effect also might be explained by the higher flexibility of the cuticle underlying the tarsal scopula, whereas the metatarsal scopula has a more rigid base. Also possible differences in material properties of setae and spatulae, which remain largely unknown, may contribute to this effect.

Our data show that desiccation has an effect on the friction force of legs of both spider species. Drying can result in the reduction of the flexibility of the underlying cuticle, setae, and spatulae and thus may suppress the adaptability of the scopula to the substrate. Possible changes in the chemistry of dried cuticle and in the thickness of the adsorbed water layer at the seta surface may contribute to this result. However, these effects are not experimentally supported to date.

Although scopulae of *A. seemanni* generated friction forces as high as those in *C. salei*, representatives of the first species were hardly able to hold on to smooth vertical surfaces. This might be due to the relatively high body weight of *A. seemanni*, which is about five times higher than *C. salei*. Moreover, *C. salei* not only relies on its scopulae, but additionally on the dragline used as a safety thread, while walking and resting on vertical substrata. Dragline-supported locomotion was not observed in *A. seemanni*.

Acknowledgments Sincere thanks to F.G. Barth (Universität Wien, Vienna, Austria) for supplying *C. salei* spiders and to P. Walter (Universität Ulm, Ulm, Germany) for his help with the cryo-SEM technique. H. Schwarz and J. Berger (MPI für Entwicklungsbiologie, Tübingen, Germany) helped with conventional SEM preparations. V. Kastner (MPI für Entwicklungsbiologie, Tübingen, Germany) prepared sections with the cryomicrotome and provided linguistic corrections of an early version of the manuscript.

F.G. Barth, C. Schaber (Universität Wien, Vienna, Austria), and J. Barnes (University of Glasgow, UK) provided valuable suggestions on the manuscript. This work was supported by the Federal Ministry of Science of Germany (BMBF) grant BioFuture 0311851 to S.N.G.

References

Arzt E, Gorb S, Spolenak R (2003) From micro to nano contacts in biological attachment devices. Proc Natl Acad Sci USA 100(19):10603–10606

Autumn K, Liang YA, Hsieh ST, Zesch W, Chan WP, Kenny TW, Fearing R, Full RJ (2000) Adhesive force of a single gecko foot-hair. Nature 405:681–685

Barth FG (2002) A spider's world: senses and behavior. Springer, Berlin Heidelberg New York

Bauchhenß E (1979) Die Pulvillen von *Calliphora erythrocephala* Meig. (Diptera, Brachycera) als Adhäsionsorgane. Zoomorphologie 93:99–123

Chung JY, Chaudhury MK (2005) Roles of discontinuities in bio-inspired adhesive pads. J R Soc Interface 2:55–61

Eisner T, Aneshansley DJ (2000) Defence by foot adhesion in a beetle (*Hemisphaerota cyanea*). Proc Natl Acad Sci USA 97:6568–6573

Foelix RF (1996) Biology of spiders. Oxford University Press, New York

Foelix R, Jackson R, Henksmeyer A, Hallas S. (1984) Tarsal hairs specialized for prey capture in the salticid *Portia*. Rev Arachnol 5:329–334

Gao H, Wang X, Yao H, Gorb SN, Arzt E (2005) Mechanics of hierarchical adhesion structures of geckos. Mech Mater 37:275–285

Geim AK, Dubonos SV, Grigorieva IV, Novoselov KS, Zhukov AA, Shapoval SY (2003) Microfabricated adhesive mimicking gecko foot-hair. Nat Mater 2:461–463

Gillett JD, Wigglesworth VB (1932) The climbing organ of an insect, *Rhodnius prolixus* (Hemiptera; Reduviidae). Proc R Soc Lond B 111:364–376

Gorb SN (1998) The design of the fly adhesive pad: distal tenent setae are adapted to the delivery of an adhesive secretion. Proc R Soc Lond B 265:747–752

Gorb SN (2001) Attachment devices of insect cuticle. Kluwer Academic Publishers, Dordrecht

Gorb SN, Beutel RG (2001) Evolution of locomotory attachment pads of hexapods. Naturwissenschaften 88:530–534

Gorb SN, Scherge M (2000) Biological microtribology: anisotropy in frictional forces of orthopteran attachment pads reflects the ultrastructure of a highly deformable material. Proc R Soc Lond B 267:1239–1244

Hiller U (1968) Untersuchungen zum Feinbau und zur Funktion der Haftborsten von Reptilien. Z Morphol Tiere 62:307–362

Hiller U (1975) Comparative studies on the functional morphology of two gekkonid lizards. J Bombay Nat Hist Soc 73:278–282

Homann H (1957) Haften Spinnen an einer Wasserhaut?. Naturwissenschaften 11:318–319

Huber G, Gorb SN, Spolenak R, Arzt E (2005a) Resolving the nanoscale adhesion of individual gecko spatulae by atomic force microscopy. Biol Lett 1:2–4

Huber G, Mantz H, Spolenak R, Mecke K, Jacobs K, Gorb SN, Arzt E (2005b) Evidence for capillarity contributions to gecko adhesion from single spatula: nanomechanical measurements. Proc Natl Acad Sci USA 102:16293–16296

Irschick DJ, Austin CC, Petren K, Fisher RN, Losos JB, Ellers O (1996) A comparative analysis of clinging ability among pad-bearing lizards. Biol J Linn Soc 59:21–35

Ishii S (1987) Adhesion of a leaf feeding ladybird *Epilachna vigintioctomaculata* (Coleoptera: Coccinellidae) on a vertically smooth surface. Appl Entomol Zool 22:222–228

Johnson KL, Kendall K, Roberts AD (1971) Surface energy and the contact of elastic solids. Proc R Soc Lond A 324:301–313

Kesel AB, Martin A, Seidl T (2003) Adhesion measurements on the attachment devices of the jumping spider *Evarcha arcuata*. J Exp Biol 206:2733–2738

Kesel AB, Martin A, Seidl T (2004) Getting a grip on spider attachment: an AFM approach to microstructure adhesion in arthropods. Smart Mater Struct 13:512–518

Kosaki A, Yamaoka R (1996) Chemical composition of footprints and cuticula lipids of three species of lady beetles. Jpn J Appl Entomol Zool 40:47–53

Langer MG, Ruppersberg JP, Gorb SN (2004) Adhesion forces measured at the level of a terminal plate of the fly's seta. Proc R Soc Lond B 271:2209–2215

Niederegger S, Gorb S (2003) Tarsal movements in flies during leg attachment and detachment on a smooth substrate. J Insect Physiol 49:611–620

Niederegger S, Gorb S, Jiao Y (2002) Contact behaviour of tenent setae in attachment pads of the blowfly *Calliphora vicina* (Diptera, Calliphoridae). J Comp Physiol A 187:961–970

Peressadko AG, Gorb SN (2004a) When less is more: experimental evidence for tenacity enhancement by division of contact area. J Adhes 80:1–15

Peressadko AG, Gorb SN (2004b) Surface profile and friction force generated by insects. In: Boblan I, Bannasch R (eds) First international industrial conference bionik 2004. VDI Verlag, Berlin, pp 257–263

Persson BNJ (2003) On the mechanism of adhesion in biological systems. J Chem Phys 118:7614–7621

Persson BNJ, Gorb SN (2003) The effect of surface roughness on the adhesion of elastic plates with application to biological systems. J Chem Phys 119:11437–11444

Richards AG, Richards PA (1979) The cuticular protuberances of insects. Int J Insect Morphol Embryol 8:143–157

Roscoe DT, Walker G (1991) The adhesion of spiders to smooth surfaces. Bull Br Arachnol Soc 8:224–226

Rovner JS (1978) Adhesive hairs in spiders: behavioral functions and hydraulically mediated movement. Symp Zool Soc Lond 42:99–108

Ruibal R, Ernst V (1965) The structure of the digital setae in lizards. J Morphol 117:271–279

Scherge M, Gorb SN (2001) Biological micro- and nanotribology. Springer, Berlin Heidelberg, New York

Stadler H, Mondon M, Walentin J, Jiao Y, Ziegler C, Gorb SN (2001) Viscosity force of the fly's pad secretion measured by atomic force microscopy. In: Wisser A, Nachtigall W (eds) Technische Biologie und Bionik. 5. Bionik—Kongress, Dessau 2000. Gustav Fisher Verlag, Stuttgart, pp. 340–344

Stork NE (1983) The adherence of beetle tarsal setae to glass. J Nat Hist 17:583–597

Stork NE, Evans MEG (1976) Tarsal setae in Coleoptera. Int J Insect Morphol Embryol 5:219–221

Williams EE, Peterson JA (1982) Convergent and alternative designs in the digital adhesive pads of scincid lizards. Science 215:1509–1511

J Comp Physiol A (2004) 190: 575–580
DOI 10.1007/s00359-004-0520-9

ORIGINAL PAPER

Frictional properties of contacting surfaces in the hemelytra-hindwing locking mechanism in the bug *Coreus marginatus* (Heteroptera, Coreidae)

P. J. Perez Goodwyn · S. N. Gorb

Received: 24 November 2003 / Revised: 26 March 2004 / Accepted: 27 March 2004
/ Published online: 28 April 2004

Abstract The structure and function of the hemelytra-to- hindwing locking mechanism of the bug *Coreus marginatus* were analysed. The system consists of a cuticular protrusion in the ventral side of the hemelytra, which locks the subcostal border of the hindwing in flight. The speed and distance slid by both surfaces against one another during flight were assessed using a combination of high-speed video recordings and a 2D geometrical model. The friction coefficient between sliding surfaces was assessed using a micromanipulator, coupled with force transducers. This was done under three experimental conditions: freshly dissected, air dried and rehydrated ethanol preserved samples. The results showed a high speed of sliding, approximately 0.18 m s^{-1}), with a relatively low friction coefficient (0.2 μ). There was no evident difference in the friction measured under the various treatments, with the exception of the rehydrated condition, which was lower. The surface morphology of the wing locking mechanism, namely outgrowths of one part having rounded edges, and completely flat surface on the counterpart, effectively aids in the reduction of friction at the microscopic level. The structure is effective even dry, and after being preserved in ethanol, suggest- ing that no cuticle secreted lubrication substance is responsible for its effectiveness. The ultrastructure pre- sumably confers mechanical stability to the system under the high load it is subjected to in flight.

Keywords Attachment devices · Biomechanics · Friction · Insect cuticle · Insect flight

Introduction

In Heteroptera, forewings have two different functions: flight and protection. The forewings (hemelytra hereafter) are folded over the body, protecting both the hindwings and the body.

P. J. Perez Goodwyn · S. N. Gorb (✉)
Evolutionary Biomaterials Group, Dept. Arzt, Max-Planck-Institut für Metallforschung,
Heisenbergstr. 3, 70569 Stuttgart, Germany
e-mail: s.gorb@mf.mpg.de

P. J. Perez Goodwyn
e-mail: Perez-Goodwyn@mf.mpg.de

For this reason wings in this group are highly specialized, the forewings being scler- otized and the hindwings membranous. At rest, the wings are kept tightly against the body, forming a stable unit to protect the animal's body (Schuh and Slater 1995). In flight, the wings are coupled and beat simultaneously. In both circumstances, wing-locking mechanisms are involved. Morphologically, wing-locking mechanisms are extremely complex, and in most cases the cuticle is covered with microtrichia (small cuticular outgrowths) (Nachtigall 1986; Gorb and Perez Goodwyn 2003). Functionally, the wing-locking mechanisms of Heteroptera can be divided into those active at rest and those active in flight (Gorb and Perez Goodwyn 2003).

The membranous hindwings of Heteroptera are the only pair bearing an aerofoil for flight; however, they lack flight muscles. The whole muscle power is restricted to the hemelytra, and the force generated is transferred to the hindwing through the hemelytra-to-hindwing- locking mechanism, the only mechanism active during flight. This mechanism is deemed to be extremely resis-tant to friction, since it must resist load forces surpassing the body weight on a relatively small contacting area and at a very high beating frequency (Schneider and Bohne 1977; Gorb and Perez Goodwyn 2003).

In this work, we analysed the morphology and ultrastructure of the hemelytra-to-hindwing "double-wave" locking mechanism and we experimentally approached the assessment of the friction generated in this mechanism. The questions asked were: what is the speed at which this mechanism is working? What is the distance the contacting surfaces effectively slide during each wing beat? Which is the friction coefficient between the interacting surfaces? Is the ultrastructure of the cuticle of this mechanism different from the general surface of the insect?.

Materials and methods

Coreus marginatus (L.) (Coreidae) a plant-feeding bug, was chosen for the experiments. This insect is common in southern Germany from early Spring to early Autumn, being prone to fly away when disturbed.

In order to study the cuticle's layer morphology and fibre orientation, samples were frozen with liquid N_2 and cracked with a scalpel. These were then placed in a 1 mol l^{-1}) NaOH solution at 90°C for 1 h, rinsed with distilled water, placed in frozen acetone, dehydrated with ethanol and critical-point dried. The samples were then placed on aluminium pegs, sputtered with gold- palladium, and observed under SEM (Hitachi S-800) (Gorb 1997).

For the assessment of flapping frequency, tethered flight was analysed. A pin attached to a mobile arm support was glued with wax to the specimen's pronotum. High-speed recordings of flight-tethered specimens were analysed, using a Kodak Motion Analyser PS 220 at 1,000 frames s^{-1}) The position of the camera was perpendicular to the flapping plane, which is slightly tilted with respect to the body's sagittal plane. Also, the position of several points of the wings of a preserved specimen was measured with a measuring microscope (Mitutoyo measuring microscope MF510TA, ± 1 lm precision).

Using these last data, a 2D geometrical model was built using Corel Draw 9 software (Fig. 3C). In the 2D model, it was considered that each wing rotates virtually about one single axis. The 2D co-ordinates of wing points used for model design were measured under a measuring microscope in two extreme positions of wings (forward and backward ones). The plane of wing's rotation was aligned parallel to the horizontal plane; that is why the 2D model describes the sliding distance closest to the real one. Since the model is 2D, the actual sliding distance may be slightly larger due to the fact that the z-axis was not considered during the measure- ments. However, the contribution of the movement in the third dimension to the total slid distance seems to be negligible.

This model made it possible to virtually manipulate wing position in order to calculate the sliding distance of the wing-locking mechanism. This model assumed a fixed rotation axis for the wings, following the natural movements of the preserved specimen. A linear regression fit function was obtained for this data, $y = -1.115 + 9.45x$(f1); where y is the hemelytra nodus position, and x is the distance slid by the mechanism.

Since the wing-locking mechanism itself is very small, it is not possible to observe it directly in high-speed vi- deo-recordings. In order to overcome this difficulty, a small dot was painted over the nodus of the hemelytra. Single frames of the high-speed recording were used to trace this point, which was visible throughout the flap cycle (Fig. 3A). With the obtained function (f1), sliding distance data were calculated, imputing the position of the nodus in the video recordings.

For the friction measurements, a device combining a three-axial motorized microma-nipulator together with two cell force transducers was used (Fig. 4A). A small square piece of each wing (ca. 2.5×2.5 mm) containing the wing-locking mechanism of each coun-terpart were cut off. The depicted device in Fig. 4A was built using Plexiglas, in which two screw-driven clamps each held one piece of the wing. Each clamp was attached to a force measuring sensor, which was connected to a Biopac MP100 force measuring system. Two perpendicu- lar sensors were used, one to measure the load and the other the frictional force of the system (Fig. 4A). These sensors were driven by the micromanipulator, which allowed positioning the samples and applying normal force to the contact area between hemelytra and hind- wing. Under a binocular microscope, the wings were coupled into a natural position.

The locked wings were slid 10 μm in either direction, approximately 10–12 times. The load was then increased and the procedure repeated until natural uncoupling occurred (Fig. 4B). The average point of the normal force curve for each 10 μm slid was registered, as well as the average load for that portion of the curve. The two measurements were plotted as force versus load, and as l (friction coefficient) versus load.

For these friction measurements three protocols were followed: using freshly dissected wings (two specimens $N = 2$, over each pair, procedure repeated ten times until uncou-pling, $n = 10$), alcohol preserved then dried ($N = 2$, $n = 7$), and alcohol-preserved samples rehydrated with phosphate buffer solution (pH = 7.3) ($N = 3$, $n = 10$), and were compared with an ANOVA test, with a Tukey post test (Origin 7 SR2).

Results

Morphology

On the ventral side of the hemelytra, two overlapping wave-shaped outgrowths are located (anal side wave, ASW; and costal side wave CSW) (Figs. 1, 2A, C) (Gorb and Perez Goodwyn 2003). The tile-like microtrichia of the ASW are flattened and blunt, whereas the setae of the CSW are conical and elongated. The clamp-like structure of the hemelytra is complementary to the cylindrical-shaped subcostal margin of the hindwing (Fig. 1), which is covered by blunt and thick microtri- chia with smooth borders (Fig. 2D).

In general, the insect procuticle bears several layers of chitin microfibrils oriented anti-parallel to the surface (Gorb 2001). In this mechanism the orientation of the fibrils also

Fig. 1 Diagram of the hemelytra to hindwing wing-locking mechanism. *ASW* anal side wave; *CSW* costal side wave; *LR* locking ridge

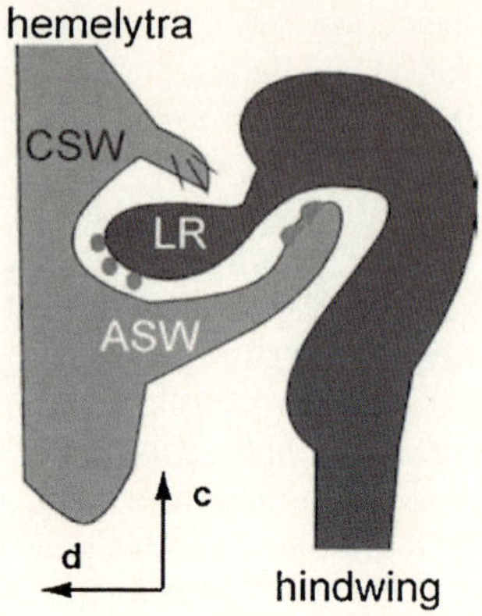

seems to be parallel (Fig. 2E, F). The exoc- uticle thickness is 1.14–3.70 μm, whereas the endocuticle thickness is 2.40–3.40 μm. However, the endocuticle is thicker on the dorsal surface under the ASW (7.80 μm). In the hindwing, the orientation of the microfibrils is also parallel to the surface in the whole wing without differentiation between dorsal and ventral surfaces. However, in the subcostal region, there is a clear distinction between the dorsal and ventral surfaces (Fig. 2B, D).

Wing beating and mechanism sliding velocity

The wing beating frequency calculated was 88–90 s^{-1}). The mechanism's maximum sliding distance calculated from the geometrical model was 0.86 mm in each half cycle. The velocity variation in a complete flapping cycle is depicted in Fig. 3B.

From these calculations, mean velocity is (0.86 mm × 2) × 90 s^{-1} = 154.8 mm s^{-1} = 0.154 m s^{-1}. In 1 s the mechanism must slide 154.8 mm, but this movement is accelerated to a maximum from 0 velocity twice per cycle. This estimation then would be inaccurate. If the acceleration has a sinusoidal character in the beating cycle, the estimated value would be the mean velocity, which in turn should be equal to the effective velocity of a sinusoidal movement. Maximum velocity may then be calculated as follows: effective velocity$\cdot 2/\sqrt{2} = 0.213\,\mathrm{m\,s^{-1}}$.

Friction experiments

Force versus load plotting showed a clear linear increase (dry treatment $b = 0.02$, $r = 0.62$; fresh $b = 0.14$, $r = 0.67$; dry $b = 0.15$, $r = 0.55$). The friction coefficient (μ) at loads under 1 mN was highly variable (0.0001–0.75), whereas it was stable at higher loads. There is an approximate stable value for μ ranging between 0.1 and 0.2 (Fig. 4C). The friction coefficients varied, depending on the treatments (ANOVA $P < 0.05$); however, it was significantly lower for the rehydrated samples but equal for the dry and fresh ones (Tukey post-test $P < 0.05$). The highest load achieved for each experimental condition was limited to the unlocking resistance of the mechanism itself.

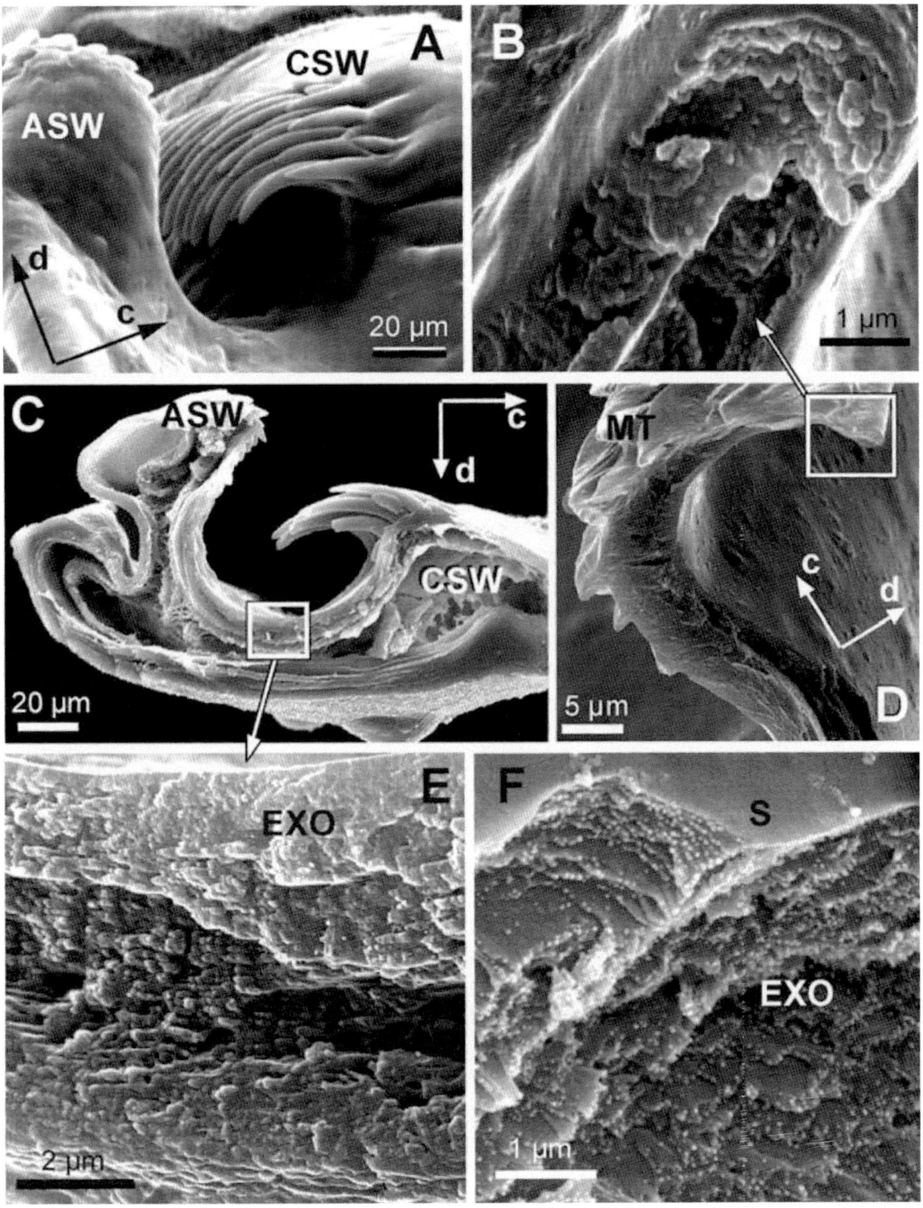

Fig. 2 A–F SEM details of the wing locking mechanism in *Coreus marginatus*. A Hemelytra wing-locking mechanism (''double wave'') overall view. B Detail of hindwing subcostal border. Note differentiation between dorsal and ventral surfaces. C Section through hemelytra wing-locking mechanism. D Hindwing subcostal border. E Section through cuticle of the wing-locking mechanism (hemelytra). F Diagonally broken portion of cuticle showing layer orientation and microfibrils in the hemelytra wing-locking mechanism. *ASW* anal side wave; *CSW* costal side wave; *EXO* exocuticle; *MT* microtrichia; S surface

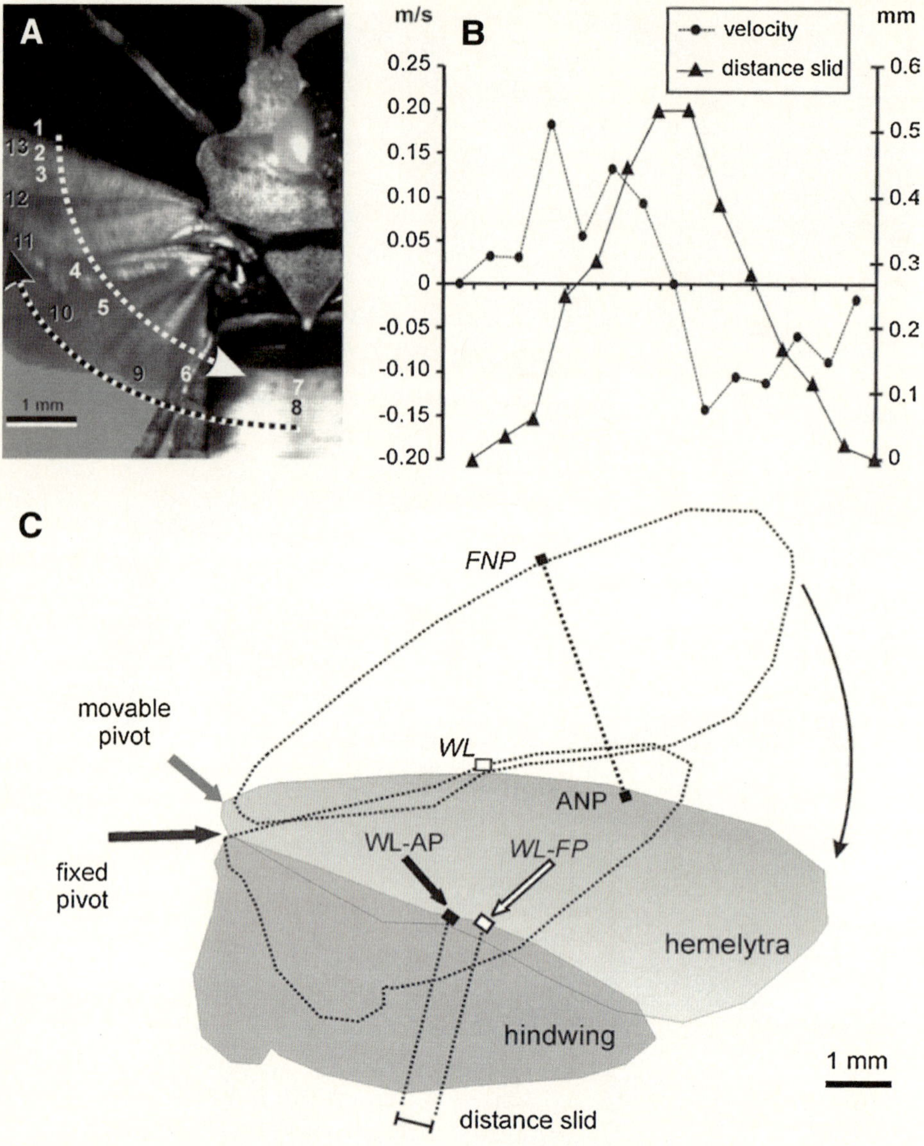

Fig. 3 A Dorsal view of *Coreus marginatus* in tethered flight. *Numbers* indicate position of wing during a complete flap cycle (1 s-0,001 between each point). *White numbers and arrow*: upstroke; *black numbers and arrow*: downstroke. B Sliding velocity and distance slid by wing-locking mechanism during a complete flapping cycle (points correspond to those of Fig. 3A). C Geometrical model used for calculating the distance slid by the wing-locking mechanism's counterparts in flight, before and after an example movement. *Dotted wings contour*: forward position of wings; *filled wings drawing*: wings after backward movement. *FNP* position of hemelytra node in forward position of wings; *ANP* node position after the movement; *WL* hemelytra wing-locking mechanism ''double-wave'' in forward position of wings; *WL-AP* ''double-wave'' mechanism after the movement; *WL-FP* point where the ''double-wave'' mechanism was before the movement, and from which it slid through the hindwing locking rail

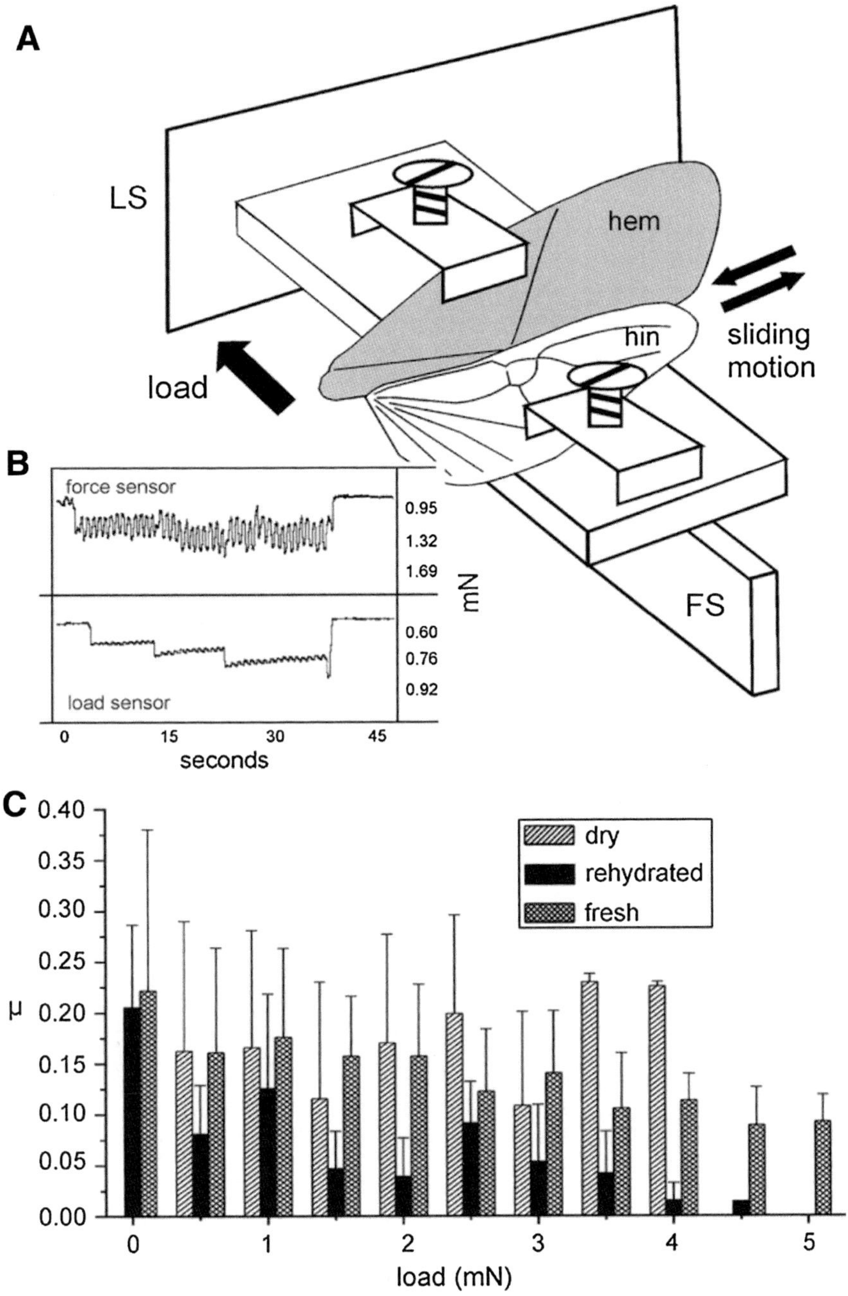

Fig. 4 A Experimental setup for friction measurements. The coupled wings portions (complete wings drawn for the sake of clarity) were clamped to two perpendicular force transducers. The system was slid to either side several times, and then the load increased and the procedure was repeated until natural uncoupling occurred. FS force sensor; LS load sensor; *hem* hemelytra; *hin* hindwing. B Example of original recording of the experimental setup; note increasing load. C Friction coefficient (μ) versus load as average columns and SD (0.5 mN pooled interval), under three experimental conditions: Fresh, dry and rehydrated

Discussion

In flight, the subcostal rail of the hindwing slides through the double wave clamp of the hemelytra, and transmits the force from the hemelytra to the hindwing, synchronizing both wing pairs (Schneider and Bohne 1977). The whole muscular force generated by the mesothoracic (hemelytra) muscles is the minimum load of this system, whereas the wing beating frequency lies in the range of medium to high values compared to other insects (Dudley 1999). Such a load-transmitting device could cause energy loss and even damage of the surfaces, if the friction in the sliding area is high.

The microtrichia of the subcostal rail of the hindwing have a smooth, rounded shape. Assuming that both surfaces are stiff, this structure reduces the contact area between this structure and the inner surface of the double wave clamp (Figs. 1, 2D). The morphological basis for friction reduction consists of a very basic principle, by which spherical or rounded outgrowths slide over a flat surface, reducing the contact points and completely avoiding mechanical interlocking of the interacting surfaces (Bowden and Tabor 1986).

The maximum experimentally estimated velocity is 0.186 m s^{-1} (Fig. 3B), somewhat lower than the estimation based on a sinusoidal assumption. This cycle might not be strictly sinusoidal, since the speed in the upstroke (negative in the graph) is slower, and the velocity variation is asymmetrical (Fig. 3A).

The friction coefficient measured is similar to that of Babbitt alloy sliding over steel (Bowden and Tabor 1986), but the variation at low loads and at microscopical dimensions of the measured structures makes such a comparison difficult. The lack of difference between the fresh and dry experimental setups is surprising. Because of the absence of significant differences in the frictional coefficient between rehydrated and fresh samples, one may conclude that no secreted lubricant is involved in this mechanism. The lower friction coefficient measured on the rehydrated samples may be due to the lubricant action of the buffer. Presumably, the arrangement of fibrils is responsible for the stiffness of the structure of the whole mechanism.

The gula of the beetle *Geotrupes stercorarius* functions as a condyle in the head-prothorax articulation, and even though the frequency of the movement is very low, it is subject to a high load. The cuticle's ultra- structure of this sclerite seems to be different from the rest of the body (Arzt et al. 2002). The orientation of the microfibrils in the exocuticle is actually perpendicular to the surface. Even though the wing-locking mechanism is also deemed to be working under a high load, no correspondence to the ultrastructure of the beetle's gula was found.

Our data show that the combination of the material structure, surface sculpture, and presumably the hydrophobic surface, makes this mechanism flexible and stable, and at the same time reduces the friction to a minimum, being effective for the double function of interlocking and sliding.

Acknowledgements Valuable help from Dr. Andrei Peressadko, Vicky Kastner and Jürgen Berger is greatly acknowledged. This work was supported by a Max-Planck-Fellowship to P.J.P.G. and by the Federal Ministry of Education, Science and Technology, Germany to S.N.G. (BioFuture 0311851). The experiments comply with the current laws of Germany.

References

Arzt E, Enders S, Gorb S (2002) Towards a micromechanical understanding of biological surface devices. Z Metallkunde 93:345–351

Bowden FP, Tabor D (1986) The friction and lubrication of solids. Clarendon Press, Oxford

Dudley R (1999) The biomechanics of insect flight: form, function, evolution. Princeton University Press, Princeton, NJ

Gorb SN (1997) Ultrastructural architecture of the microtrichia of the insect cuticle. J Morphol 234:1–10

Gorb SN (2001) Attachment devices of insect cuticle. Kluwer, The Netherlands

Gorb SN, Perez Goodwyn PJ (2003) Wing locking mechanisms in aquatic Heteroptera. J Morphol 257:127–146

Nachtigall W (1986) Konstruktionen: Biologie und Technik.: VDI, Düsseldorf

Perez Goodwyn PJ, Gorb SN (2002) Attachment forces of the hemelytra-locking mechanisms in aquatic bugs (Heteroptera: Belostomatidae). J Insect Physiol 49:753–764

Schneider P, Bohne J (1977) The sliding mechanism of coupling front and hind wing in Heteroptera. Fortschr Zool 24:139–142

Schuh RT, Slater JA (1995) True bugs of the world (Hemiptera: Heteroptera). Classification and natural history, Comstock, Ithaca

J Comp Physiol A
DOI 10.1007/s00359-008-0408-1

ORIGINAL PAPER

Surface structure and frictional properties of the skin of the Amazon tree boa *Corallus hortulanus* (Squamata, Boidae)

R. A. Berthé · G. Westhoff · H. Bleckmann · S. N. Gorb

Received: 14 October 2008 / Revised: 12 December 2008 / Accepted: 21 December 2008

Abstract The legless locomotion of snakes requires specific adaptations of their ventral scales to maintain friction force in different directions. The skin microornamentation of the snake *Corallus hortulanus* was studied by means of scanning electron microscopy and the friction properties of the skin were tested on substrates of different roughness. Skin samples from various parts of the body (dorsal, lateral, ventral) were compared. Dorsal and lateral scales showed similar, net-like microornamentation and similar friction coefficients. Average friction coefficients for dorsal and lateral scales on the epoxy resin surfaces were 0.331 and 0.323, respectively. In contrast, ventral scales possess ridges running parallel to the longitudinal body axis. They demonstrated a significantly lower friction coefficient compared to both dorsal and lateral scales (0.191 on average). In addition, ventral scales showed frictional anisotropy comparing longitudinal and perpendicular direction of the ridges. This study clearly demonstrates that different skin microstructure is responsible for different frictional properties in different body regions.

Keywords Snake skin · Microornamentation · Friction · Biological materials · Biotribology

R. A. Berthé (✉) · S. N. Gorb
Evolutionary Biomaterials Group, Department of Thin Films and Biological Systems, Max Planck Institute for Metals Research, Heisenbergstraße 3, 70569 Stuttgart, Germany
e-mail: berthe@uni-bonn.de

R. A. Berthé · G. Westhoff · H. Bleckmann
Institute of Zoology, University of Bonn, Poppelsdorfer Schloss, 53115 Bonn, Germany
e-mail: gwesthoff@uni-bonn.de

S. N. Gorb
Department of Zoology: Functional Morphology and Biomechanics, Zoological Institute of the University of Kiel, Am Botanischen Garten 1-9, 24098 Kiel, Germany
e-mail: sgorb@zoologie.uni-kiel.de

Introduction

Snakes lack legs and use the surface of the body itself to generate propulsion on the ground during locomotion. For this purpose, some frictional grip is mandatory, in order to get the force onto the ground. Depending on the snake species, type of movement, environment and preferred substrate, different parts of the body must have different functional requirements and therefore different frictional properties. Frictional properties of the contact pair depend on the stiffness of materials of the contact pair, their physico-chemical properties, and their surface profiles (Bowden and Tabor 1986; Scherge and Gorb 2001).

The snakes' epidermis is made up of different layers with the innermost called the stratum germinativum. The outer layers, which are renewed during shedding, are, from the inside, α-, mesos-, β-layer, and *Oberhäutchen*. The *Oberhäutchen*, mainly consisting of β-keratin, is in direct contact with the environment. It is well known that the *Oberhäutchen* possesses a fine surface structure called microornamentation (Leydig 1873; Ruibal 1968), whose details were described by earlier authors with the use of electron microscopy (Hoge and Souza Santos 1953; Price 1983; Bea and Fontarnau 1986; Fontarnau and Bea 1987; Stille 1987; Chiasson et al. 1989; Chiasson and Lowe 1989; Price and Kelly 1989; Price 1990).

Functions of the lizard and snake microornamentation have been previously discussed in the literature. It was suggested that the microstructure functions as a kind of zip-fastener supporting the moulding process by holding old and new skin together until the old skin is entirely shed (Maderson 1966). Ruibal and Ernst (1965) proposed the function of surface strengthening by the surface microgrooves. Anti-contamination capability of the scales has been shown for uropeltid snakes (Gans and Baic 1977; Gower 2003). An anti-fouling effect for sea-snake scales was proposed by McCarthy (1987). In contrast to these functional interpretations of microornamentations, Price (1982) proposed that interspecific differences are independent of ecological or environmental factors.

A qualitative frictional analysis of the snake skin at the nanoscale has been performed previously by using atomic force microscopy in three species (*Boa constrictor*, *Morelia spilotes*, and *Python regius*) (Hazel et al. 1999). The authors found friction anisotropy by using the AFM-cantilever in anterior and posterior directions of the scale. They suggested that narrow, backward pointing "spikes" (referred to as microfibrils by the authors), as part of the microornamentation in ventral scales, were responsible for anisotropic properties. However, no comparison to scales from other regions of the body has been done so far. Since keratinised biological micro- and nano-structures may have specific interactions with various substrata as has been previously shown for gecko setae (Hiller 1968), we expect to reveal micro- and nanostructural adaptations to legless locomotion on the ventral and possibly lateral scale surface of the snake in a macroscopical tribological experiment.

The goal of the present study was to analyse relationships between the microstructure of scales and their frictional properties at different regions of the body in the Amazon tree boa *Corallus hortulanus* (Squamata, Boidae), which is a specialised arboreal snake. The high degree of specialisation was the reason for us to work on this particular species. Frictional properties of the skin were characterised on the smooth and rough substrata in different direction relative to the body axis of the snake.

Materials and methods

Animals and preparations

Two specimens of the Amazon tree boa *C. hortulanus* were used in this study. Specimens were killed by an overdose of isoflurane and stored at −70°C in polyethylene bags. Pieces of frozen snakes were cut off and thawed at room temperature. The skin was removed and carefully cleaned from remaining subdermal tissue. During the preparation, we avoided touching skin regions used in further experiments and microscopic analysis.

Scanning electron microscopy

Small pieces of the skin were dehydrated for 15 min in absolute methanol and critical point dried using a critical point drying apparatus (E3000 Series, Quorum Technologies, UK). Dried samples were mounted on aluminum stubs, sputter coated with 6 nm of gold–paladium (SCD 500 Sputter Coater equipped with QSG 100 Quartz Film Thickness Monitor, BAL-TEC, Liechtenstein) and viewed using a Hitachi S-4800 (Hitachi, Japan) scanning electron microscope at an acceleration voltage of 3 kV. Surface structures were measured from SEM micrographs using the software SigmaScan Pro 5 (Systat Software Inc., USA). The length and the width of the grooves as well as distance between grooves were measured. The "length" is defined as the maximum dimension of a groove, which was always oriented in approximately rostro-caudal direction. The "width" of a groove is defined as the maximum dimension orthogonal to the length (see Fig. 1).

Substrate preparation for friction measurements

Plates, made of the polymerised epoxy resin (61.3% NSA, 23.6% ERL 4221, 14.2% D.E.R. 736, 0.9% DMAE; for details see Spurr 1969) were used as substrate surfaces in measurements of the snake skin friction. The surfaces of the epoxy resin were positive replicas of glass and different polishing papers. In order to prepare the replicas, a two step molding technique was applied (Gorb 1999, 2007). The negative molds were made

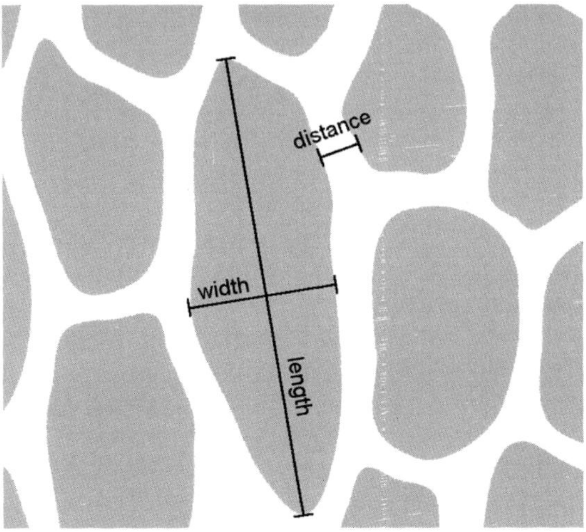

Fig. 1 Schematic drawing of microornamentation to illustrate the method of measuring surface structures. Grooves are indicated in *grey*

by coating the original surface attached to a plane wooden plate with a two-compound polymer polyvinylsiloxane (PVS) (President light body, Coltène, Switzerland). As original surfaces, smooth glass plates and polishing papers (Fibrmet Discs, Buehler, Germany) were used. Polishing papers used had grain sizes of 0.3, 1, 3, 9, and 12 μm. Additionally, the sandpapers of the grit size of P220, P100, and P60 were used. Grit sizes rougher than P60 resulted in a regular hooking of the scales edges. Thus the friction properties of the surface itself could not be measured. The PVS negatives were then used to obtain the positives by covering the negative surface with fluid epoxy resin and polymerizing it for 24 h at 70°C.

The final epoxy substrates had average surface roughnesses (R_a) of 0.08 (glass mold), 0.25, 0.42, 1.11, 2.25, 2.75, 12.67, 13.94 μm (the mold of P60 sandpaper was too rough to measure with the instruments available). These values are the means obtained from the measurements using three surface profiling instruments: surface perthometer Dektak 8 (Veeco Instruments Inc., USA), white light interferometer New View 5000 (Zygo Corporation, USA), and Surftest 301 (Mitutoyo Corporation, Japan). The scanned distances per sample were 3 mm for Dektak (all surfaces except P60) and three times 2.5 mm for Surftest (molds of 0.3 μm polishing paper to P220 sandpaper). For Niew View the examined areas were 0.07 × 0.05 mm (glass mold, 0.3, and 1 μm polishing paper), 0.14 × 0.11 mm (3 μm paper), and 0.36 × 0.20 mm (9 and 12 μm paper). Thus, different physical principles for evaluation of the surface roughness were used. The white light interferometer uses interferences caused by the roughness of the surface while the other two instruments scan the surface mechanically with a stylus.

Friction measurements

For friction measurements, the thawed fresh skin was wrapped around a polytetrafluoroethylen (PTFE) plate (10 × 10 × 3 mm) containing a hole of about 2 mm in diameter. The hole was situated in the centre of the plate and was not covered by the skin on one side. A few layers of cellulose were placed between the plate and skin, so that the skin could be kept moist by pouring water through the hole without wetting the surface of the scales. Otherwise, the capillary forces, caused by water, could have falsified the measurements.

The setup for the estimation of the frictional coefficient consisted of a flat motorized panel, which could be slowly inclined in a controllable way from the horizontal to the vertical position. The skin sample, prepared as mentioned above, was placed onto the epoxy resin substrates and the panel tilted at an angular velocity of 3.5° per second until the sample slid down. The sample orientation was randomly chosen but recorded in each experiment. Twelve individual samples from three different regions of the body (dorsal, ventral, lateral) were tested 14 times in four different orientations (rostro-caudad, caudo-rostrad, sinistro-dextrad, dextro-sinistrad for dorsal and ventral scales and rostro-caudad, caudo-rostrad, dorso-ventrad, ventro-dorsad for lateral scales, respectively). The data obtained for each surface for dorsal, lateral, and ventral scales were statistically compared (Kolmogorov–Smirnov test with software SPSS 16, SPSS Inc., USA) for each combination of longitudinal with transversal directions. Friction coefficients of opposite directions were pooled in the cases of sinistrad and dextrad in ventral and dorsal scales, as well as dorsad and ventrad in lateral scales.

Results

Electron microscopy

All body regions possess a regular fine microstructure (microornamentation) that is characterized by micropores and pits or indentions of various shape and diameter. The dimensions of these structures are summarized in Table 1. The microstructure of the ventral scales (Fig. 2e, f) clearly differs from dorsal (Fig. 2a, b) and lateral (Fig. 2c, d) scales in respect to pronounced longitudinal ridges at the scale areas with surface contact.

Dorsal and lateral scales are characterized by a honeycomb or net like pattern with depressions about 750 nm long and about 350 nm wide on average at the middle of the scales. There is only slight structural anisotropy in the way that the depressions are elongated along the longitudinal axis of the snake (Fig. 2a, c). Towards the edges of the scales the depressions become shorter and therefore more circular (Fig. 2b, d).

Ventral scales possess a fine structure consisting of ridges oriented along the longitudinal body axis. These ridges are separated by 300 nm wide grooves running more or less parallel to each other. The ridges build up a system of rails that presumably represent the principal contact zone with the substrate. Depressions on the ventral scales are about one-third the diameter of those found on dorsal and lateral scales. The cell borders are usually oriented along the ridges. Such geometry results in a zigzag line at the anterior and posterior ends of the cells and straight lines at their sides. The anterior part of each ventral

Table 1 Dimensions of the microornamentation of *Corallus hortulanus* scales

Location of the scale	Position on the scale	Mean (nm)	Standard deviation	Number of objects measured
Dorsal	Caudal–middle (Fig. 2a)			
	Length of the groove	740	189	29
	Width of the groove	352	61	29
	Distance between grooves	107	31	62
	Cranial (Fig. 2b)			
	Length of the groove	410	62	21
	Width of the groove	367	46	21
	Distance between grooves	133	12	19
Lateral	Middle (Fig. 2c)			
	Length of the groove	773	214	31
	Width of the groove	356	68	31
	Distance between grooves	90	15	37
Ventral	Middle (Fig. 2e)			
	Length of the groove	258	116	39
	Width of the groove	106	22	39
	Distance between grooves	285	46	66
	Distance between ridges (crest to crest)	327	45	51
	Cranial (Fig. 2f)			
	Diameter of the groove	230	48	46
	Distance between grooves	236	45	89

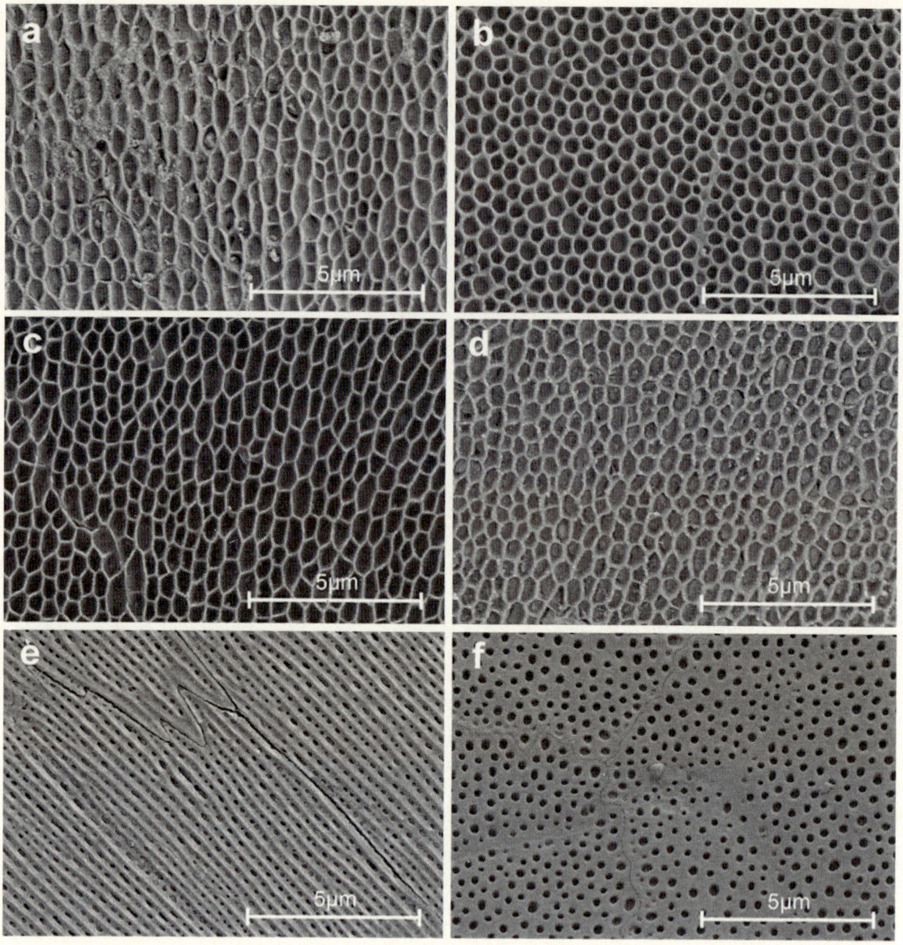

Fig. 2 SEM micrographs of scale regions of *Corallus hortulanus* at a magnification of 10,000. **a** Dorsal scale, middle part, anterior is at the *top*. **b** Dorsal scale, anterior part, anterior is at the *top*. **c** Lateral scale, middle part, anterior is at the *top*. **d** Lateral scale, anterior part, anterior is at the *top*. **e** Ventral scale, anterior is in the *upper left corner*. **f** Ventral scale, most anterior part of the scale; this surface was covered by the adjacent rostral scale; anterior is at the *top*

scale, covered by the next overlapping anterior scale, does not possess any ridges. This part is almost flat with circular depressions, which are on average 230 nm in diameter (Fig. 2f).

Friction measurements

Altogether 5,346 friction coefficient (FC) values were recorded. They ranged from 0.15 to 0.47 (Table 2). The ventral scales with a FC mean of 0.191 represent the lowest value. That is about 40% less than mean FCs of dorsal (0.331) and lateral (0.323) scales (Fig. 3). The intermediate range of roughness within the nine different roughnesses always represented the lowest friction coefficients regardless of the origin of scales (ventral, dorsal, lateral). For each sliding direction (rostrad, caudad and sideward) and each roughness, ventral scale friction differed significantly from dorsal and lateral scales ($P < 0.01$) whereas dorsal and lateral scales appeared to be similar (Fig. 4d–f).

The comparison of FCs measured in rostro-caudal versus medio-lateral direction revealed most pronounced frictional anisotropy in ventral scales. Ventral scales showed an increase of 20.6% compared to only 4.8% (dorsal scales) and 4.3% (lateral scales) in medio-lateral direction (Fig. 3).

This anisotropy was observed in all nine roughnesses tested on ventral scales (Fig. 4c). The results of testing of rostrad against sideward direction are highly significant on all roughnesses. For caudad against sideward direction, eight of nine roughnesses showed highly significant differences. Rostrad against caudad for ventral scales and all directional combinations for dorsal and lateral scales do not show such a consistency throughout the different roughnesses (Fig. 4a–c).

Table 2 Descriptive statistics of the friction measurements

Orientation	Roughness R_a (in μm)	*N*	Dorsal scales (mean ± SD)	Lateral scales (mean ± SD)	Ventral scales (mean ± SD)
Rostrad	0.08	52	0.454 ± 0.069	0.412 ± 0.093	0.207 ± 0.038
Rostrad	0.25	52	0.367 ± 0.058	0.353 ± 0.074	0.218 ± 0.042
Rostrad	0.42	52	0.325 ± 0.052	0.311 ± 0.053	0.203 ± 0.066
Rostrad	1.11	52	0.300 ± 0.043	0.268 ± 0.064	0.170 ± 0.037
Rostrad	2.26	52	0.293 ± 0.044	0.280 ± 0.075	0.153 ± 0.032
Rostrad	2.75	52	0.272 ± 0.040	0.281 ± 0.062	0.161 ± 0.022
Rostrad	12.67	52	0.261 ± 0.050	0.250 ± 0.035	0.144 ± 0.019
Rostrad	13.94	52	0.283 ± 0.056	0.301 ± 0.063	0.151 ± 0.027
Rostrad	*	52	0.299 ± 0.067	0.346 ± 0.079	0.163 ± 0.038
Caudad	0.08	52	0.472 ± 0.067	0.416 ± 0.090	0.227 ± 0.055
Caudad	0.25	52	0.384 ± 0.076	0.349 ± 0.075	0.222 ± 0.043
Caudad	0.42	52	0.344 ± 0.058	0.320 ± 0.066	0.204 ± 0.057
Caudad	1.11	52	0.308 ± 0.048	0.283 ± 0.072	0.165 ± 0.032
Caudad	2.26	52	0.301 ± 0.048	0.289 ± 0.081	0.153 ± 0.025
Caudad	2.75	52	0.271 ± 0.040	0.283 ± 0.060	0.162 ± 0.029
Caudad	12.67	52	0.266 ± 0.037	0.279 ± 0.049	0.157 ± 0.043
Caudad	13.94	52	0.303 ± 0.059	0.307 ± 0.043	0.181 ± 0.063
Caudad	*	52	0.375 ± 0.090	0.408 ± 0.092	0.188 ± 0.055
Sideward	0.08	94	0.455 ± 0.061	0.408 ± 0.086	0.233 ± 0.037
Sideward	0.25	94	0.402 ± 0.061	0.348 ± 0.075	0.242 ± 0.036
Sideward	0.42	94	0.333 ± 0.049	0.310 ± 0.046	0.220 ± 0.040
Sideward	1.11	94	0.310 ± 0.038	0.273 ± 0.051	0.191 ± 0.025
Sideward	2.26	94	0.297 ± 0.046	0.283 ± 0.065	0.185 ± 0.020
Sideward	2.75	94	0.287 ± 0.040	0.284 ± 0.046	0.194 ± 0.021
Sideward	12.67	94	0.297 ± 0.042	0.289 ± 0.039	0.212 ± 0.028
Sideward	13.94	94	0.322 ± 0.057	0.362 ± 0.061	0.221 ± 0.031
Sideward	*	94	0.360 ± 0.068	0.437 ± 0.106	0.234 ± 0.045

Shown are the mean friction coefficients for all roughnesses for dorsal, lateral and ventral scales. An asterisk in the roughness column stands for the roughness of the P60 sandpaper which was not measurable

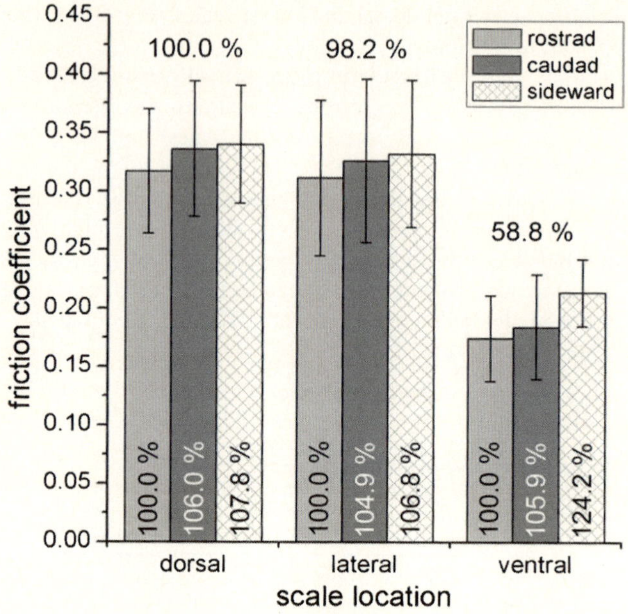

Fig. 3 Friction coefficients of scales. *Columns* show mean values of corresponding friction measurements for different roughnesses and directions. *Error bars* indicate standard deviations. The percentages indicated over column are set to 100% for dorsal scales and the percentages within the columns are set to 100% for rostrad direction within one scale location

However, lateral and dorsal scales show some significant results scattered among the three coarser roughnesses (Fig. 4a, b).

Discussion

Reptile scale microornamentation has been previously investigated in the context of phylogeny and ecological adaptations (Picado 1931; Hoge and Souza Santos 1953; Ruibal and Ernst 1965; Porter 1967; Ruibal 1968; Stewart and Daniel 1972, 1973, 1975; Gans and Baic 1977; Price 1982, 1983; Peterson 1984a, b; Peterson and Bezy 1985; Renous et al. 1985; Stille 1987; Irish et al. 1988; Price and Kelly 1989; Chiasson and Lowe 1989; Chiasson et al. 1989; Price 1990; Harvey 1993; Harvey and Gutberlet 1995; Maderson et al. 1998; Arnold 2002; Gower 2003). Microornamentation characters, such as micropits, cells size and shape, and raised posterior cell edges seem to be evolved several times independently among taxa, which is interesting from phylogenetical and ecological points of view (Arnold 2002). In phylogenetic reconstructions, microornamentation has been used as an argument to support relationships of species (in snakes: Picado 1931; Price 1983; Stille 1987; Chiasson and Lowe 1989; Chiasson et al. 1989; Price and Kelly 1989; Gower 2003; in lizards: Stewart and Daniel 1975; Peterson and Bezy 1985; Harvey 1993; Harvey and Gutberlet 1995; Arnold 2002).

Previous authors claimed that structural differences in microornamentations in different species do not represent any adaptations to different environments (Price 1982; Peterson 1984a, b; Peterson and Bezy 1985), because similar microstructures have been found in

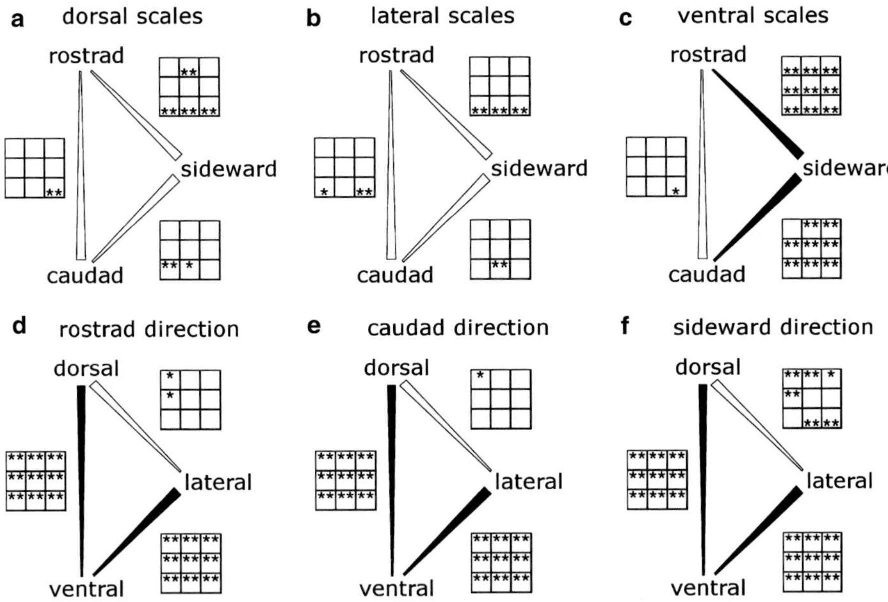

Fig. 4 Comparison of friction coefficients estimated in different sliding directions (**a–c**) and of scales from different body regions (**d–f**). *Broad end* of the *connecting line* indicates where significantly higher coefficients of friction were found more often on the nine different roughnesses. *Filled connections* are used when one direction (or body region) has significantly higher friction coefficients on the majority of the nine roughnesses (94 and 100% in these cases). *Open connections* indicate that half or less of the roughnesses have significantly higher friction coefficients. The *asterisks* in the panels next to the connecting lines show the levels of significance for the nine roughnesses. (*first row* 0.08, 0.25, and 0.42 μm; *second row* 1.11, 2.26, and 2.75 μm; *third row* 12.67 and 13.94 μm, and replication of P60 sandpaper; *no asterisk* $P \geq 0.05$, *$0.01 < P < 0.05$, **$P < 0.01$). *Asterisks* at the *top* of *box* indicate that the friction coefficient of the direction (or body region) at the upper tip of the comparison bar is significantly higher at the corresponding roughness

taxa living in different habitats and a variety of microstructures can be found in different species occupying similar ecological niches.

However, it seems to be unlikely that microornamentation is a character independent of selective pressures. Especially in snakes, the contact between the body surface and the environment demands a functional adaptation of the surface. Since there might be multiple ways to achieve similar functional tasks at the ultrastructural level of the scales, the comparison of FCs of different body regions within one species aids in understanding of functional significance of the microornamentation and provides a good basis for further interspecies comparison.

Possible functional significance of microstructure of ventral scales in different boid snakes (*B. constrictor*, *M. spilotes*, *P. regius*) has been shortly discussed in the work by Hazel et al. (1999), who demonstrated frictional anisotropy at the nanoscale in caudo-rostral direction versus rostro-caudal direction in contact with an AFM cantilever tip. How applicable are these results to real situation remains unknown. The friction anisotropy in caudo-rostral direction versus rostro-caudal direction is not supported by the present study, where friction has been estimated at various substrata in a more realistic situation. We could not find significant anisotropy in caudo-rostral versus rostro-caudal direction on the majority of roughnesses tested. However, this discrepancy between Hazel et al. (1999) and our results might be due to the different species investigated. In contrast to the species used by Hazel et al. (1999), in *C. hortulanus* it is difficult recognize to point the rostral or caudal orientation based on the appearance of the ultrastructure (Fig. 1e). Furthermore, representatives of the genera *Morelia*, *Boa* and *Python* are capable of the non-undulating rectilinear movement, whereas *Corallus* species are not (personal observations). A caudo-rostral anisotropy seems to be advantageous in rectilinear locomotion.

Snake locomotion by means of lateral undulation involves continuous movement of the ventral scales over the ground and load generation against resisting points at the side of the animal (Mosauer 1932a, b; Gray 1946; Gray and Lissmann 1950). When climbing upwards *C. hortulanus* bends its body ventrally against higher resisting points, comparable to the lateral bending during lateral undulation (personal observation). As friction impedes locomotion by lateral undulation, a reduction of friction in the longitudinal direction is beneficial for the propulsion of the snake. At the same time, high perpendicular friction or a fixed object as a counter bearing is needed to avoid slipping when forces are applied (Mosauer 1932a, b; Gray 1946). This principle has been used in constructing snake robots where wheels provide such kind of a frictional anisotropy (Hirose and Morishime 1990).

Our results show for the first time, that there is a significant anisotropy of friction coefficient for the ventral scales with higher coefficients for sideward directions. This anisotropy is constant over the broad range of different roughnesses tested. We would expect selective pressures on functional adaptations to cope with a variety of substrates and roughnesses during locomotion. Thus, we interpret our results clearly as functional adaptation of snake scale microornamentation. In contrast the dorsal and lateral scales do not show the consistency in anisotropy over the range of different roughnesses. This fact additionally supports the specialization of ventral scales.

The dorsal and lateral scales do show higher friction coefficients compared to the ventral ones (Fig. 3). Since the dorsal and lateral scales are hardly involved in surface contact, their high friction coefficients are not expected to be of functional relevance in locomotion. However, friction is not the only factor that may cause evolutionary change and may explain structural differences between dorsal and ventral scales. Dirt shedding capability and anti-reflection property are other characters of scales which depend on the microstructure (for lizards see Arnold 2002). Optimization of surface for one of these functions (dirt shedding, anti-reflection, friction) is most likely in conflict with the optimization for one another. A trade-off is to be expected depending on the selective pressure.

In the case of *C. hortulanus* the ventral scales have lower friction than dorsal and lateral scales, but are shiny, whereas the two last-mentioned scales are matt. Generally, a matt surface is less easily spotted by moving predators or prey which should be beneficial to an ambush hunter.

The basal, i.e. most anterior, regions of ventral scales revealed no anisotropic microstructures and are covered by the caudal edge of an adjacent rostral scale. This can be regarded as a further argument, supporting functional significance of the anisotropic microstructure of ventral scales in the non overlapping areas which are in contact with the substrate.

The results of friction measurements and SEM analysis revealed a correlation between microstructural pattern of the ventral scales and frictional anisotropy. Longitudinal ridges are involved in friction reduction in the sliding direction. Scales from other regions of the snakes' body show higher friction on respective surfaces (168%). Further analysis has to reveal whether the lower friction in ventral scales is to support propulsion of the snake or to avoid abrasion of the contacting scales.

Acknowledgments Cornelia Miksch (Max Planck Institute for Metals Research, Stuttgart) assisted with the microscopy techniques. This work, as part of the European Science Foundation EUROCORES Programme FANAS was supported by the German Science Foundation DFG (contract no. GO 995/4-1) and the EC Sixt Framework Programme (contract no. ERAS-CT-2003-980409) to S.G. The experiments comply with the "Principles of animal care", publication no. 86–23, revised 1985 of the National Institute of Health, and also with the current laws of Germany.

References

Arnold EN (2002) History and function of scale microornamentation in lacertid lizards. J Morphol 252:145–169

Bea A, Fontarnau R (1986) The study of the sloughing cycle in snakes by means of scanning electron microscopy. In: Roček Z (ed) Studies in herpetology. Charles University, Prague, pp 373–376

Bowden FP, Tabor D (1986) The friction and lubrication of solids. Clarendon Press, Oxford

Chiasson RB, Lowe CH (1989) Ultrastructural scale patterns in *Nerodia* and *Thamnophis*. J Herpetol 23:109–118

Chiasson RB, Bentley DL, Lowe CH (1989) Scale morphology in *Agkistrodon* and closely related crotaline genera. Herpetologica 45:430–438

Fontarnau R, Bea A (1987) A quick, simple method of replicating for scanning electron microscopy applied to the oberhautchen micro-ornamentation study. J Herpetol 21:366–369

Gans C, Baic D (1977) Regional specialisation of reptilian scale surface: relation of texture and biological role. Science 195:1348–1350

Gorb SN (1999) Ultrastructure of the thoracic dorso-medial field (TDM) in the elytra-to-body arresting mechanism in tenebrionid beetles (Coleoptera: Tenebrionidae). J Morphol 240:101–113

Gorb SN (2007) Visualisation of native surfaces by two-step molding. Micros today 15:44–46

Gower DJ (2003) Scale microornamentation of uropeltid snakes. J Morphol 258:249–268

Gray J (1946) The mechanism of locomotion in snakes. J Exp Biol 23:101–120

Gray J, Lissmann HW (1950) The kinetics of locomotion of the grass-snake. J Exp Biol 26:354–367

Harvey MB (1993) Microstructure, ontology, and evolution of scale surface in xenosaurid lizards. J Morphol 216:161–177

Harvey MB, Gutberlet RL (1995) Microstructure, evolution, and ontogeny of scale surface in cordylid and gerrhosaurid lizards. J Morphol 226:121–139

Hazel J, Stone M, Grace MS, Tsukruk VV (1999) Nanoscale design of snake skin for reptation locomotions via friction anisotropy. J Biomech 32:477–484

Hiller U (1968) Untersuchungen zum Feinbau und zur Funktion der Haftborsten von Reptilien. Z Morphol Tiere 62:307–362

Hirose S, Morishime A (1990) Design and control of a mobile robot with an articulated body. Int J Robot Res 9:99–114

Hoge AR, Souza Santos P (1953) Submicroscopic structure of "stratum corneum" of snakes. Science 118:410–411

Irish FJ, Williams EE, Seling E (1988) Scanning electron microscopy of changes in epidermal structure occurring during the shedding cycle in squamate reptiles. J Morphol 197:105–126

Leydig F (1873) Über die äusseren Bedeckungen der Reptilien und Amphibien. Archiv für mikroskopische Anatomie 9:753–794

Maderson PFA (1966) Histological changes in the epidermis of the Tokay (*Gekko gecko*) during the sloughing cycle. J Morphol 116:39–50

Maderson PFA, Rabinowitz T, Tandler B, Alibardi L (1998) Ultrastructural contributions to an understanding of the cellular mechanisms involved in lizard skin shedding with comments on the function and evolution of a unique lepidosauran phenomenon. J Morphol 236:1–24

McCarthy CJ (1987) Sea snake puzzles. In: van Gelder JJ, Strijbosch H, Bergers PJM (eds) In: Proceedings of the fourth ordinary general meeting of the Scocietas Europaea Herpetologica, Nijmegen, pp 279–284

Mosauer W (1932a) On the locomotion of snakes. Science 76:583–585

Mosauer W (1932b) Über die Ortsbewegung der Schlangen. Zoologisches Jahrbuch (Abteilung für allgemeine Zoologie und Physiologie der Tiere) 52:191–215

Peterson JA (1984a) The scale microarchitecture of *Sphenodon punctatus*. J Herpetol 18:40–47

Peterson JA (1984b) The microstructure of the scale surface in iguanid lizards. J Herpetol 18:437–467

Peterson JA, Bezy RL (1985) The microstructure and evolution of scale surfaces in xantusiid lizards. Herpetologica 41:298–324

Picado C (1931) Epidermal microornaments of the crotalinae. Bull Antivenin Inst Am 4:104–105

Porter WP (1967) Solar radiation through the living body walls of vertebrates with emphasis on desert reptiles. Ecol Monogr 37:273–296

Price RM (1982) Dorsal snake scale microdermatoglyphics: ecological indicator or taxonimical tool? J Herpetol 16:294–306

Price RM (1983) Microdermatoglyphics: the *Liodytes-Regina* problem. J Herpetol 17:292–294

Price RM (1990) Microdermatoglyphics: an appeal for standardization of methodology and terminology with comments on recent studies of North American natricines. J Herpetol 24:324–325

Price RM, Kelly P (1989) Microdermatoglyphics: basal patterns and transition zones. J Herpetol 23:244–261

Renous S, Gasc J, Diop A (1985) Microstructure of the tegumentary surface of the Squamata (Reptilia) in relation of their spatial position and their locomotion. Forts Zool 30:478–479

Ruibal R (1968) The ultrastructure of the surface of lizard scales. Copeia 1968(4):698–703

Ruibal R, Ernst V (1965) The structure of the digital setae of lizards. J Morphol 117:271–294

Scherge M, Gorb SN (2001) Biological micro- and nanotribology. Springer, Berlin

Spurr RA (1969) A low-viscosity epoxy resin embedding medium for electron microscopy. J Ultra Mol Struct R 26:31–43

Stewart GR, Daniel RS (1972) Scales of the lizard Gekko gecko: surface structure examined with the scanning electron microscope. Copeia 1972:252–257

Stewart GR, Daniel RS (1973) Scanning electron microscopy of scales from different body regions of three lizard species. J Morphol 139:377–388

Stewart GR, Daniel RS (1975) Microornamentation of lizard scales: some variation and taxonomic correlation. Herpetologica 31:117–130

Stille B (1987) Dorsal scale microdermatoglyphics and rattlesnake (*Crotallus* and *Sisturus*) phylogeny (Reptilia: Viperidae: Crotalinae). Herpetologica 43:89–104

J Comp Physiol A
DOI 10.1007/s00359-010-0556-y

ORIGINAL PAPER

Material properties of the skin of the Kenyan sand boa *Gongylophis colubrinus* (Squamata, Boidae)

Marie-Christin G. Klein · Julia K. Deuschle · Stanislav N. Gorb

Received: 11 May 2010 / Revised: 21 June 2010 / Accepted: 22 June 2010

Abstract On the basis of structural data, it has been previously assumed that the integument of snakes consists of a hard, robust, inflexible outer surface (*Oberhäutchen* and β-layer) and soft, flexible inner layers (α-layers). The aim of this study was to compare material properties of the outer and inner scale layers of the exuvium of *Gongylophis colubrinus*, to relate the structure of the snake integument to its mechanical properties. The nanoindentation experiments have demonstrated that the outer scale layers are harder, and have a higher effective elastic modulus than the inner scale layers. The results obtained provide strong evidence about the presence of a gradient in the material properties of the snake integument. The possible functional significance of this gradient is discussed here as a feature minimizing damage to the integument during sliding locomotion on an abrasive surface, such as sand.

Keywords Effective elastic modulus · Nanoindentation · Gradient materials · Biological materials · Wear resistance

Introduction

Snakes are limbless reptiles that use their entire body for sliding locomotion. This supposes that the epidermis, at least that of the ventral body side, has to permanently endure friction forces. One can expect some specializations in the snake epidermis, against abrasion. Even

M.-C. G. Klein (✉) · S. N. Gorb
Department of Functional Morphology and Biomechanics, Zoological Institute of the University of Kiel, Am Botanischen Garten 1-9, 24098 Kiel, Germany
e-mail: mklein@zoologie.uni-kiel.de

S. N. Gorb
e-mail: sgorb@zoologie.uni-kiel.de

M.-C. G. Klein · J. K. Deuschle · S. N. Gorb
Evolutionary Biomaterials Group, Max Planck Institute for Metals Research, Heisenbergstr. 3, 70569 Stuttgart, Germany
e-mail: deuschle@mf.mpg.de

though snakes from different lineages have different body size, scale dimension, body shape and body mass (Mattison 1995) their epidermis basically consists of six layers overlying the dermis (Baden and Maderson 1970; Landmann 1979) (Fig. 1c). The following epidermis layers from the outer scale surface towards the dermis are known (Fig. 1c): (1) the *Oberhäutchen*, (2) β-layer, (3) mesos-layer, (4) α-layer, (5) lacunar tissue and (6) clear layer.

During maturation of the epidermis, the relatively thin *Oberhäutchen* fuses together with the thick β-layer resulting in a homogeneous *stratum* without cell boundaries. In the lowermost region of the β-layer, cell boundaries are still recognizable (Landmann 1979). When combined together these two layers compose the outer surface of the snake epidermis, which interacts with the environment. The next layers, underlying this relatively thick outer shell, are so-called α-layers composed of α-keratin (Gregg and Rogers 1986; Sawyer et al. 2000; Alibardi and Sawyer 2002). The mesos-layer consists of 2–10 layers of flattened cells containing α–filaments, which do not form a matrix keratin pattern. The α-layer consists of several keratinized cell layers (number of layers can vary according to species and seasonal influence, perhaps even in one species) and contains α-keratin (mammalian-type) with a coiled-coil conformation rather than β-keratin (bird-type) with a pleated sheet conformation present in the β-layer (Landmann 1979, 1986; Irish et al. 1988). The lacunar tissue and the clear layer make up the inner most part of the epidermis. The clear layer lies directly above the *stratum germativum*. The lacunar tissue consists of 1–4 cell layers. The clear layer consists of one layer of flattened cells. Both layers are similar and their cells resemble immature α-cells (Landmann 1979, 1986).

Snakes shed their entire skin (Irish et al. 1988) 4–8 times a year depending on the species. Moulting provides the renewal of the epidermis, but is not likely to be the only adaptation against wear because frequent moulting is costly from an energetic point of view. In addition, scratches in the outermost layer of the skin may affect frictional properties of the skin and lead to further increase in abrasion. Both the exuvium surface geometry is not different from that of an intact animal, since the new epidermis is developed as a copy of an underneath old epidermis. In a living snake, each scale has two sides exposed to the environment [both of them are here called the outer scale layer (OSL)] (1) the outer scale surface (OSS) is in contact with the substrate and (2) the inner scale surface (ISS) is in contact to the following scale. Both the outer and inner scale surfaces are

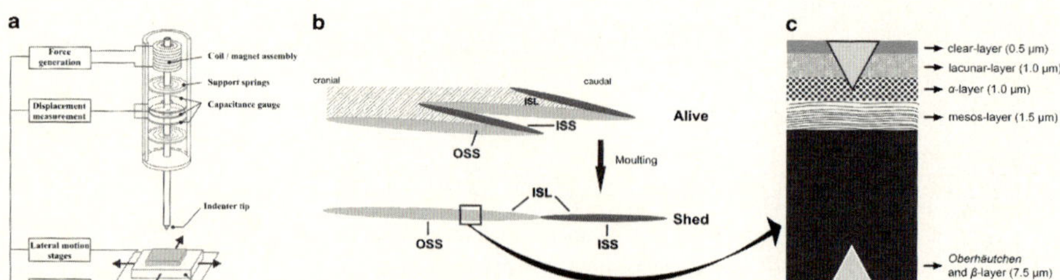

Fig. 1 **a** Schematic drawing of the indentation system setup with the coil-magnet assembly for load application, the capacitive displacement sensor and the indenter shaft support springs (Barbakadze et al. 2006). **b** Schematic drawing of a shedding cycle on scale level. The *black square* illustrates the skin part which was tested. **c** Diagram of the cross section of the snake integument (after Landmann 1986; Maderson 1964, Toni and Alibardi 2007 and the present study). *Triangles* indicate the indentation depth undertaken in this study. *ISL* inner scale layers, *ISS* inner scale surface, *OSS* outer scale surface

composed of the six above-mentioned cell layers. The epidermis side facing the *stratum germativum* is here called the inner scale layer (ISL) the ISLs, which becomes unfolded and exposed after the skin has been shed (Fig. 1b).

Previous authors have hypothesized that the presumably softer α-keratin (Maderson 1964; Alexander 1970; Baden and Maderson 1970; Matoltsy 1976; Wyld and Brush 1979; Furch and Marchuk 1983; Landmann 1986; Carver and Sawyer 1987; O'Guin et al. 1987; Alibardi and Toni 2006a, b, c) is responsible for the flexibility of the α-layer, whereas β-keratin makes the β-layer inflexible (Mercer 1961; Maderson 1964; Matoltsy 1976; Landmann 1986). The β-layer was hypothesized to be a strengthening layer against mechanical and chemical injuries (Landmann 1979). This hypothesis is based, among others, on the data of Baden and Maderson (1970), who showed that matrix proteins of β-keratin contain 7.4% half cystine linked by disulfide bonds, whereas the matrix proteins of α-keratin only contains 6.5%. In addition, the *Oberhäutchen* contains a large amount of matrix material (Landmann 1979). Furthermore, the number of stable disulfide bonds tends to increase from the inside outward through the keratinized layers of the epidermis (Goslar 1958; Banerjee and Mittal 1978). The cell boundaries in squamata *Oberhäutchen* and β-layers are dissolved completely, in contrast to β-keratin in feathers (Filshie and Rogers 1962). In addition, both the mesos- and clear layer contain a keratin network (Landmann 1979), which in combination with a reduced matrix is suggested to be rather flexible (Matoltsy 1976). In the α-layers, the marginal layer of α-keratin containing cells decreases in thickness from without inwards (Landmann 1979), which could be associated with a gradient. The dermis, which lies beneath the epidermis, consists among other components, such as capillaries, of various fibers, and is less stiff as compared to the epidermis (Landmann 1979).

The aim of this study was to compare local material properties of the outer and inner scale layers of the exuvium to test previous hypotheses about the contribution of the structure and composition of the snake integument to its material properties. The exuvium of the Kenyan sand boa *Gongylophis colubrinus* (Squamata, Boidae) (Fig. 2a) was used in the experiment. *G. colubrinus* inhabits desert and moves through the sand or on it. Thus, its ventral epidermis is under pressure, and is continuously in contact with highly abrasive

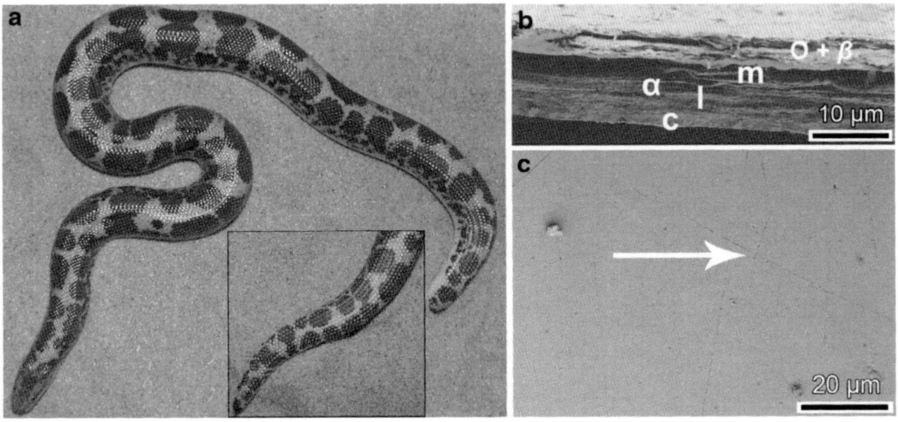

Fig. 2 Body (**a**), tail region (**a** *inset*) and SEM images of the ventral scales of *G. colubrinus*. Fracture of a ventral scale of an exuvium showing different epidermis layers (outer scale surface is on the *top* of the *image*) (**b**). Caudal part of a ventral scale of an exuvium illustrating abrasion marks on the *Oberhäutchen* (*arrow*) (**c**). *O* + *β Oberhäutchen* + *β*-layer, *m* mesos layer, *α* α-layer, *l* lacunar tissue, *c* clear layer

sand particles. This snake, occupying a habitat similar to the sand skink (*Scincus scincus*), has been previously suggested to have a very robust epidermis, which exhibits low friction and high abrasion resistance (Rechenberg and El Khyari 2004).

Using the nanoindentation technique, both the effective elastic modulus and hardness of the outer and inner epidermis layers were characterized and compared. This method, allowing characterization of material properties at the local level, has been previously used to evaluate mechanical properties on small volumes of various biological materials (Arzt et al. 2002; Enders et al. 2004; Barbakadze et al. 2006). The results obtained in the present study provide strong evidence about the presence of the gradient in material properties of the snake integument.

Materials and methods

The nanoindentation measurements were conducted using exuvia from an adult *G. colubrinus* male (Fig. 2a). Cryo scanning electron microscopy was conducted using snake samples, obtained from the University of Bonn. The samples were kept in a −20°C deep freezer subsequently after death.

Nanoindentation

Nanoindentation is a technique to measure mechanical properties of thin and heterogeneous samples with small volumes (Oliver and Pharr 1992; Xu et al. 1998; Ebenstein and Pruitt 2006). Through the force–displacement curves obtained using this technique, hardness and modulus of the material can be determined (Oliver and Pharr 1992) for penetration depths as small as several hundred nanometers (Enders et al. 2004). The equations used to extract hardness and elasticity information from force–displacement curves according to Oliver and Pharr (1992; 2004) are as follows:

$$H = \frac{p_{\max}}{A_c}, \tag{1}$$

where H is hardness, $P_{\max}$ is the maximal load and A_c is the contact area,

$$E_{\text{eff}} = \frac{\sqrt{\pi}}{2}\frac{S}{\beta\sqrt{A_c}} \tag{2}$$

with E_{eff} being the reduced or effective elastic E modulus, β the correction factor for the indenter form, and S the contact stiffness. In contrast to the Young's modulus, the effective elastic modulus can be determined by means of nanoindentation. Furthermore, since the Poisson coefficient is not known the elastic modulus could not be determined. Owing to the variation of material structure depending on the depth in biological samples, the mechanical properties might fluctuate considerably with depth (Enders et al. 2004). Therefore, the continuous stiffness measurement method was applied, which allows a continuous measurement of mechanical properties as a function of depth (Oliver and Pharr 1992; Enders et al. 2004). Although the Oliver–Pharr method is mainly meant for isotropic materials and therefore is not ideal for anisotropic biological material, it is the best method to date for studies of material properties. In our previous work, we successfully applied this method for mechanical characterization of biological samples (Arzt et al. 2002; Enders et al. 2004; Barbakadze et al. 2006). Furthermore, this method is valid for our anisotropic material because we are comparing both sides of a biological material in relation to one another using relative and not absolute values and the comparability with previous results from former studies is given.

The Nano Indenter SA2 (MTS Nano Instruments, Oak Ridge, TN, USA) (Fig. 1a) was used in this study. This system allows testing of materials with low contact stiffness, such as soft biological tissues or gels (Enders et al. 2004; Deuschle et al. 2007). A standard Berkovich indenter tip was used in all experiments. This tip is a three-sided pyramid with a face opening angle of 65.3° (Enders et al. 2004). The tests were conducted under normal room conditions (relative humidity 40–45%, temperature 22–25°C). During each test, the relative humidity was kept constant and the temperature fluctuations were recorded to be below ±0.5°C.

Sample preparation for nanoindentation

Exuvia pieces (5 × 5 mm) of *G. colubrinus* were cut from the ventral scales of the mid-body region and attached to an aluminum sample holder with 5925 cyanacrylat superglue (Kisling AG, Schwäbisch Hall, Germany). The samples were fixed in one of two ways: (1) the outer scale layers, the *Oberhäutchen*/*β*-layer, of the outer scale surface were facing the indenter tip; (2) the inner scale layers, clear layer, of the outer scale surface was facing the tip (Fig. 1b). This way the mechanical properties into the direction perpendicular to the surface and therefore out-of-plane regarding the keratin fibers were obtained. The results acquired from this experimental setup, in contrast to in-plane indentation, are relevant for our study since the forces exerted on the snake skin are mainly out-of-plane. For each surface, three different individual scales were prepared. 12–19 indentation tests, on the posterior edge of the scale, were carried out on each individual scale.

Nanoindentation test procedure

All tests on one sample were performed within 1 h to minimize noise in the data caused changes in the environmental conditions of the laboratory. The samples were loaded under constant strain rate conditions, with a rate of 0.05 s^{-1} to a maximum penetration depth of 2 μm. Before the sample was fully unloaded the maximum load was reduced to 95% and the thermal drift was measured. Thereafter, the indenter tip was moved to the next preselected test location. To ensure proper surface detection, the procedure described in Deuschle et al. (2007) was used. After the tests were completed, each indentation curve was manually processed to verify a correct surface approach.

Preparation for scanning electron microscopy of the cross sections

5 × 10 mm pieces of exuvia and scale margins of fresh skin (kept frozen at −20°C) were cut from the ventral scale region and mechanically fixed on a specimen holder in a vertical position relative to the holder plane, so that the fractured integument could be examined in the cross section. The mounted samples were blown with a jet of compressed air, to remove dust particles, frozen in liquid nitrogen and then fractured using a cold metal knife at −140°C in the Gatan ALTO 2500 Cryo preparation system. After coating with gold palladium (6 nm thickness and a rate of 1:9), the samples were observed in a Hitachi-S4800 (Tokyo, Japan) scanning electron microscope at the temperature of −120°C and accelerating voltage of 2–3 kV. Exuvia were also examined using the Hitachi-S4800 (Tokyo, Japan) SEM, although at room temperature.

Preparation for scanning electron microscopy of the scale structures

5 × 10 mm pieces of ventral, dorsal and lateral scale regions of a *G. colubrinus* exuvia were cut and mechanically fixed on a specimen holder, so that the outer scale surface of the integument could be examined. The mounted samples were then prepared as described above and observed in the SEM at an accelerating voltage of 2–3 kV.

Results

The epidermis of *G. colubrinus* shows a variation in scale microstructure, when comparing the ventral and lateral sides with the dorsal sides of the body (Fig. 3). Dorsal, ventral and lateral scales have in common that the outer scale surface bears strap-shaped *Oberhäutchen* cells, which exhibit pits. The cells have caudal elevations and random denticulations from the cranial to central part of the scale (Fig. 3a). However, in contrast to the dorsal scales, which bear denticulations aligned in comb-like rows running perpendicular to the longitudinal body axis on the caudal scale part and have a central ridge, the lateral and ventral scales are relatively smooth from the central to caudal part of the scale (Fig. 3b).

The ventral epidermis of a *G. colubrinus* exuvium is about 12 μm thick. The layered organization of the epidermis layers is clearly visible in fractures of both exuvia and fresh scales (Figs. 2b, 4a, b). Abrasion marks were often seen on the outer scale surface (*Oberhäutchen*) of a ventral exuvium scale (Fig. 2c).

Taking into consideration the indentation depth, applied in this study (Fig. 1c), the information about mechanical response was collected from the outer regions of the *Oberhäutchen* + β-layer (outer surface) and from the α-layers (inner surface). The hardness and effective elastic modulus values for the exuvia samples were studied as a function of the indentation depth (Fig. 5). No significant difference was found, when comparing the measurement series separately for either the inner or outer scale surface for each scale tested. The load–displacement curves revealed a significant difference in the mechanical response, and thus in the modulus values obtained for the outer (4.1 GPa) and inner (3.2 GPa) scale layers (Table 1; $p < 0.001$, one-way RM ANOVA and *t* test). There was also a statistically significant difference between hardness values of the outer (0.28 GPa) and inner (0.14 GPa) scale layers (Table 1; $p < 0.001$, one-way RM ANOVA and *t* test).

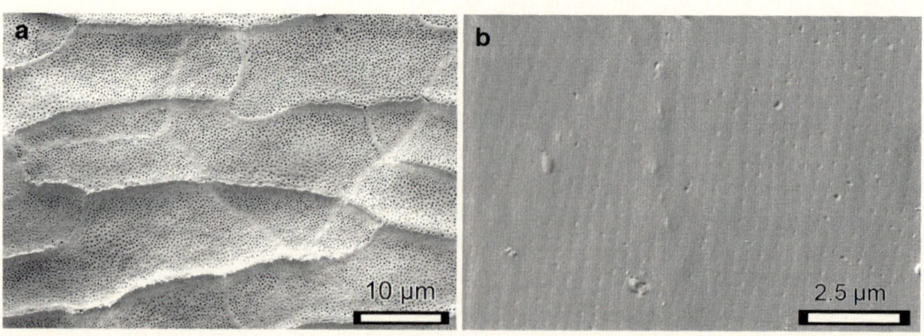

Fig. 3 *G. colubrinus*, SEM images illustrating some of the scale microstructure. Cranial part of a ventral **a** scale and of the caudal part of a dorsal **b** scale. *Image top* indicates caudal direction

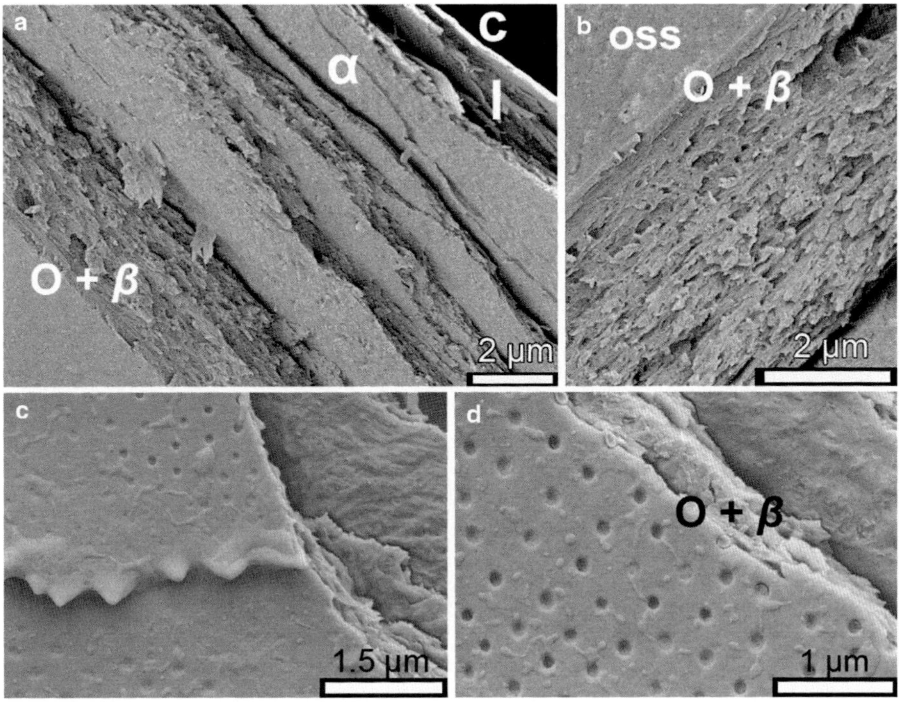

Fig. 4 *G. colubrinus*, Cryo-SEM of ventral scales. Profile of a fresh scale with an overview (**a**) of the epidermis layers and a detailed view (**b**) of the β-layer. The outer scale surface (*OSS*) is in the bottom *left* (**a**) or *top left* (**b**). Rostral *OSS* part of a fresh scale shown from above (**c**). Caudal *OSS* part of a fresh scale shown from above (**d**). $O + \beta$ *Oberhäutchen* + β-layer, *m* mesos-layer, α α-layer, *l* lacunar tissue, *c* clear layer, *OSS* outer scale surface

Figure 5b, c shows that the modulus and hardness values change with indentation depth for both the inner and the outer scale layers. Table 2 shows the modulus and hardness values for the inner and outer scale layers within three intervals (400–500, 1,000–1,100 and 1,600–1,700 nm). When comparing data obtained within these depth intervals, the modulus values increase with indentation depth for the successive inner scale layers (3.30, 3.43 and 3.48 GPa) and decrease for the successive outer scale layers (5.02, 4.73 and 4.46 GPa), whereas the hardness values do not change significantly (Table 2).

Tables 3 and 4 present differences between modulus and hardness results measured at various depths for the indentation from inner (Table 3) and outer (Table 4) directions. The comparison within the inner scale layers shows that the only significant difference for the modulus was found between ranges of depth of 400–500 and 1,600–1,700 nm. This would suggest a fine transition of the modulus from the inside to the outside. When looking at the hardness values for the inner scale layers, only hardness measured in the depth ranges 400–500 and 1,600–1,700 nm shows no significant difference. This would suggest that inner layers start out hard, become less hard at a depth of 1,000 nm and then become harder again, when the indenter reaches the α-layer.

When comparing modulus values for outer scale layers, all three depth intervals show significant differences between each other. In hardness, there is only a significant difference between depth intervals 400–500 and 1,000–1,100 nm.

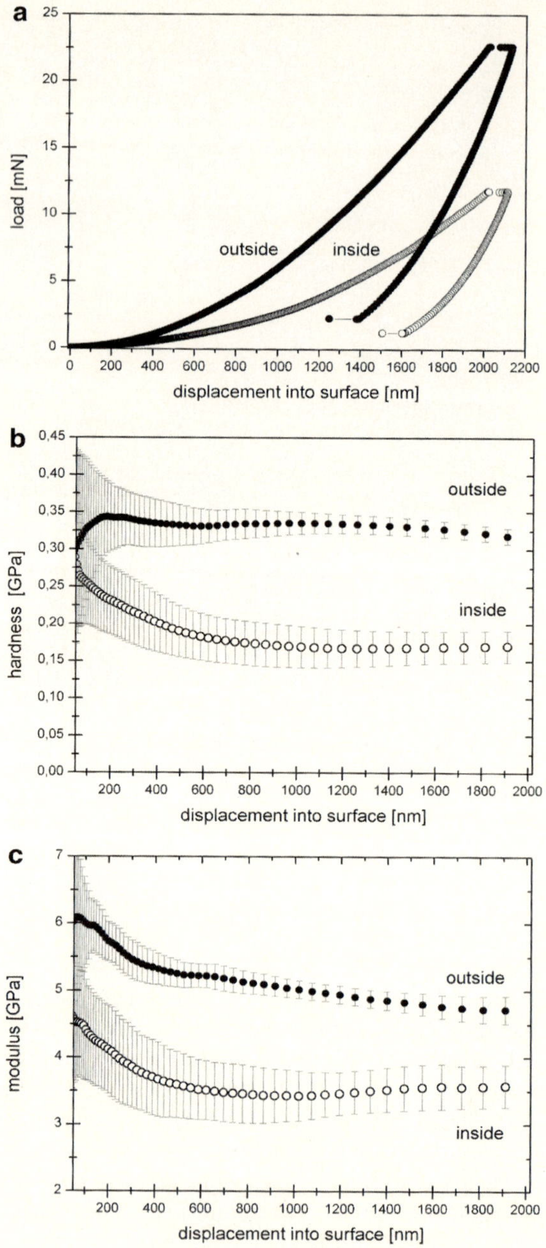

Fig. 5 Results of nanoindentation measurements for both the outer and inner scale layers. **a** Load–displacement curves. **b** Hardness versus displacement. **c** Effective elastic modulus versus displacement. *Each curve* represents the average of at least 12 tests and the *error bars* denote standard deviations

Discussion

Our results of the nanoindentation tests on ventral *G. colubrinus* scales provide strong evidence for the existence of a gradient in material properties across the ventral epidermis of the sand boa. Because the snake epidermis consists of dead cells or cellular derivatives (Irish et al. 1988), we assume no discrepancy in material properties between fresh and shed

Table 1 Mean and standard deviation values of hardness and effective elastic modulus measurements on the inner scale layers (ISL) and outer scale layers (OSL) of three different ventral scales of *G. colubrinus* exuvium

ISL			OSL			
Measurement	Mean (GPa)	SD (GPa)	Measurement	Mean (GPa)	SD (GPa)	
Effective elastic modulus						
1 ($n = 19$)	3.15	0.31	1 ($n = 16$)	4.86	0.13	$p < 0.01$, *t* test
2 ($n = 12$)	3.30	0.28	2 ($n = 20$)	4.98	0.15	$p < 0.01$, Tukey's test
3 ($n = 14$)	3.55	0.35	3 ($n = 16$)	4.91	0.25	$p < 0.01$, *t* test
Pooled ($n = 45$)	3.16	0.30	Pooled ($n = 52$)	4.13	0.34	$p < 0.01$, Tukey's test
Hardness						
1 ($n = 19$)	0.13	0.02	1 ($n = 16$)	0.27	0.03	$p < 0.01$, *t* test
2 ($n = 12$)	0.16	0.02	2 ($n = 20$)	0.35	0.02	$p < 0.01$, *t* test
3 ($n = 14$)	0.18	0.03	3 ($n = 16$)	0.34	0.01	$p < 0.01$, Tukey's test
Pooled ($n = 45$)	0.14	0.02	Pooled ($n = 52$)	0.28	0.02	$p < 0.01$, Tukey's test

Statistical comparison was carried out using the *t* test and one-way RM ANOVA

Table 2 Comparison of the mean and standard deviation values of hardness and effective elastic modulus measured at three different indentation depth intervals (400–500, 1,000–1,100 and 1,600–1,700 nm) of the inner scale layers (ISL) and outer scale layers (OSL) of all three *G. colubrinus* scales tested

ISL			OSL			
Depth range (nm) ($n = 45$)	Mean (GPa)	SD (GPa)	Depth range (nm) ($n = 52$)	Mean (GPa)	SD (GPa)	
Effective elastic modulus						
400–500	3.30	0.36	400–500	5.02	0.59	$p < 0.01$, Tukey's test
1,000–1,100	3.43	0.56	1,000–1,100	4.73	0.23	$p < 0.01$, Tukey's test
1,600–1,700	3.48	0.36	1,600–1,700	4.46	0.18	$p < 0.01$, Tukey's test
Hardness						
400–500	0.15	0.03	400–500	0.31	0.07	$p < 0.01$, Tukey's test
1,000–1,100	0.17	0.05	1,000–1,100	0.32	0.04	$p < 0.01$, Tukey's test
1,600–1,700	0.15	0.03	1,600–1,700	0.32	0.03	$p < 0.01$, Tukey's test

Statistical comparisons were carried out using one-way RM ANOVA

epidermis. The outer scale layers are harder and exhibit a higher effective elastic modulus than the inner scale layers containing only a reduced keratin matrix. The data correlates well with antibody labeling results showing the distribution of α- and β-keratin within

Table 3 Comparison of the mean and standard deviation values measured at three different depth intervals (400–500, 1,000–1,100 and 1,600–1,700 nm) of hardness and effective elastic modulus on the inner (ISL) scale layers of all three *G. colubrinus* scale tested

Depth range (nm) ($n = 45$)	Mean (GPa)	SD (GPa)	Depth range (nm) ($n = 45$)	Mean (GPa)	SD (GPa)	
Effective elastic modulus						
400–500	3.30	0.36	1,000–1100	3.43	0.56	NS
1,000–1,100	3.43	0.56	1,600–1,700	3.48	0.36	NS
1,600–1,700	3.48	0.36	400–500	3.30	0.36	$p = 0.01$, Tukey's test
Hardness						
400–500	0.15	0.03	1,000–1,100	0.17	0.05	$p < 0.01$, Tukey's test
1,000–1,100	0.17	0.05	1,600–1,700	0.15	0.03	$p < 0.01$, Tukey's test
1,600–1,700	0.15	0.03	400–500	0.15	0.03	NS

Statistical comparisons were carried out using one-way RM ANOVA

Table 4 Comparison of the mean and standard deviation values measured at three different depth intervals (400–500, 1,000–1,100 and 1,600–1,700 nm) of hardness and effective elastic modulus on the outer scale layers (OSL) of all three *G. colubrinus* scale tested

Depth range (nm) ($n = 52$)	Mean (GPa)	SD (GPa)	Depth range (nm) ($n = 52$)	Mean (GPa)	SD (GPa)	
Effective elastic modulus						
400–500	5.02	0.59	1,000–1,100	4.73	0.23	$p < 0.01$, Tukey's test
1,000–1,100	4.73	0.23	1,600–1,700	4.46	0.18	$p < 0.01$, Tukey's test
1,600–1,700	4.46	0.18	400–500	5.02	0.59	$p < 0.01$, Tukey's test
Hardness						
400–500	0.31	0.07	1,000–1,100	0.32	0.04	$p = 0.05$, Tukey's test
1,000–1,100	0.32	0.04	1,600–1,700	0.32	0.03	NS
1,600–1,700	0.32	0.03	400–500	0.31	0.07	NS

Statistical comparisons were carried out using one-way RM ANOVA

molts of juveniles of the python *Lasius fuscus* and the colubrid snake *Natrix natrix* (Toni and Alibardi 2007), whereas the hard β-keratin is only present in the β-layer. In addition, variations in the measured moduli at different depths may also be explained by recent studies (Dalla Valle et al. 2005; Alibardi and Toni 2006a, b, c; Toni and Alibardi 2007), which have shown that there are at least three different types of β-keratins. For instance some that corresponds to glycine–proline-rich proteins that associate to α-keratin bundles and are responsible for the hardening of the reptilian epidermis. These proteins are temporarily synthesized at high levels in the *Oberhäutchen* and β-layer. Synthesis decreases and stops in layers underlying these (mesos- and α-layers) (Landmann 1986). Moreover, Toni and Alibardi (2007) differentiate, among others, between β-keratins that appear as basic proteins and β-keratins associate with cytokeratins as matrix proteins. However, our nanoindentation method did not have enough in-depth resolution to mechanically characterize such fine sub-layers.

The mean values of the modulus and hardness obtained for the outer (4.9 and 0.27 GPa) and inner scale layers (3.1 and 0.13 GPa) are comparable with the results, obtained in other studies on the mechanical properties of keratinous materials (Bonser and Purslow 1995; Moran et al. 2007; Huber et al. 2008). The hardness and modulus values for the inner scale layers are relatively high close to the surface (Fig. 5b, c). This can be explained by the fact that the tip oscillation (independent of the amplitude) affects the contact formation as well as contact stabilization for displacements between 0 and approximately 300 nm and thus results obtained at these shallow depths are not reliable to make any conclusion (Deuschle 2008).

Modulus values for β-keratin range between 1.0 and 8.0 GPa (Bonser and Purslow 1995; Bonser 2001; Huber et al. 2008) and the values for α-keratin are also quite variable (modulus 3.0–7.5 GPa; hardness 0.1–0.4 GPa) (Farren et al. 2004; Wei et al. 2005; Moran et al. 2007). This means material properties of biological materials preferably containing either β- or α-keratin vary more than would be expected. One can assume that it is not only the keratin type which influences hardness and elasticity, but also the arrangement filaments and the architecture of the keratin matrix, for instance the thickness of the β- and α-keratin layers. The material ultrastructure of the snake's integument (relatively tight packaging of layers) or the bird feather (hollow feather shaft) influences material properties according to the specific function of the material. For instance, Bonser (2001) found that the rachis of two different feather types of Ostrich revealed two significantly different moduli. The keratin composite of the wing feather had a modulus of 1.71 GPa and the body contour feather 2.42 GPa. Bonser (2001) argues that the difference in moduli might tell us something about the mechanical loads to which the feathers are subjected. The mechanical environment of contour feathers might be more demanding than that of flight feathers. It was also suggested that strain in feathers will depend on both rachis shape and keratin properties (Bonser 2001). Thus, the feather geometry and the material ultrastructure are important features in determining the mechanical behavior of a whole feather (Astbury and Bell 1939). Furthermore, the method used in order to gain information on material properties (tensile, bending and nanoindentation tests) may also plays a role in varying values (Huber et al. 2008). This would explain the variation of values for β- keratin and for α-keratin and between the two.

Experiments carried out by Licht and Bennett (1972) on scaleless snakes showed that the β-layer contrary to previous data does not function as a permeability barrier. On the basis of these results and his own ones, Landmann (1979) suggested that the β-layers' main function is to protect the epidermis against abrasion. Toni and Alibardi (2007) showed that the α- and β-keratin differ in chemical structure and therefore presumably also in material properties (Matoltsy 1976). The β-keratin molecule is stiff and inelastic, which presumably makes materials containing it more robust and therefore protecting the underlying layers against damage, whereas the α-keratin presumably provides flexibility to the material of the skin (Matoltsy 1976; Landmann 1979).

The epidermis of both the sand boa *G. colubrinus* and the sand skink *S. scincus* possess high durability and friction reducing properties (Rechenberg and El Khyari 2004). These authors suggested that these specific properties of the skink skin, and perhaps sand swimming reptiles in general are due to the microstructure on the scale surface. They supposed that microstructures effect in the reduction of the contact area to the sand particles and may serve as sites for electrostatic charge concentration. The charges may enhance repulsive force between skin and substrate that may lead to the wear reduction. Baumgartner et al. (2007) suggested that the low friction and high abrasion resistance of the sand skink skin is due to specific material properties that rely on the glycosylation of keratins.

Data of our study show for the first time hardness and effective elastic modulus values at different depths of the snake skin and therefore give rise to an additional hypothesis that may explain high abrasion resistance due to a gradient in material properties of the integument from a hard and inflexible outside to a soft and elastic inside. There are other biological materials, such as cartilage, tooth, bone and wood which have to cope with abrasion and friction. Some of these systems have a similar layered structure with gradient material properties. It has previously been shown that in the tooth, such layered organization with the gradient of material properties is a key mechanism of the wear resistance (Wang and Weiner 1998; Fong et al. 2000). The visible part of the tooth, above the gum, is enveloped by the enamel, which is mechanically hard (Zheng et al. 2003) and highly resistant to wear (Fong et al. 2000). Its primary function is to protect the underlying dentin (the tooth's major component) and the more flexible and soft pulp (Zheng et al. 2003). Nanoindentation tests have shown that enamel is five times harder than dentin and possesses a three times higher elastic modulus (Fong et al. 2000). Using indentation tests on pulp and pulpless dog teeth with the Knoop indentor, Fusayama and Maeda (1968) showed that the dentin hardness in pulpless teeth is lower than that of vital teeth. The authors believe this to be due to absence of the pulp functioning as a damping layer for the stiffer dentin (Fusayama and Maeda 1968).

A system which has to endure high amounts of stress under pressure is more effective against abrasion wear, if it has a gradient in structure and properties, since such design leads to more uniform stress distribution and thus to the minimization of probability of local stress concentration (Gibson and Ashby 1988; Wang and Weiner 1998), which lead to the materials fatigue and failure. Since a hard, inflexible system will break easily under pressure and a flexible system will be easily worn off, the gradient material will improve wear resistance by combining advantages of both stiffness (outside) and flexibility (inside).

The β-layer of the snake skin can also be compared by analogy with some technical surface developments, such as diamond-like carbon (DLC) coatings. Nanoindentation tests showed that DLC layers have much higher hardness values (14–18 GPa) than those of coated materials (Xu and Rowcliffe 2002). That is why DLC coatings have low friction coefficients (Lettington 1998), which in combination with more compliant underlying materials make them useful for protecting objects from scratches.

Conclusion and outlook

The microscopy data provide evidence for the presence of a structural gradient in the ventral epidermis. The nanoindentation results demonstrate the presence of the gradient in the material properties of the snake skin. Taken together, these two facts may suggest that the structural gradient is responsible for gradient of mechanical properties, which might represent one of the mechanisms reducing abrasion of the snake skin. Further comparative data on different body regions of different snakes, as well as the local experimental testing of the skin wear, might aid in the understanding of the role of structural and mechanical features of the skin in abrasion reduction.

Acknowledgments Dr. Guido Westhoff (University of Bonn, Germany) provided frozen snakes and valuable comments on snake biology. S.G. was supported in this work by the Federal Ministry of Education, Science and Technology, Germany (BMBF project Biona 01RB0812A). Animal care for the live *G. colubrinus* was provided by M.-C. Klein. The experiments comply with the "Principles of animal care", publication no. 86–23, revised 1985 of the National Institute of Health and also with the current laws of Germany.

References

Alexander NJ (1970) Comparison of α and β keratin in reptiles. Z Zellforsch 110:153–165

Alibardi L, Sawyer RH (2002) Immunocytochemical analysis of beta keratins in the epidermis of chelonians, lepidosaurians, and archosaurians. J Exp Zool 293:27–38

Alibardi L, Toni M (2006a) Cytochemical, biochemical and molecular aspects of the process of keratinization in the epidermis of reptilian scales. Prog Histochem Cytochem 40:73–134

Alibardi L, Toni M (2006b) Immunological characterization and fine localization of lizard beta-keratin. J Exp Zool 306B:528–538

Alibardi L, Toni M (2006c) Immunolocalization and characterization of beta-keratins in growing epidermis of chelonians. Tissue Cell 38:53–63

Arzt E, Enders S, Gorb S (2002) Towards a micromechanical understanding of biological surface devices. Z Metallkunde 93:345–351

Astbury WT, Bell FO (1939) X-ray data on the structure of natural fibres and other bodies of high molecular weight. Tabulae Biol 17:90–112

Baden HP, Maderson PFA (1970) Morphological and biophysical identification of fibrous proteins in the amniote epidermis. J Exp Zool 174:225–232

Banerjee TK, Mittal AK (1978) Histochemistry of the epidermis of the checkered water snake *Natrix piscator* (Colubridae, Squamata). J Zool 185:415–435

Barbakadze N, Enders S, Gorb S, Arzt E (2006) Local mechanical properties of the head articulation cuticle in the beetle *Pachnoda marginata* (Coleoptera, Scarabaeidae). J Exp Zool 209:722–730

Baumgartner W, Saxe F, Weth A, Hajas D, Sigumonrong D, Emmerlich J, Singheiser M, Böhme W, Schneider JM (2007) The sandfish's skin: morphology, chemistry and reconstruction. J Bionic Eng 4:1–9

Bonser RHC (2001) The elastic properties of wing and contour feather keratin from the Ostrich *Struthio camelus*. Ibis 143:144–145

Bonser RHC, Purslow PP (1995) The Young's modulus of feather keratin. J Exp Bio 198:1029–1033

Carver WE, Sawyer RH (1987) Development and keratinization of the epidermis in the common lizard, *Anolis carolinensis*. J Exp Zool 243:435–443

Dalla Valle L, Toffolo V, Belvedere P, Alibardi L (2005) Isolation of mRNA encoding a glycine-proline-rich β-keratin expressed in the regenerating epidermis of lizard. Dev Dynam 234:934–947

Deuschle J (2008) Mechanics of soft polymer indentation. Dissertation, Max-Planck-Institute for Metals Research Stuttgart and Institute for Metallography of the University of Stuttgart

Deuschle J, Enders S, Arzt E (2007) Surface detection in nanoindentation of soft polymers. J Mater Res 22:3107–3119

Ebenstein DM, Pruitt LA (2006) Nanoindentation of biological materials. Nano Today 1:26–33

Enders S, Barbakadze N, Gorb SN, Arzt E (2004) Exploring biological surfaces by nanoindentation. J Mater Res 19:880–887

Farren L, Shayler S, Ennos AR (2004) The fracture properties and mechanical design of human fingernails. J Exp Biol 207:735–741

Filshie BK, Rogers GE (1962) An electron microscope study of the fine structure of feather keratin. J Biophys Biochem Cytol 13:1–12

Fong H, Sarikaya M, White SN, Snead ML (2000) Nano-mechanical properties profiles across dentin–enamel junction of human incisor teeth. Mat Sci Eng 7:119–128

Fraser RDB, MacRae TP (1980) Molecular structure and mechanical properties of keratin. In: Vincent JFV, Currey JD (eds) The mechanical properties of biological materials. Symposia of the Society for Experimental Biology

Furch E, Marchuk D (1983) Type I and type II keratins have evolved from lower eukaryotes to from the epidermal intermediate filaments in mammalian skin. Proc Natl Acad Sci USA 80:5857–5861

Fusayama T, Maeda T (1968) Effect of pulpectomy on dentin hardness. J Dent Res 48:452–460

Gibson LJ, Ashby MF (1988) Cellular solids: structures and properties. Pergamon Press, New York

Goslar HG (1958) Beiträge zum Häutungsvorgang der Schlangen. Acta Histochem 5:182–212

Gregg K, Rogers GE (1986) Feather keratins: composition, structure and biogenesis. In: Bereither-Hahn J, Matoltsy GA, Sylvia-Richards K (eds) Biology of the integument: vertebrates, vol 2. Springer, New York, pp 666–694

Herter K (1960) Das Tierreich VII/4. Kriechtiere. Walter de Gruyter & Co, Berlin

Huber G, Orso S, Spolenak R, Wegst UGK, Enders S, Gorb SN, Arzt E (2008) Mechanical properties of a single gecko seta. Int J Mater Res 2008:1113–1118

Irish FJ, Williams EE, Seling E (1988) Scanning electron microscopy of changes in the epidermal structure occurring during the shedding cycle in squamate reptiles. J Morphol 197:105–126

Landmann L (1979) Keratin formation and barrier mechanisms in the epidermis of *Natrix natrix* (Reptilia: Serpentes): an ultrastructural study. J Morphol 162:93–126

Landmann L (1986) Biology of the integument. In: Bereither-Hahn J, Matoltsy GA, Sylvia-Richards K (eds) The skin of reptiles, Chap 9: Epidermis and dermis. Springer, Heidelberg, pp 150–185

Lettington AH (1998) Applications of diamond-like carbon thin films. Carbon 36:555–560

Licht P, Bennett AF (1972) A scaleless snake: tests of the role of reptilian scales in water loss and heat transfer. Copeia 4:702–707

Maderson PFA (1964) The skin of lizards and snakes. B J Herpet 3:151–154

Matoltsy AG (1976) Keratinization. J Invest Dermatol 67:20–25

Mattison C (1995) The encyclopedia of snakes. Cassel & Co, London

Mercer EH (1961) Keratin and keratinization. Pergamon Press, Inc, Oxford

Moran P, Towler MR, Chowdhury S, Saunders J, German MJ, Lawson NS, Pollock HM, Pillay L, Lyons D (2007) Preliminary work on the development of a novel detection method for osteoporosis. J Mater Sci 18:969–974

O'Guin WM, Galvin S, Schermer A, Sun TT (1987) Patterns of keratin expression define distinct pathways of epithelial development and differentiation. Curr Top Dev Biol 22:97–125

Oliver WC, Pharr GM (1992) An improved technique for determining hardness and elastic modulus using load and displacement sensing indentation experiments. J Mater Res 7:1564–1583

Oliver WC, Pharr GM (2004) Measurement of hardness and elastic modulus by instrumented indentation: advances in understanding and refinements in methodology. J Mater Res 19:3–20

Rechenberg I, El Khyari AR (2004) Der Sandskink der Sahara—Vorbild für Reibungs- und Verschleißminderung. http://www.bionik.tuberlin.de/institute/festo04.pdf,Berlin

Sawyer RH, Glenn T, French JO, Mays B, Shames RB, Barnes GL, Rhodes W, Ishikawa Y (2000) The expression of beta keratins in the epidermal appendages of reptiles and birds. Am Zool 40:530–539

Toni M, Alibardi L (2007) Alpha- and beta-keratin of the snake epidermis. Zoology 110:41–47

Wang RZ, Weiner S (1998) Strain-structure relations in human teeth using Moire fringes. J Biomech 31:135–141

Wei G, Bhushan B, Torgerson PM (2005) Nanomechanical characterization of human hair using nanoindentation and SEM, vol 105. Elsevier, Amsterdam, pp 248–266

Wyld JA, Brush AH (1979) The molecular heterogeneity and diversity of reptilian keratins. J Mol Evol 12:331–347

Xu Z, Rowcliffe D (2002) Nanoindenation on diamond-like carbon and alumina coatings. Surf Coat Technol 161:44–51

Xu HHK, Smith DT, Jahanmir S, Romber E, Kelly JR, Thompson VP, Rekow ED (1998) Indentation damage and mechanical properties of human enamel and dentin. J Dent Res 77:472–480

Zheng J, Zhou ZR, Zhang J, Li H, Yu HY (2003) On the friction and wear behaviour of human tooth enamel and dentin, vol 255. Elsevier, Amsterdam, pp 967–974

3. Attachment Devices

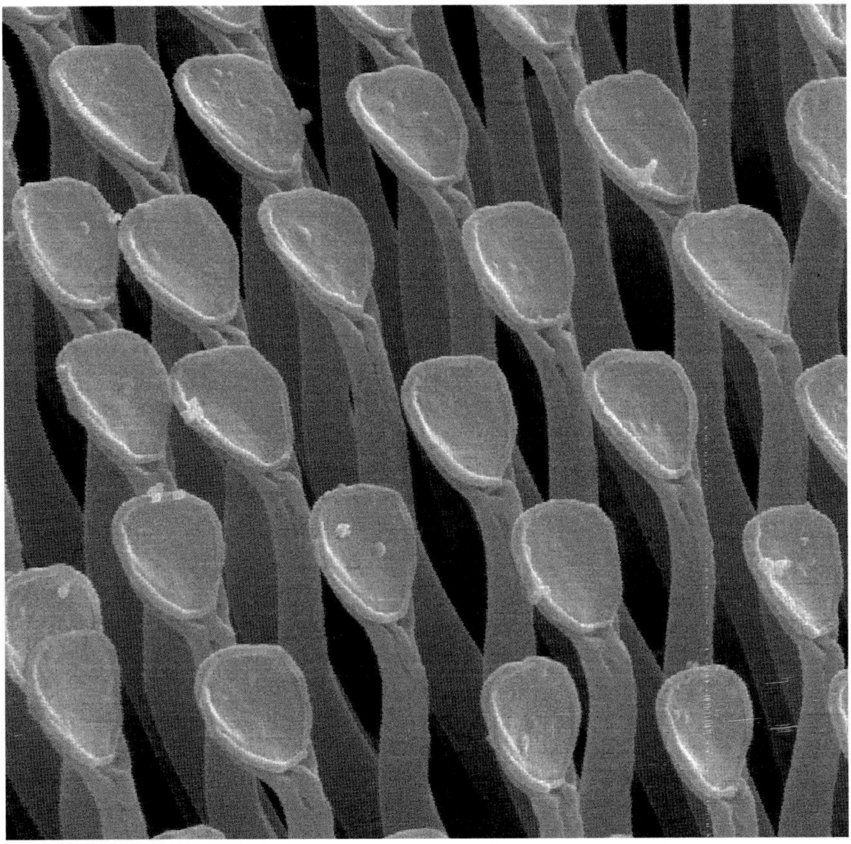

Naturwisswnschaften (2001) 88:530–534
DOI 10.1007/s00114-001-0274-y

SHORT COMMUNICATION

Evolution of locomotory attachment pads of hexapods

S. N. Gorb · R. G. Beutel

Received: 2 July 2001 / Accepted in revised form: 10 September 2001 / Published online: 31 October 2001

Abstract This study shows that, in their evolution, hexapods have convergently developed two distinctly different mechanisms to attach themselves to a variety of substrates during locomotion. The first mechanism is provided by hairy surfaces and the second one by smooth flexible pads. The main similarity of both mechanisms is that the structured pad surface or particular properties of pad materials guarantee a maximum real contact with diverse substrata, regardless of their microsculpture. Ten characters of the two alternative designs were coded and analyzed together with a data matrix containing 105 additional morphological characters of different stages and body parts. The analysis demonstrates that similar structures (arolium, euplantulae, hairy tarsomeres) have evolved independently in several hexapod lineages. The evolution of flight and the associated necessity of being able to cling to vegetation or other substrates are suggested to be major triggers for the evolution of attachment structures.

Introduction

Mechanisms of animal attachment have fascinated scientists over the centuries, but many details of functions of locomotory attachment devices still remain unclear. Such releasable attachment mechanisms have been discussed recently in the literature (Autumn et al. 2000; Pain 2000), but the evolutionary aspects of an animal's ability to walk on the ceiling still remain unexplored. Hexapods possess attachment devices, which diversify enormously in the evolution of this group, and represent, therefore, an interesting example to illustrate the evolution of biomechanical systems of attachment pads connected to locomotion.

S. N. Gorb (✉)
Biological Microtribology Group, Biochemistry Department, MPI für Entwicklungsbiologie, Spemannstrasse 35, Tübingen 72076, Germany
e-mail: stas.gorb@tuebingen.mpg.de

R. G. Beutel
Institut für Spezielle Zoologie und Evolutionsbiologie mit Phyletischem Museum, Friedrich-Schiller-Universität, Erbertstrasse 1, Jena 07743, Germany

Regarding hexapod pretarsal structures, the first comparative accounts were given by (De Meijere 1901) and (Holway 1935) and their terminology was presented by (Dashman 1953). Specific information about the general design of hexapod structures adapted to attachment is scattered throughout the vast systematic literature. To date, both the morphological and ultrastructural bases of the ability of hexapods to walk on vertical surfaces have been studied in detail only in representatives of selected groups including Orthoptera, Blattodea, Thysanoptera, Hemiptera, Hymenoptera, Diptera, and Coleoptera (Beutel and Gorb 2001). It is remarkable that these highly specialized structures are not restricted to one particular area of the leg. They may be located on different parts, such as claws, derivatives of the pretarsus, tarsal apex, tarsomeres, or tibia. Therefore, it appears obvious that structures adapted to attachment have evolved independently several times. However, in the past, authors have discussed this phenomenon considering only members of a few hexapod taxa, and have not provided a phylogenetic analyses of the characters in question.

Here, we summarize our ultrastructural studies of the cuticle of hexapod adhesive pads and additional data from the literature in representatives of most hexapod orders. Our principal goal was to review structures adapted to substrate attachment during locomotion and to provide their evolutionary interpretation based on a cladistic analysis of characters of attachment structures. Ten characters of the attachment structures were coded and analyzed, together with a data matrix containing 105 additional morphological characters of different stages and body parts. Leg structures specialized for prey-capture and/or copulation were excluded.

Material and methods

Light microscopy and scanning electron microscopy

Specimens were fixed in 2.5% glutaraldehyde (0.01 M phosphate buffer at pH 7.3), postfixed for 1 h in 1% osmium tetroxide in phosphate buffer at 2°C, washed, dehydrated and embedded in a low-viscosity resin. Semi-thin sections (0.5-2.0 μm) of pads embedded in Spurr resin were sectioned using a diamond knife, were stained with Toluidine Blue or with triple stain after Cason and were examined by light microscopy. An additional technique was used to obtain information about inner structures. Sections were picked up on pioloform-covered cover-slips, treated with Maxwell's solution (Maxwell 1978) for 2–5 min in order to remove the resin, washed in absolute ethanol and critical-point dried. All dried preparations were mounted on holders, sputter-coated with gold-palladium (8–10 nm) and examined in a Hitachi S-800 or in a JEOL JSM-5400 scanning electron microscope at 20 kV.

Pad characters

The following ten characters of the two alternative designs were coded and analyzed together with a data matrix containing 105 additional morphological characters of different stages and body parts: (1) *arolium*: 0–absent, 1–present not eversible, 2–present eversible; (2) *pulvilli*: 0–absent, 1–smooth, 2–hairy; (3) *euplantulae*: 0–absent, 1–present; (4) *hairy adhesive soles of tarsomeres*: 0–absent, 1–present; (5) *fossula spongiosa*: 0–absent, 1–present; (6) *eversible structure between tibia and tarsus*: 0–absent, 1–present; (7) *claw pad*: 0–absent, 1–present; (8) *tarsal thorns transformed into adhesive structures*: 0–absent,

Fig. 1 **a** Scheme of action of the "hairy" and "smooth" pad attachment systems on the smooth and ▶ structured substrata. Both systems are able to adapt to the surface profile. **b** Diversity of the leg attachment devices in hexapods. *Blue areas* indicate preferably smooth systems; the *orange areas* indicate preferably hairy systems. *Upper part of the panel*: arolium; pulvilli (*pu*); empodial pulvillus (*ep*); hairy soles of tarsomeres; eversible pretarsal bladder. *Lower part of the panel*: eversible structure between tibia and tarsus; fossula spongiosa; euplantulae (*eu*) and claw pad (*cp*); tarsal thorns transformed into adhesive structures (*th*) and claw pad (*cp*); claw setae. **c** "Smooth" systems: arolium of *Tipula* sp. (Diptera); pulvilli of *Coreus marginatus* (Heteroptera); arolium surface of *Tipula* sp. (Diptera), arolium surface of *Panorpa communis* (Mecoptera); euplantula surface of *Tettigonia viridissima* (Ensifera); structure of the arolium material of *Apis mellifera* (Hymenoptera), structure of the arolium material of *Cercopis vulnerata* (Auchenorrhyncha). *Arrow* indicates distal direction, *AR* arolium, *PU* pulvilli, *TAR* tarsal segments, *UN* claw. **d** "Hairy" systems: pulvilli of *Episyrphus balteatus* (Diptera); hairy tarsomere of *Cantharis fusca* (Coleoptera, Polyphaga); hairy tarsomere of *Forficula auricularia* (Dermaptera); tenent hairs of *F. auricularia* (Dermaptera); tenent hairs of *E. balteatus* (Diptera); tenent hairs of *Sialis lutaria* (Megaloptera); tenent hairs of *C. fusca* (Coleoptera, Polyphaga). *HS* Hairy soles, *PT* hair plate, *PU* pulvilli, *SH* hair shaft, *TAR* tarsal segments, *UN* claw. **e** Strict consensus tree of 36 minimal length trees. *Blue and yellowish areas* indicate clades with preferably "smooth" or "hairy" types of attachment pads, respectively

1–present; (9) *empodium plate-like*: 0–no, 1–yes; (10) *adhesive claw setae*: 0–absent, 1–present.

Cladistic analysis

Computer software (PAUP version 3.1) was used to calculate minimum length trees. Analysis of character evolution was conducted in MacClade (version 3). Chilopoda and Symphyla were included as outgroup taxa and treated as were all other groups in the analysis (simultaneous analysis; Nixon and Carpenter 1993). The traditional concept of a monophyletic group comprising myriapods and hexapods seems to be supported by a recent, detailed study of head structures. Even if tracheates may not be monophyletic, as indicated in some studies based on developmental and DNA sequence data, myriapods are certainly members of a monophyletic taxon Mandibulata and therefore still suitable as an outgroup. Hypotheses proposing a closer relationship between Hexapoda and Crustacea or crustacean subgroups (Malacostraca or Branchiopoda) are not consistent (Adoutte and Phillippe 1993; Friedrich and Tautz 1995; Regier and Schultz 1997) and the choice of crustacean outgroup taxa would be problematic. The subgroups are very heterogeneous morphologically and specialized in various ways differing strongly from terrestrial mandibulates. As a first step, phylogenetic relationships between groups of Hexapoda were calculated using characters not related to attachment. Then all characters were pooled and re-analyzed. Additional analyses were carried out after successive reweighting. Finally, evolution of selected attachment structures was analyzed in the context of relationships between different groups of Hexapoda.

Results

An analysis of the full set of 115 characters resulted in 72 minimal length trees with 303 steps (CI: 0.677, HI: 0.531, RC: 0.535; Beutel and Gorb 2001). Finally, 54 minimal length trees with 267 steps were obtained, when adhesive pad characters had been excluded (CI: 0.670, HI: 0.498, RC: 0.540). These are largely identical with the previous cladograms. However, Plecoptera are the sistergroup of all remaining neopteran orders in all trees. Only three minimal length trees were obtained after successive reweighting.

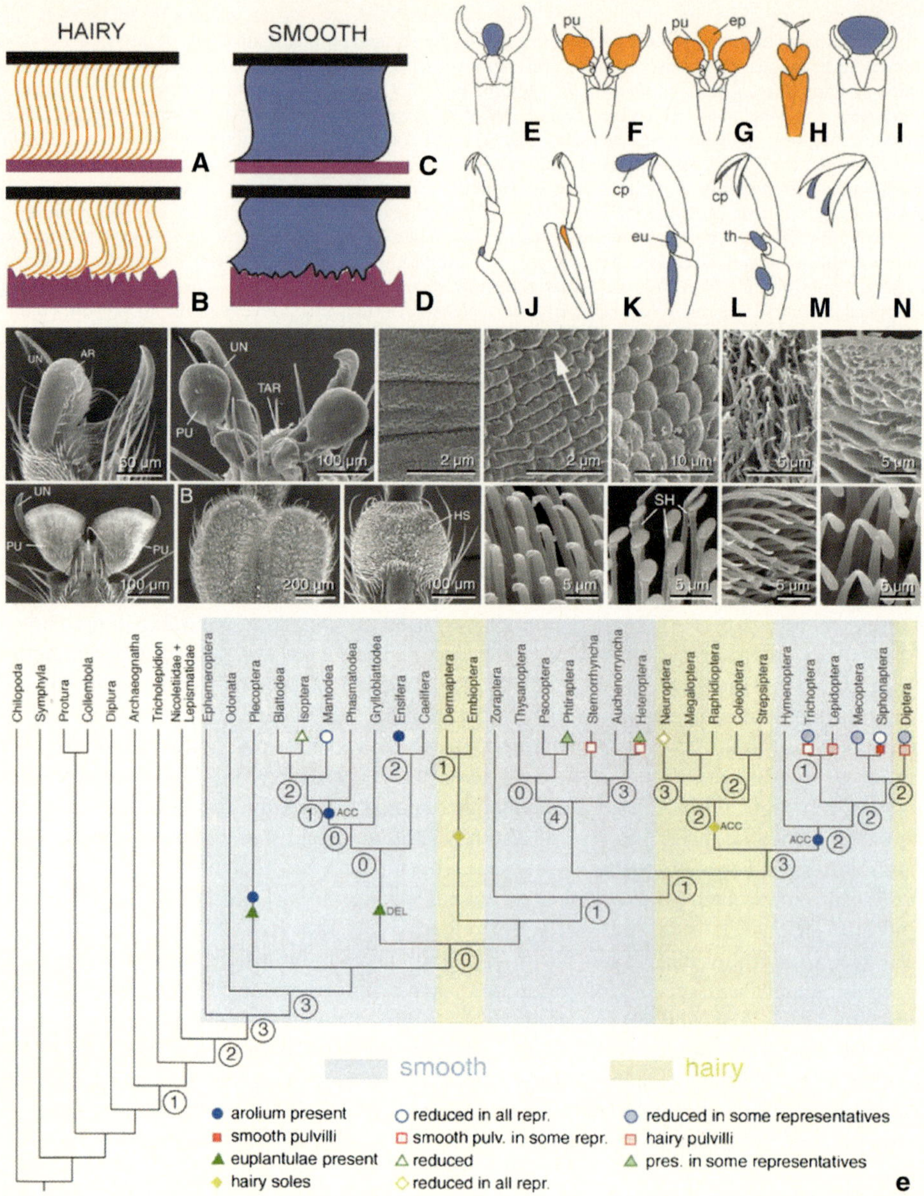

Surface characteristics

Results of the ultrastructural studies show that, in their evolution, hexapods have developed two distinctly different mechanisms to attach themselves to a variety of substrates during locomotion (Fig. 1a, b). The first mechanism is provided by hairy surfaces and the second by smooth flexible pads. The main similarity of both mechanisms is that the structured pad surface or particular properties of pad materials guarantee a maximum real contact with diverse substrata, regardless of their microsculpture.

In general, hairy systems always contain cuticle protuberances on their surfaces. However, protuberances on the hairy pads of Coleoptera, Dermaptera and Diptera belong to different types (Fig. 1c). Representatives of the first two lineages have socketed setae. Presumably, in these pads, the sensory cells degenerate in the trichoid complex during development. Dipteran protuberances are acanthae: single sclerotized protuberances originating from a single cell (Richards and Richards 1969). The key morphological characteristic of acanthae is their lack of a socket and a sensory cell (Richards 1965; Hepburn 1969). Both types of structures can be covered with additional minute outgrowths referred to as microtrichia.

The surface of smooth systems appears smooth in the light microscope. However, it may consist of a pattern of hexagonal structures as in representatives of some Orthoptera (Fig. 1d). The hymenopteran and mecopteran arolium may be patterned in lines running perpendicular to the longitudinal axis of the pretarsus. A similar pattern occurs on the surface of attachment thorns in some species of non-apocritan Hymenoptera. The surface of a tipulid arolium exhibits a complex pattern of microfolds. Microfolds found on the surface of attachment thorns in some species of non-apocritan Hymenoptera are much larger.

Ultrastructural architecture of pad material

Hairs of the fly pulvillus are hollow acanthae, and some of them contain pores under the end plate (Gorb 1998). The pores deliver an epidermal secretion directly in the contact area. Additionally, porous channels at the base of the shaft may transport secretion to the surface (Bauchhenss 1979).

The cuticle of smooth pads is composed of naturally friction-active materials with a specific inner structure (Fig. 1d). Tiny threads are located just under the epicuticle. These filaments are branches of thicker threads located more deeply in the cuticle (Henning 1974; Kendall 1970). It has been shown that these threads in the orthopteran euplantulae can change their shape under load and adapt the pad surface on the substrate profile (Gorb et al. 2000). Internal architecture of the arolium of cicadas resembles foam, consisting of fluid-filled cells surrounded by flexible cuticle.

Discussion

Evolution of pad structures and hexapod phylogeny

Cladistic analysis clearly demonstrated that the two alternative designs of leg attachment systems have convergently evolved (Fig. 1e). A plausible explanation for the existence of such a restricted number of design principles is that leg pads are primarily involved in and adapted to facilitate locomotion. This means that ability to detach from a substrate quickly is of equal importance to fast attachment on a variety of substrates. Other releasable attachment mechanisms, such as suckers, hooks, or glue cannot work as well, because of their inability to detach rapidly. Additionally, attachment pads are made of the same basic material, which possibly has some constraints for design of such systems.

In spite of the adaptability of pad characters, the phylogenetic analysis performed demonstrates some general evolutionary trends in the structures studied. Structures, such as the arolium, euplantulae, and hairy tarsomeres, have evolved independently in several lineages. Nevertheless, some of them support monophyletic lineages (e.g. Embioptera +

Dermaptera; Paurometabola excl. Dermaptera + Embioptera; Isoptera + Blattodea; Neuropterida + Strepsiptera + Coleoptera; Hymenoptera + Mecopterida). Other structures, such as claw pads (Ephemeroptera), balloon-shaped eversible pads (Thysanoptera), or fossulae spongiosae (Reduviidae), are unique to members of larger or smaller monophyletic units. Groups with the potential to evolve a great variety of designs in adhesive pads are Hemiptera and Diptera.

Ephemeropterans are characterized by an unusual, autapomorphous condition of the distal leg: the presence of claw pads. Monophyly of Paraneoptera and the branching pattern within this group is consistent with other studies (Kristensen 1991). An interesting point is the evolution of a broad variety of attachment structures in members of this lineage. A balloon-shaped eversible structure between the claws is a unique autapomorphy of Thysanoptera. Distally broadened adhesive setae on the claws have evolved in Psocoptera. A claw pad is present in phthirapterans, probably as an adaptation to ectoparasitism. An arolium and smooth pulvilli have evolved independently in several lineages of Hemiptera. Euplantulae are present in some mallophagans and mirines (Heteroptera, Miridae), but are absent from other heteropterans. A *fossula spongiosa* is present in piratines (Heteroptera, Reduviidae).

Endopterygota, the sistergroup of Paraneoptera, is divided into two large monophyletic groups, Neuropterida and Coleoptera + Strepsiptera on one hand and Hymenoptera + Mecopterida on the other. Both lineages show distinctly different trends concerning attachment structures. Members of the former group are characterized by hairy adhesive soles (reversal in Neuroptera). An arolium and hairy or smooth pulvilli are characteristic for representatives of the latter monophyletic taxon.

The clades Mecopterida, Antliophora and Amphiesmenoptera are not supported by adhesive pad characters. However, it is evident that development of different types of adhesive structures plays an important role in the diversification of several orders. Hymenopterans have retained an arolium, which is ascribed to the groundplan of a clade comprising Hymenoptera + Mecopterida, and representatives of non-apocritan families ("Symphyta") possess an additional unique type of attachment structure derived from tarsal thorns (plantar lobes). Most Mecopterida have also retained the arolium; however, it is secondarily absent from the ground-living boreids and from notiothaumids. Most dipterans possess a plate-like empodium (Tabanoidea, Nematocera excl. Tipulidae) and hairy pulvilli. The latter character state was found in all taxa examined with the exception of Tipulidae. The presence of an arolium in members of this family is probably a plesiomorphic condition.

It is reasonable to assume that the evolution of flight and the correlated necessity of being able to cling to vegetation or other substrates was a major trigger for the evolution of adhesive pads. Attachment structures are not found in entognathous hexapods, which live mainly in soil and leaf litter. However, rather complicated claws and associated structures have evolved in diplurans, collembolans, and proturans. The rise of Insecta (Ectognatha) is a crucial event in hexapod evolution, which is reflected in distinct changes of leg structure. Subdivision of the tarsus and reduction of the ring-shaped pretarsus, still more or less distinct in Diplura and Ellipura, are autapomorphies of this large clade. These changes, especially adaptive flexibility of the tarsus, are probably preconditions for further modification and the appearance of different attachment devices.

Acknowledgements This project was partly supported by the Federal Ministry of Education, Science and Technology, Germany to S.N.G. (Project BioFuture 0311851) and partly by the Deutsche Forschungsgemeinschaft (R.G.B., grant No. BE 1789/1-2).

References

Adoutte A, Phillippe H (1993) The major lineages of metazoan evolution: summary of traditional evidence and lessons from ribosomal sequence analysis. In: Pichon Y (ed) Comparative molecular neurobiology. Birkhäuser, Basel, pp 1–30

Autumn K, Liang YA, Hsieh ST, Zesch W, Chan WP, Kenny TW, Fearing R, Full RJ (2000) Adhesive force of a single gecko foot-hair. Nature 405:681–685

Bauchhenss E (1979) Die Pulvillen von *Calliphora erythrocephala* Meig. (Diptera, Brachycera) als Adhäsionsorgane. Zoomorphologie 93:99–123

Beutel R, Gorb SN (2001) Ultrastructure of attachment specializations of hexapods (Arthropoda): evolutionary patterns inferred from a revised ordinal phylogeny. J Zool Syst Evol Res 39

Dashman T (1953) Terminology of the pretarsus. Ann Entomol Soc Am 46:56–62

De Meijere JC (1901) Über das letzte Glied der Beine bei den Arthropoden. Zool Jahrb Anat 14:417–476

Friedrich M, Tautz D (1995) Ribosomal DNA phylogeny of the major extant arthropod classes and the evolution of myriapods. Nature 376:165–167

Gorb SN (1998) Origin and pathway of the epidermal secretion in the damselfly head-arresting system (Insecta: Odonata). J Insect Physiol 44:1053–1061

Gorb SN, Jiao Y, Scherge M (2000) Ultrastructural architecture and mechanical properties of attachment pads in *Tettigonia viridissima* (Orthoptera Tettigoniidae). J Comp Physiol A 186:821–831

Henning B (1974) Morphologie und Histologie der Tarsen von *Tettigonia viridissima* L. (Orthoptera, Ensifera). Z Morphol Tiere 79:323–342

Hepburn HR (1969) The proventriculus of Mecoptera. J Ga Entomol Soc 4:159–167

Holway TR (1935) Preliminary note on the structure of the pretarsus and its possible phylogenetic significance. Psyche 42:1–124

Kendall UD (1970) The anatomy of the tarsi of *Schistocerca gregaria* Forskål. Z Zellforsch 109:112–137

Kristensen NP (1991) Phylogeny of extant hexapods. In: CSIRO (eds) The insects of Australia. Cornell University Press, Ithaca, N.Y., pp 125–140

Maxwell MH (1978) Two rapid and simple methods used for the removal of resins from 1.0 μm thick epoxy sections. J Microsc 112:253–255

Nixon KC, Carpenter JM (1993) On outgroups. Cladistics 9:413–426

Pain S (2000) Sticking power - want to walk on the ceiling? All it takes is a bit of fancy footwork. New Sci 168:62–67

Regier JC, Shultz JW (1997) Molecular phylogeny of the major arthropod groups indicates polyphyly of crustaceans and a new hypothesis for the origin of hexapods. Mol Biol Evol 14:902–913

Richards AG (1965) The proventriculus of adult Mecoptera and Siphonaptera. Entomol News 76:253–256

Richards PA, Richards AG (1969) Acanthae: a new type of cuticular process in the proventriculus of Mecoptera and Siphonaptera. Zool Jahrb Anat 86:158–176

J Comp Physiol A (2000) 186: 821–831
DOI 10.1007/s003590000135

ORIGINAL PAPER

Ultrastructural architecture and mechanical properties of attachment pads in *Tettigonia viridissima* (Orthoptera Tettigoniidae)

Stanislav Gorb · Yuekan Jiao · Matthias Scherge

Accepted: 4 July 2000 / Published online: 19 September 2000

Abstract Natural releasable attachment systems of insect legs, where attachment-detachment performances are often very fast, seem to be optimized to get a maximum of real contact to the substratum. Tarsi of *Tettigonia viridissima* bear flexible attachment pads with unusual ultrastructural architecture of the cuticle. The indentation of the attachment pads was measured under different loads using a force-tester. Since the mechanical properties are influenced by material structure, the freeze-substitution experiments were undertaken to investigate the influence of loads on material structure. Both profile changes of the surface and the orientation of cuticle microfibrils were visualized by means of scanning electron microscopy followed by fracturing of the frozen material. The results show that the flexible pad material deforms replicating the substrate profile down to the micrometer roughness. The pad material showed both elastic and viscous behavior under loads. Elastic deformation of the pad occurred under normal force applied for 4–6 s (elastic modulus 27.2 ± 11.6 kPa). Two viscous relaxation processes were found, of time constants $\tau_1 = 1.88 \pm 0.616$ s and $\tau_2 = 41.2 \pm 9.95$ s. Low stiffness of material studied here aids in surface replication and increase of area of real contact between the pad and the underlying substrate.

Keywords Ultrastructure · Cuticle · Material properties · Elasticity · Viscosity

S. Gorb (✉) · Y. Jiao
Biological Microtribology Group, Biochemistry Department, Max-Planck-Institute of Developmental Biology, Spemannstr. 35, D-72076 Tübingen, Germany
e-mail: Stas.Gorb@tuebingen.mpg.de

M. Scherge
Microtribology Group, Ilmenau Technical University, Institute of Physics, Weimarer Str. 32, D-98684 Ilmenau, Germany
e-mail: Matthias.Scherge@physik.tu-ilmenau.de

Introduction

Attachment forces mediated by friction or adhesion are usually proportional to the area of real contact between two surfaces (Persson 1998). Attachment systems, where attachment-detachment performances are often very fast, have to be optimized to achieve a contact to the substratum in the shortest time (Radhakrishnan 1998). There are two alternative designs of such natural releasable systems used in locomotion. Pads of flies, beetles, and earwigs are covered by relatively long deformable setae (Bauchhenss and Renner 1977; Stork 1980), which presumably can bend and thus repeat the surface profile. The second type of pads, so called arolia and euplantulae, occur in cockroaches, grasshoppers, bees, and bugs (Snodgrass 1956; Arnold 1974; Henning 1974; Ghasi-Bayat et al. 1980a). These organs are smooth and soft, but have an exceptional flexibility (Slifer 1950; Roth and Willis 1952).

Insects with smooth type of attachment pads can walk on a variety of plant surfaces and always deal with an unpredictable roughness of the substrate. In spite of this, the attachment system allows an insect to attach to the surface and to detach from it easily. It has been previously shown that attachment of a great green bush-cricket *Tettigonia viridissima* (Orthoptera: Tettigoniidae) onto a surface is strongly dependent on the deformation of the pads (Jiao et al. 2000). However, the ability of leg attachment pads to adapt to a variety of surface profiles remained a hypothesis which needed to be proven experimentally.

Despite numerous literature sources devoted to the microsculpture and ultrastructural architecture of tanned cuticle of insect sclerites (Hepburn 1985), very little work has been done on the structure of the flexible cuticle (Vincent and Prentice 1973; Carruthers and Davey 1983; Hackman and Goldberg 1985). Previous authors have shown that material of a smooth pad differ from common structure of the cuticle. Pad cuticle is not layered and consists of rod-like fibers oriented at some angle to the cuticle surface (Slifer 1950; Roth and Willis 1952; Kendall 1970). Since mechanical properties of a biological material are determined by its structure, the ultrastructural study of the surface and material fractures by means of scanning electron microscopy followed by freezing-substitution experiments was undertaken to visualize material behavior of the pad under loads. We have hypothesized that pads, due to their flexibility, are able to replicate the surface profile down to the microscale roughness.

Normal forces ranging from nano- to micronewtons have been used in studies of the mechanical deformation of cells (Bausch et al. 1998, 1999). To study the mechanical properties of attachment pads, a force tester, capable of generating and detecting greater forces, was applied (Scherge and Schaefer 1999; Gorb and Scherge 2000).

Materials and methods

Microscopy and freeze-substitution experiments

Living specimens of *T. viridissima* were captured in Ilmenau, Germany. The distal euplantula of a living insect was lightly pressed against smooth and rough surfaces and frozen in this condition with liquid N_2. As rough surfaces, silicon plates with cut-cone shaped outgrowths were used. The outgrowths were 3.983 μm (SD = 0.178, $n = 20$) high and 7.122 μm (SD = 0.279, $n = 20$) in diameter at their bases. The distance between outgrowths from center to center was 11.955 μm (SD = 2.822, $n = 20$) (Fig. 5A, B). In a control experiment, we used euplantulae not contacting the substratum.

The freeze-substitution technique is, meanwhile, a standard method in cell science (Schwarz and Humbel 1988; Meissner and Schwarz 1990). In this study, we tried to apply this method to understand behavior of biological material under loads. Usually, frozen biological material can recover to a certain extent after being thawed. The OsO_4 substitution was necessary to fixate all material deformations in the frozen condition and keep such a deformation in the course of the whole microscopic procedure. The shock-frozen pads from different experiments were fractured using a razor blade and transferred to 0.5% OsO_4 solution in absolute acetone at −80°C for 48 h, washed in absolute acetone at −20°C, transferred to absolute ethanol and critical-point dried.

An additional technique was used to obtain information about orientation of inner structures. Semi-thin sections (0.5–2.0 μm) of pads, embedded in Spurr-resin (Spurr 1969), were sectioned using a diamond knife. The sections were picked up on pioloform-covered glass cover-slips, treated with Maxwell's solution (Maxwell 1978) for 2–5 min in order to remove the resin, washed in absolute ethanol, and critical-point dried.

All preparations were mounted on holders, sputter-coated with gold-palladium (10 nm) and examined in a Hitachi S-800 scanning electron microscope at 20 kV.

Force tester

The force tester (Tetra, Ilmenau, Germany; Fig. 1A) included three main parts, a platform, a glass spring, and a fiber optical sensor. The lower sample was mounted on the platform, the upper sample was attached to the glass spring with a spring constant of 106.9 N m^{-1}. Driven by a motor, the platform was capable of moving the lower sample up and down. Spring deflection was detected by the fiber-optic sensor connected to a computer used for data acquisition. The resulting force between two samples was recorded.

The approach used in this study aids to estimate dependence of indentation with respect to applied force. The mechanical properties studied may not represent the local properties responsible for intimate contact at submicroscopic level. The local properties of the cuticle surface and chitin rods will be subject of next study using an atomic force microscopy.

Sample preparation

A living animal was immobilized on the platform using sticky tape (Fig. 1A); one leg was firmly fixed with wax, the distal pad was used as the lower sample. A smooth silicon chip was cleaned in an ultrasonic bath in acetone, and then in ethanol was attached on the spring as the upper sample. All experiments were carried out at room temperature (22–24°C) at a relative humidity of 47–56%.

Acquisition of data

The pad on the platform was rapidly brought in direct contact with the silicon surface (approaching process) to set a certain normal force. The pad was kept in contact with the silicon surface for a defined period of time (resting time) to record the response of the pad material under load (Fig. 1B). Typically, the approaching process took about 4–6 s to apply a force of 0.6–0.7 mN. The recorded force-distance curve (Fig. 2A) was used to calculate the elasticity of the pad. The force recorded in the resting time was transformed into stress to calculate viscosity. Force curve fluctuations, obtained in experiments with living animals, were presumably caused by haemolymph pulsation in the pad.

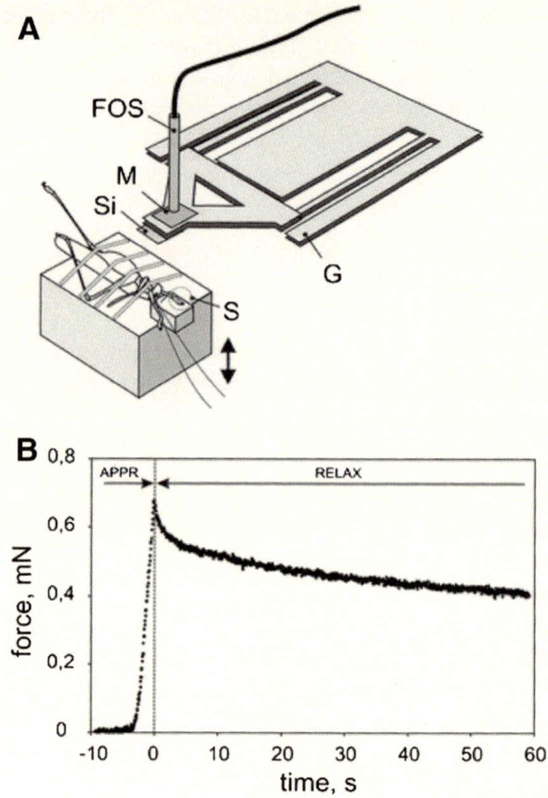

Fig. 1 A, B. Experimental setup. **A** Micro-force tester. The lower sample (*S*), the pad of a living insect, was connected to the platform, and the upper sample (*Si*), a silicon chip, was attached to the spring. Driven by a motor, the platform moved the lower sample to approach to and retract from the upper sample. The deflection of the glass spring (*G*) was monitored by the fiber optical sensor (*FOS*) through the mirror (*M*). **B** A typical force curve acquired for the evaluation of the mechanical properties of the pad. The force curve includes two distinct parts, *approaching process* (APPR) and *remaining period* (RELAX). In the approaching process, the pad was rapidly approached to the hard silicon surface to obtain an interacting force *F* (*applied force*), Generally it took 4–6 s to reach an applied force of 0.6–0.7 mN. After that the pad was kept in contact with the silicon surface, the interacting force slowly relaxed

As a control experiment, the deformation of wax under load by a sapphire ball of 1.47 mm diameter was measured. Wax turned out to be much stiffer than the pad and the influence on the pad was therefore ignored.

Calculation of elasticity

The pad deformed when pressed against the silicon surface as a result of the applied force. The indentation of the pad was obtained by comparing the spring deflections caused by the pad and a hard lower sample (for example the sapphire ball; Fig. 2A). The deflection of the spring pushed by the hard sample was equal to the traveling distance of the lower sample (no indentation). The spring, pushed by the pad, deflected less because of the indentation into the soft material of the pad. This means that the lower sample had to travel longer in order to obtain the same deflection of the spring. The indentation of the pad was obtained by subtracting the displacements of pad and hard surface for the same traveling distance of the lower sample (Fig. 2B).

To calculate pad elasticity, Hertz theory was applied (Hertz 1881). The pad shape was assumed to be part of a sphere (Fig. 3). Hertz theory describes the deformation of two elastic bodies in contact under applied force. It gives the relationship between the external applied force and the deformation as

Fig. 2 **A, B.** The force-distance curve, derived from the approach part of a force curve, was used to calculate the elastic modulus of the pad. **A** The indentation of the pad, which was obtained from the comparison of force-distance curves of the pad (*dotted line*) and hard sample (*solid line*). The deflection of the spring, caused by a hard sample, was approximately equal to the displacement of the sample. The spring, pushed by the pad, deflected less because of the indentation of the pad. The indentation of the pad was calculated by subtracting displacements of the pad and the hard surface at the same spring deflection. The material indentation δ under force F corresponding to the deflection was obtained. **B** The indentation δ of the pad versus applied force F (*dotted line*). The *solid line* is the indentation data fitted according to the Hertz theory

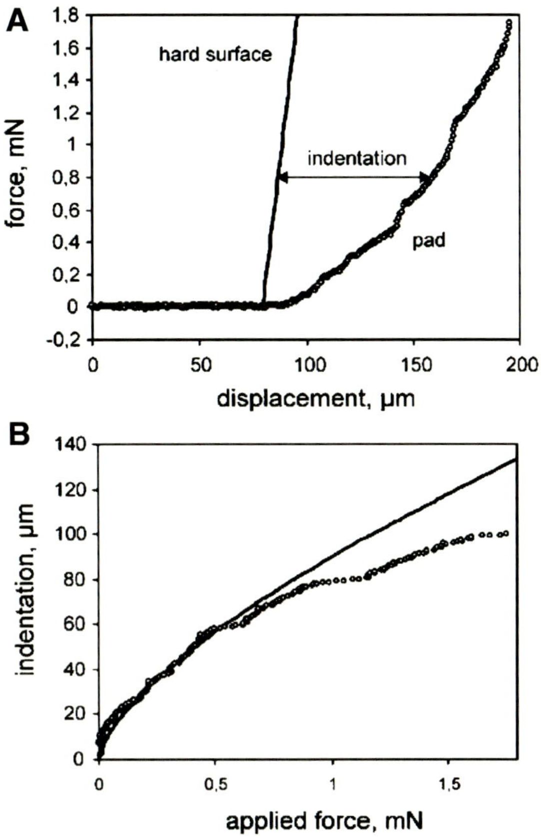

Fig. 3 A–C. Scheme of the attachment pad of the *Tettigonia viridissima*. **A** Ventral view of the tarsus of the third pair of legs; pad width. *TI*, tibia, *T1-T4*, leg tarsomeres. The pad of the most distal tarsomere (*gray-colored area*) was used in experiments. **B** Course scheme of the pad (*gray-colored area*) outlined by a semicircle. *R*, radius of pad curvature; *r*, pad width. **C** Cross section of the pad (strongly schematized after (Henning 1974). *AS*, air sacks; *CL*, epidermal cell layer; *EXO*, rod-containing exocuticle of the pad; *HM*, haemolymph; *TD*, tendon of the claw flexor muscle; *TK*, tanned cuticle

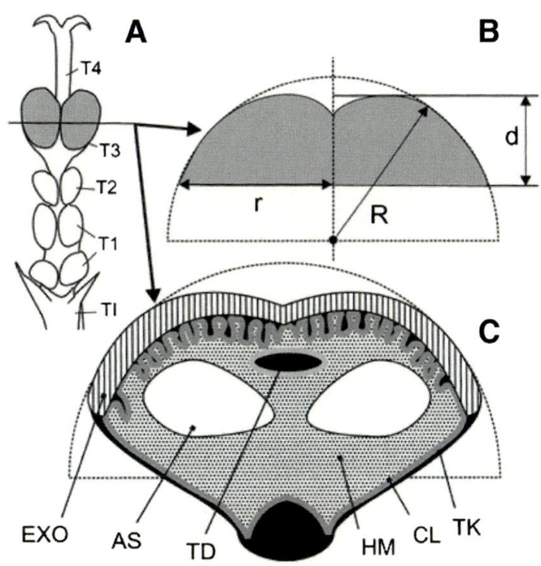

$$F = \frac{4}{3}\frac{E}{1 - V^2} a\delta \quad (1)$$

where E is Young's modulus (elastic modulus), v is the Poisson ratio of the pad, a and δ are the contact radius and the indentation caused by the applied force F. Contact radius a is related to indentation δ as

$$a^2 = R\delta \quad (2)$$

where R, the radius of pad curvature, was obtained from the pad cross sections. Equations 1 and 2 imply that the indentation δ can be expressed with applied force F as

$$\delta = \left(\frac{3}{4}\frac{1 - v^2}{E}\right)^{2/3} \frac{1}{R^{1/3}} F^{2/3} \quad (3)$$

Calculation of viscosity

At the beginning of the resting period, the interacting force showed a rapid decrease, after which it decreased slowly. The decrease of the interacting force indicates relaxation of the pad material. This behavior reflects the viscoelastic properties of the pad material. To evaluate viscosity in the remaining period, a Kelvin model was used. This model includes two sets of serial dashpot and spring (μ_1, E_1, and μ_2, E_2, respectively) and an additional spring E_0 (Fig. 7B) (Fung 1981). The two sets of serial dashpot and spring are responsible for both the rapid and slow force relaxation. Three springs together, E_0, E_1 and E_2, determine an elastic modulus of the whole system. The spring E_0 determines the stress of the pad at time $+\infty$. Let us approximate that the value ε is basically constant in the remaining period, as the strain is generated by an applied force at the end of the approaching process. Under this assumption we have stress $\sigma(t)$:

$$\sigma(t) = \varepsilon\left(E_0 + E_1 e^{-t/\tau_1} + E_2 e^{-t/\tau_2}\right) \quad (4)$$

where two time-constants are given as $\tau_1 = \mu_1/E_1$ and $\tau_2 = \mu_2/E_2$.The stress curve (Fig. 7A) was obtained from the applied force in the resting period divided by the contact area of pad and silicon surface. By fitting with the Kelvin model, the constants τ_1, τ_2, spring constants E_0, E_1, E_2, viscous constants μ_1, and $\mu_2$were obtained.

Results

Pad structure

The dorsal surface of the *T. viridissima* tarsus is hard and consists of the tanned layered cuticle. On its ventral side, each tarsus bears three to four attachment pads called euplantulae (Fig. 3A). A single pad is composed of two halves. The preterminal tarsomere has the biggest euplantula (1.7–2.5 mm wide) mainly used in experiments.

In the light microscope, the pad surface appeared smooth. In reality, it is composed of hexagonal structures (area 14.7 μm^2; SD = 1.96; $n = 22$) with underlying tiny fibers of 0.08 μm in diameter (SD = 0.01; $n = 25$) (Fig. 4A,C). These fibers are branches of thicker principal fibers or rods of 1.12 μm in diameter (SD = 0.09; $n = 20$) located deeper in the cuticle. The principal fibers, which correspond to the exocuticle layer, originate from the underlying endocuticle. The endocuticle is layered in the horizontal plane. The space between fibers is filled with a liquid which can be transported to the cuticle surface through terminal channels. The fibers seem to be interconnected by a non-fibrous matrix (Fig. 4). However, the nature of these cross links between the fibers remains unclear. In sections and fractures, the uppermost cuticular layer resembles a thin film (180 nm thick; SD = 39; $n = 20$). The thickness and the non-fibrous structure of this layer correspond to the epicuticle, the outermost layer of the insect integument.

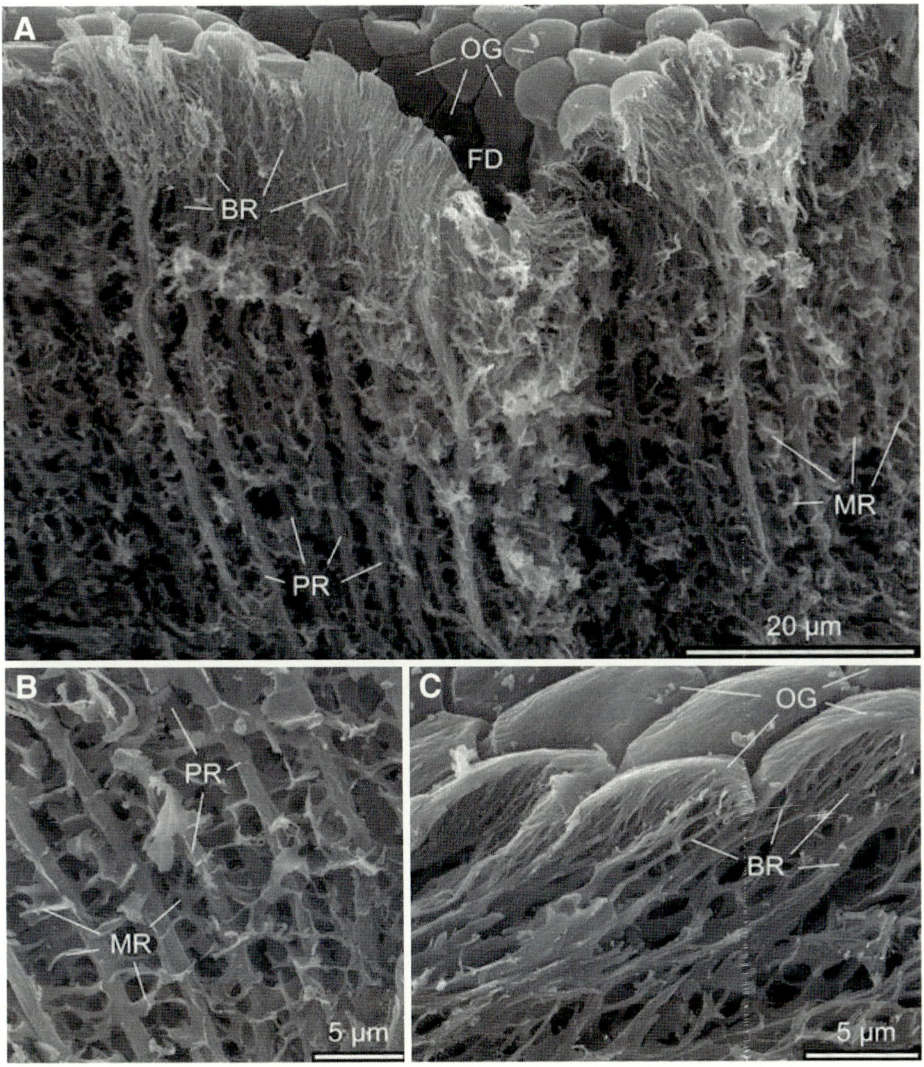

Fig. 4 A–C. Ultrastructure of the pad material, freeze-substitution preparations. **A** Overview of the fractured pad. **B** Area of the principal rods. C Material fractured closely to the surface. *BR*, branched rods; *FD*, fold; *MR*, matrix residuals; *OG*, outgrowths; *PR*, principal rods

Material behavior

Two series of freeze-substitution experiments were carried out. In the first series, pads were pressed with different loads against a structured silicon surface (Fig. 5A, B) and frozen in this condition. Further SEM studies revealed that the indentation patterns of the surface correspond to the pattern of the silicon surface. Under higher loads, the indentation depth corresponded to the height of silicon structures (approx. 4 μm). The epicuticle deformed to this extent without any damage (Fig. 5C, D). Under lower loads very weak deformations occurred. The epicuticle, however, perfectly replicated silicon pattern (Fig. 5D, F). The indentation was in the range 50–500 nm. In the fractured preparation, deformation of the branched fibers was revealed (Fig. 5G). The principal fibers were not involved in this deformation (Fig. 4A).

In the second series of the experiments, pads were pressed with high loads against a smooth silicon surface and frozen in this condition in order to visualize deformation of principal fibers. The principal fibers are usually sloped at some angle in the distal direction of the pad (60–80°; Fig. 6A). Under load, the principal fibers sloped more in the same direction (Fig. 6B). The slope angle ranged from 5° to 45°, presumably, depending on the load.

Elasticity

Experiments revealed a gradual deformation of the pad occurred under the applied force during the approaching period (4–6 s). The deformation was fitted well with the Hertz model at an applied force lower than 0.6–0.7 mN (Fig. 2B), although the top of the pad was not exactly spherical, and the pad material was not isotropic (Fig. 3B,C). The deformation could not be described by Hertz theory anymore at larger forces (Fig. 2B). The corresponding indentation caused by applied forces of 0.6–0.7 mN was found to be 35–65 μm. Ultrastructural data show that the material studied is anisotropic. However, there are no experimental data on the anisotropy of mechanical properties of the material. Since the Poisson ratio is unknown for the most of biological material, an average one (0.5) was used for calculations. Similar approach is used also in studies on mechanical properties of cells. By applying the Hertz theory for applied forces lower than 0.6–0.7 mN with an average radius of the pad curvature R of 2.86 mm and the Poisson ratio of the pad material of 0.5, the elastic modulus of 27.2 ± 11.6 kPa was obtained (the pad thickness d was measured around 100 μm, and pad radius τ 1.5 mm, see Fig. 3B).

Viscosity

To obtain stress from the interacting force, the contact area is required. The previous study of pad adhesion on silicon surface has shown, that the retracting curve was basically linear. The separation of the pad and the silicon surface was a one-step action, and the adhesive force was proportional to the contact area (Jiao et al. 2000). The adhesive force, measured with various remaining times (2–60 s) under a certain applied force, basically remained constant (Fig. 8A). Therefore, it is concluded that the contact area did not change in the remaining period (at least during the first 60 s), and in the retracting process as well. Force curves with an applied force of 0.6–0.7 mN were used to calculate viscosity. The contact area A, caused by the applied force, could be derived from the Hertz theory (Eq. 2) as

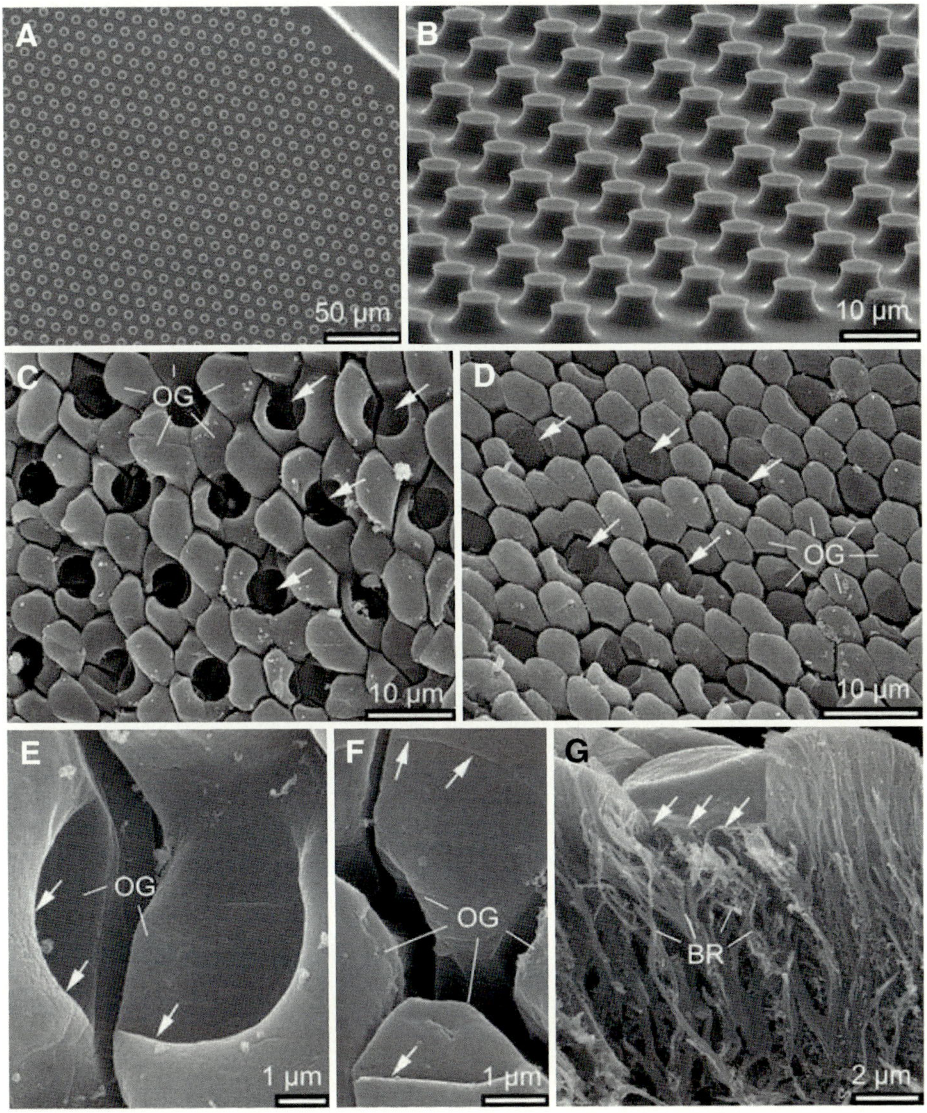

Fig. 5 A–F. Microindentation experiments. **A** Structured silicon surface, view from above. **B** Same, view from the side. **C, E** Pad surface deformation under strong load. **D, F** Pad surface deformation under moderate load. **F** Fracture of the pad material in the area of deformation. *Arrows* indicate material deformations. *BR*, branched rods; *OG*, outgrowths

$$A = \pi a^2 = \pi R\delta \tag{5}$$

For various pads we found a corresponding contact area ranging from 0.31 mm^2 to 0.58 mm^2 (full area of pad was estimated to be 1.77 mm^2).

Considering Eqs. 1, 2, and 3, the assumed constant strain ε, caused by the applied force of 0.6–0.7 mN and lower, was obtained as

Fig. 6 **A, B.** Deformation of the pad material under load, freeze-substitution preparations. **A** Control. **B** Material frozen in deformed state. *BR*, branched rods; *MR*, matrix residuals; *OG*, outgrowths; *PR*, principal rods

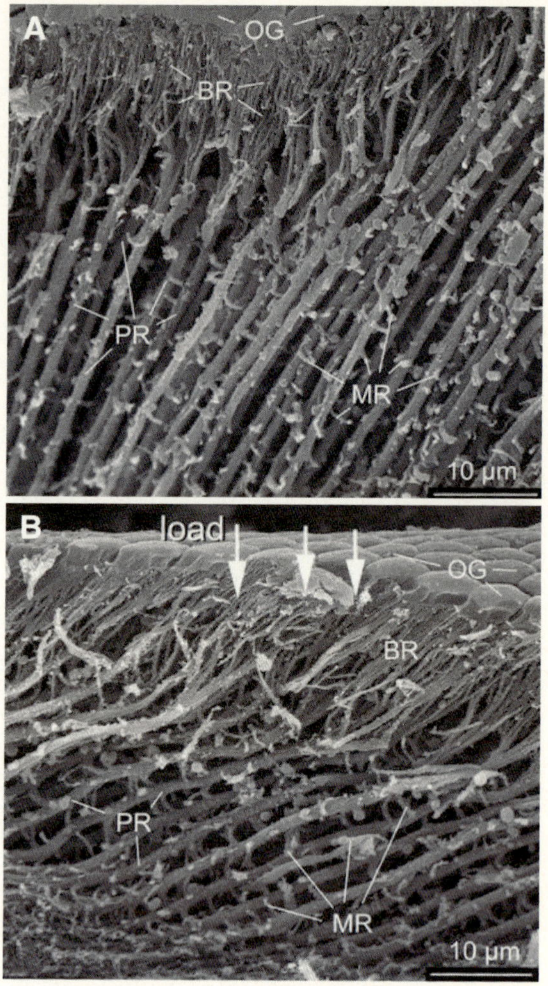

$$\varepsilon = \frac{\sigma}{E} = \frac{F/A}{E} = \frac{4}{3}\frac{1}{1-\nu^2}\frac{1}{\pi}\left(\frac{\delta}{R}\right)^{1/2} \tag{6}$$

The strain ε was found in the range 0.062–0.085 ($\nu = 0.5$), corresponding to the applied force of 0.6∼0.7 mN. Because the strain ε was obtained from Eq. 6 of the Hertz theory, the elastic modulus, determined from the viscoelasticity model ($E_0 + E_1 + E_2$), is the same as the one E obtained by applying the Hertz theory to indentation versus applied force in the elastic part of the curve.

The stress curve, obtained from the force curve of the remaining period, was fitted to the viscoelasticity model of Kelvin (Fung 1981) (Fig. 7). The parameters obtained were $E_0 = 12.0 \pm 6.73$ kPa, $E_1 = 3.94 \pm 1.06$ kPa, $E_2 = 8.05 \pm 1.98$ kPa, $\mu_1 = 7.18 \pm 2.03$ kPa s^{-1} and $\mu_2 = 310 \pm 58.2$ kPa s^{-1}. Relaxation times were found $\tau_1 = 1.88 \pm 0.616$ s for the rapid relaxation and $\tau_2 = 41.2 \pm 9.95$ s for the slow one.

Fig. 7 **A, B.** The viscoelasticity model. **A** The stress-versus-time curve. These data, derived from the remaining period of a force curve, were used for calculation of viscoelastic properties of the pad. There are two kinds of relaxation, a rapid one (at the beginning) and a slow one. The curve was fitted (*line*) to a Kelvin viscoelastic model. **B** The Kelvin viscoelastic model. It has two sets of dashpot and spring E_1, μ_1, E_2, μ_2 responsible for both rapid and slow relaxation processes. E_0 determines the stress at equilibrium

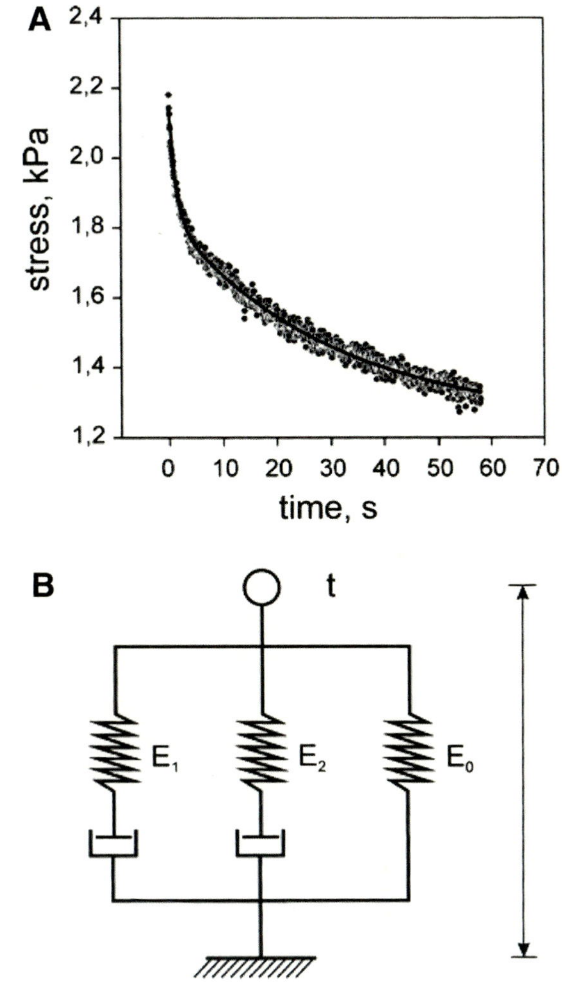

Discussion

Viscoelasticity seems to be the general property of natural fibrous composites (Sanjeevi 1982; Mak 1986; Flaud and Quemada 1988; Neville 1993). The mechanics properties of such fibrous materials, such as biopolymer solutions, cells or tissues of cell aggregates and muscles, have been extensively studied using various techniques. Measurements with a micro-pipette (Evens et al. 1980; Evens and Yeung 1989), cell poker (Petersen et al. 1982), atomic force microscope (Hoh and Schönenberger 1994; Radmacher et al. 1996), magnetic bead (Ziemann et al. 1994), and micro needle (Kolega 1991) aided in obtaining local properties of the cell surface and/or cell organelles. Other techniques, such as micro-plate manipulation (Thoumine and Ott 1997) and compression plate instrument (Forgacs et al. 1998), aided in measurements on whole cell or cell aggregates, and provided a description of bulk properties of biological materials. The force tester used in our experiments is similar to the two plate techniques above but is capable of applying and detecting much greater forces (μN–mN range). Although elastic properties of a cell are generally attributed to the actin cortex, and viscosity to the viscous cytoplasm fluids (Janmey 1991), measured mechanical properties diverge strongly depending on the technique used.

Fibrous composite nature of the pad material

Insect cuticle is a fibrous composite material (Neville 1975, 1993). The microfibrils consist of chitin rods set in a protein matrix (Blackwell and Weih 1980). The microfibrils are organized in a lamella, which is a monolayer of microfibrils. Within the lamella, single microfibrils usually lie parallel to one another. Sometimes lamellae rotate relative to one another resulting in a so-called helicoidal cuticle. Two types of membranous deformable cuticle have been previously reported in insects. The first type is a highly extensible membrane found in the locust abdomen that can extend up to ten times its length (Vincent and Wood 1972; Vincent 1975, 1981). This cuticle is highly specialized in its protein composition (Hackman and Goldberg 1987). The second type is a folding laminated membranous cuticle with a somewhat lower stretching capacity reported on the abdomen of the tsetse fly, *Glossina morsitans* (Hackman and Goldberg 1987), and the bug *Rhodnius* (Hackman 1975; Reynolds 1975).

The pad cuticle of *T. viridissima* cannot be included in either of these types. It is a third type of cuticle, which also has a fibrous architecture, but is extremely specialized according to both structure and orientation of fibers. First, the principal fibers are much thicker than common cuticle microfibrils and lie more separated from each other than fibers in arthrodial membranes of insects (Gorb 1997, 2000). They consists of a higher number of single chitin rods. The second important structural property is that fibers branch in the vicinity to the cuticle surface. Presumably the number of branches corresponds to the number of single chitin rods composing a principal fiber. Fiber orientation, almost perpendicular to the surface, is different from common parallel orientation to the cuticle surface. Such an orientation of fibers can presumably provide flexibility at two levels: (1) the local one, when preferably branched fibers deform, and (2) the global one, when the main fibers also deform. The first level of deformation is responsible for adapting the pad surface to the substrate microroughness, whereas the second one can fit the pad to the macrosculpture of natural substrata. Similar architecture of the pad material was found in "smooth systems" of other insect groups, such as Blattaria, Plecoptera, Heteroptera, Homoptera, Hymenoptera, etc. (Beutel and Gorb 2000).

Elastic behavior

The pad of *T. viridissima* behaved elastically in the approaching process, which implies that the stress σ and the strain ε with the deformation followed the relationship of $\sigma = E\varepsilon$ The bulk elastic properties of a cell rely on the fiber structure of cytoskeleton and actin cortex. For a pad we would like to attribute the elasticity to the fiber structure of the pad. To clarify this point, we considered firstly that the elasticity of a cell is mainly due to the fiber structure of the cell. Secondly, the pad of a leg freshly cut off the body almost had the same elastic modulus as a leg on the body. Air sacks left the shell when the leg was broken off the body. The loss of air sacks did not affect the elastic properties of the pad. Thirdly, the elastic modulus decreased with the time of the leg had been separated from the body. Once the leg was cut off from the body, the fiber structure, including the dendrites of the surface, the shell and the possible fiber network inside the shell, lost water (Jiao et al. 2000). Usually, biological structures that have dried out become stiffer. However, an initial phase of the water loss in the pad may lead to a decrease of haemolymph amount in the same volume of the pad. Taking into account softness of the pad material, it can be expected that the pad will lose its inner tension (elasticity) and become softer. Again, such a mechanical behavior may be compared to the cell.

Viscous behavior

The model used for calculation of viscous behavior is very similar to the one previously applied for living embryonic tissues (Forgacs et al. 1998). We need to explain how the parameters obtained from the viscoelasticity model are related to the structure of the pad. Both models consist of two sets of dashpot and spring for two viscoelastic processes, the difference between them is how the force or stress at equilibrium is determined.

A living embryonic tissue experienced two processes under compression. One was more elastic with a short relaxation time; the other was more viscous with a long relaxation time. During the more elastic response, individual cells appeared relaxed, while the cells reorganized themselves during the more viscous process. Finally, the cells completely recovered their original shapes. Hence, the stress of cell aggregates at equilibrium was determined by the surface or interfacial tensions possessed by the cells. It was independent of both the magnitude of compressive force and the size of the aggregates.

Actually, the two relaxation processes happened to cells studied with many rheological techniques. Both cell surface and cytoplasm showed two distinct behaviors, relaxation regime and flow regime, after a fast elastic response (Bausch et al. 1998, 1999). Under traction, chicken fibroblasts showed a steep relaxation and a slow one, but the slow one was blurred by an active response of the cells.

Force curves, normalized to their applied forces, showed that the stress relaxation in the remaining period was influenced by applied forces (Fig. 8B). Time-constants τ_1, τ_2 and equilibrium stress σ_0 changed with different applied forces. The dependence of time-constant τ_1, τ_2 on the applied force shows a nonlinearly of viscoelastic response of the pad material. The dependence of equilibrium stress on the applied force, was manifested with the viscoelasticity model, in which $\sigma_0 = E_0\varepsilon$. The strain ε was the assumed constant caused by the applied force F, expressed as $\varepsilon = \left(\frac{4}{3}\frac{1}{1-\nu^2}\right)^{2/3}\frac{1}{\pi R^{2/3}E^{1/3}}F^{1/3}$ from Eqs. 3 and 6.

Since the average weight of the insect is approximately 1 g, each leg has a load force of 1.6 mN while the insect is standing on a horizontal surface. The whole pad gets about half of the force, 0.8 mN. In the experiment we used forces ranging from 0.6 mN to 0.7 mN, where the contact area of the pad with the silicon surface could be estimated with the Hertz theory. The applied force of 1.4 mN may correspond to the load on the pad while the insect walks, and only three legs remain in contact with the substratum.

For the two relaxation processes, there are two alternative explanations. First, the rapid relaxation ($\tau_1 = 1.322$ s) was attributed to the structure of the endocuticle underlying the fibrous part of the pad cuticle, and the slow relaxation ($\tau_2 = 30.7$ s) to the surface structure. Second, the rapid relaxation was the property of the fiber structure of the pad only, including the dendrites of the surface structure and the endocuticle. The slow relaxation was due to the interaction of these fibers with the fluid inside the fibrous cuticle, and due to the haemolymph and air sacks inside the pad (Fig. 3C). In either of these two explanations, the slow relaxation was related with the fluid or air, but not with the solid phase of the pad material.

The functional significance of elastic properties of the pad can be easily recognized. However, it is more difficult to find an explanation for the viscous response. When pads demonstrate elastic properties only, one would expect saving elastic energy during landing after a jump, as other cuticle-related material do (Andersen and Weis-Fogh 1964;

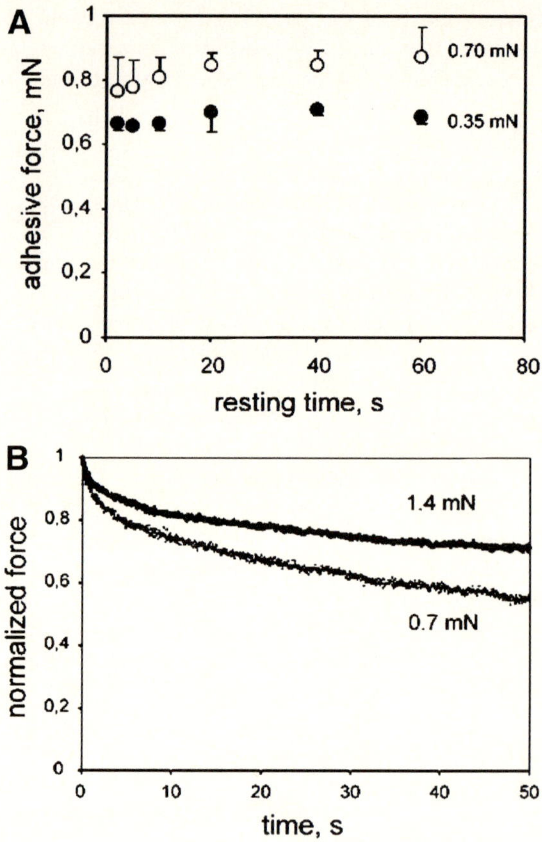

Fig. 8 A, B. Viscoelasticity of the pad material. **A** Adhesive force versus remaining time, measured at an applied force of 0.6 mN. The adhesive force did not show dependence on the remaining time. According to that the adhesive force was determined by the contact area caused by the applied force. **B** Comparison of two force curves normalized to applied forces. They were recorded on the same pad at applied forces of 0.7 mN, and 1.4 mN, respectively. The difference between them shows the nonlinearity of viscoelasticity response of the pad. The time constants τ_1, τ_2 and equilibrium stress $_0$ changed with different applied forces

Bennet-Clark 1976; Gorb 1999). In this case, an insect would be accelerated again just after landing. In other words, an elastic material would prevent getting contact to the surface in such a behavioral situation. A viscoelastic material serves as a kind of damper in which relaxation response aids in surface replication and optimization of a real contact between the pad and the underlying substrate.

Pad deformation mediates adhesion

It seems that both "hairy" and "smooth" types of attachment pads are supplemented by epidermal secretion. Such a secretion has been previously reported from pads of reduviid bugs (Edwards and Tarkanian 1970), flies (Bauchhenss and Renner 1977; Bauchhenss 1979a, 1979b; Walker et al. 1985), coccinellid beetles (Ishii 1987), cockroaches (Roth and Willis 1952), aphids (Lees and Hardie 1988; Dixon et al. 1990), bugs (Hasenfuss 1977a, 1977b, 1978; Ghasi-Bayat and Hasenfuss 1980a, 1980b, 1980c). Pad secretion is an essential factor for attachment to smooth substrata. For example, the bug *Rhodnius prolixus* with pads treated with lipid solvents, had poorer attachment performance compared with non-treated pads (Edwards and Tarkanian 1970). Aphids (*Aphis fabae*) after some period of walking on silica gel, lost their ability to attach to smooth surfaces (Dixon et al. 1990).

Our previous study showed that the pad secretion released in the contact area between pad and substrate is necessary but not sufficient for adhesion (Jiao et al. 2000). The second important factor is deformation of material. The indentation increased rapidly at smaller applied forces, and slowed down with greater forces applied. This fact can be explained by the shape of the pad: because of the hemispherical shape of the pad, both indentation and contact area increased very fast at small applied forces. With increased contact area, however, it was harder to get the pad surface indented. As a result, both the indentation and the contact area increased slower with increasing applied force. When the pad is fully indented, it behaved almost like a hard sample. In this condition, the indentation almost did not increase further with increasing applied force, and a maximum contact was reached. At applied forces, which caused the maximum contact between the pad and the silicon surface, saturated adhesive force was reached (Jiao et al. 2000). The saturated adhesive force of 1.1 mN was obtained at an applied force of 0.8 mN. The corresponding indentation at this load was about 75 μm, which was comparable with the thickness of the pad (ca. 100 μm). Before the maximum contact was reached, both the contact area and adhesive force increased with an increasing applied force. This fact shows that the adhesive force is related to the contact area caused by the applied force. It is well known from tribological literature that material deformation is one of several factors responsible for the area of real contact between solids (Bowden and Tabor 1986). The area of real contact may directly influence both adhesion and friction. In this study, data on material deformation of the pad aided in understanding factors influencing tribological properties of the pad.

Active response?

Some of the creep curves acquired with a magnetic bead inside the cytoplasm of J774 macrophages exhibited a direct crossover from elastic into the flow regime. The amplitude may be strongly enlarged or diminished (Bausch et al. 1999). The chick fibroblasts pulled by two microplates, showed a steep force relaxation. However, the force did not continue to decrease, instead it is increased to a certain value (Thoumine and Ott 1997). Cells exhibit remarkable mechanical activity even without an external force applied.

There were two interesting phenomena occurring in our experiments. First, some force curves obtained with the pads of living animals did not show rapid relaxation, whereas the two kinds of relaxation, a rapid one and slow one, always occurred on the pads of severed from the body. Second, on pads of living animals, the interacting force went up a little after the pad experienced an initial phase of the slow relaxation. In contrast, the interacting force of the pads of severed legs, always decreased. This difference might be due to the active response of the pad, although we do not have enough evidence for the final conclusion. The possible active response of pads needs to be further investigated.

Acknowledgements Discussions with Prof. Dr. U. Schwarz (MPI für Entwicklungsbiologie, Tübingen, Germany) and with Dr. R. Hilpert and Dr. J. Ritter (DaimlerCrysler Research Center, Ulm, Germany) on structure and properties of biomaterials are greatly acknowledged. We thank Mrs V. Kastner, who made valuable suggestions on an early version of the manuscript. This work is supported by the Federal Ministry of Education, Science and Technology, Germany to SG (project BioFuture 0311851).

References

Andersen SO, Weis-Fogh T (1964) Resilin, a rubber-like protein in arthropod cuticle. Adv Insect Physiol 2:1–65

Arnold JW (1974) Adaptive features on the tarsi of cockroaches (Insecta: Dictyoptera). Int J Insect Morphol Embryol 3:317–334

Bauchhenss E (1979a) Die Pulvillen von *Calliphora erythrocephala* Meig. (Diptera, Brachycera) als Adhäsionsorgane. Zoomorphologie 93:99–123

Bauchhenss E (1979b) Die Pulvillen von *Calliphora erythrocephala* Meig. als Adhäsionsorgane. München

Bauchhenss E, Renner M (1977) Pulvillus of *Calliphora erythrocephala* Meig. (Diptera; Calliphoridae). Int J Insect Morphol Embryol 6:225–227

Bausch AR, Ziemann F, Boulbitch AA, Jacobson K, Sackmann E (1998) Local measurements of viscoelastic parameters of adherent cell surfaces by magnetic bead microrheometry. Biophys J 75:2038–2049

Bausch AR, Möller W, Sackmann E (1999) Measurement of local viscoelasticity and forces in living cells by magnetic tweezers. Biophys J 76:573–579

Bennet-Clark HC (1976) Energy storage in jumping insects. In: Hepburn HR (ed) The insect integument. Elsevier, Amsterdam, pp 421–443

Beutel R, Gorb S (2000) Evolution of tarsal and pretarsal attachment structures of insects (Arthropoda). J Zool Syst Evol Res (in press)

Blackwell J, Weih MA (1980) The structure of chitin-protein complexes: the ovipositor of the Ichneumon fly *Megarhyssa*. J Mol Biol 137:49–60

Bowden FP, Tabor D (1986) The friction and lubrication of solids. Clarendon Press, Oxford

Carruthers CB, Davey KG (1983) Does cuticular elasticity regulate the size of the bloodmeal imbibed by female *Glossina austeni*? Can J Zool 61:1888–1891

Dixon AFG, Croghan PC, Gowing RP (1990) The mechanism by which aphids adhere to smooth surfaces. J Exp Biol 152:243–253

Edwards JS, Tarkanian M (1970) The adhesive pads of Heteroptera: a re-examination. Proc R Entomol Soc Lond A 45:1–5

Evens EA, Kwok R, McCown T (1980) Calibration of beam deflection produced by cellular forces in the 10^{-9} to 10^{-6} gram range. Cell Biophys 2:99–112

Evens EA, Yeung A (1989) Apparent viscosity and cortical tension of blood granulocytes determined by micropipet aspiration. Biophys J 56:151–160

Flaud P, Quemada D (1988) A structural viscoelastic model of soft tissues. Biorheology 25:95–105

Forgacs G, Foty RA, Shafrir Y, Steinberg MS (1998) Viscoelastic properties of living embryonic tissues: a quantitative study. Biophys J 74:2227–2234

Fung YC (1981) Biomechanics: mechanical properties of living tissues. Spring, Berlin Heidelberg New York

Ghasi-Bayat A, Hasenfuss I (1980a) Die Oberflächenstrukturen der Prätarsus von *Elasmucha ferrugata* (Fabricius) (Acanthosomatidae, Heteroptera). Zool Anz 205:76–80

Ghasi-Bayat A, Hasenfuss I (1980b) Über den Transportweg der Haftflüssigkeit der Pulvilli bei *Coptosoma scutellatum* (Geoffr.) (Plataspididae, Heteroptera). Nachrichtenblatt der Bayerischen Entomologen 1980:58–58

Ghasi-Bayat A, Hasenfuss I (1980c) Zur Herkunft der Adhäsionsflüssigkeit der Tarsalen Haftlappen bei den Pentatomidae (Heteroptera). Zool Anz 204:13–18

Gorb SN (1997) Ultrastructural architecture of the microtrichia of the insect cuticle. J Morphol 234:1–10

Gorb SN (1999) Serial elastic elements in the damselfly wing: mobile vein joints contain resilin. Naturwissenschaften 86:552–555

Gorb S (2000a) Ultrastructure of the neck membrane in dragonflies (Odonata). J Zool Lond 250:479–494

Gorb S, Scherge M (2000b) Biological microtribology:anisotropy in frictional forces of orthopteran attachment pads reflects the ultrastructure of a highly deformable material. Proc R Soc Lond Ser B 267:1239–1244

Hackman RH (1975) Expanding abdominal cuticle in the bug *Rhodnius* and in the tick *Boophilus*. J Insect Physiol 21:1613–1623

Hackman RH, Goldberg M (1985) The expanding alloscutal cuticle in adults of the argasid tick *Argas* (*Persicargas*) *robertsi* (Acari: Ixodoidea). Int J Parasitol 15:249–254

Hackman RH, Goldberg M (1987) Comparative study of some expanding arthropod cuticles: the relation between composition, structure and function. J Insect Physiol 33:39–50

Hasenfuss I (1977a) Die Herkunft der Adhäsionsflüssigkeit bei Insekten. Zoomorphology 87:51–64

Hasenfuss I (1977b) Woher stammt die Adhesionsflüssigkeit, die bei Insekten ein Haften an glatten Flächen ermöglicht? Verh Dtsch Zool Ges 277:277–277

Hasenfuss I (1978) Über das Haften von Insekten an glatten Flächen- Herkunft der Adhäsionsflüssigkeit. Zool Jb Anat 99:115–116

Henning B (1974) Morphologie und Histologie der Tarsen von *Tettigonia viridissima* L. (Orthoptera, Ensifera). Z Morphol Tiere 79:323–342

Hepburn HR (1985) Structure of the integument. In: Kerkut GA, Gilbert LI (eds) Comprehensive insect physiology biochemistry and pharmacology. Pergamon Press, Oxford, pp 1–58
Hertz H (1881) Über den Kontakt elastischer Körper. J Reine Angew Math 92:156–156
Hoh JH, Schoenenberger C-A (1994) Surface morphology and mechanical properties of MDCK monolayers by atomic force microscopy. J Cell Sci 107:1105–1114
Ishii S (1987) Adhesion of a leaf feeding ladybird *Epilachna vigintioctomaculata* (Coleoptera: Coccinellidae) on a vertically smooth surface. Appl Entomol Zool 22:222–228
Janmey PA (1991) Mechanical properties of cytoskeletal polymers. Curr Opin Cell Biol 2:4–11
Jiao Y, Gorb S, Scherge M (2000) Adhesion measured on the attachment pads of *Tettigonia viridissima* (Orthoptera, Insecta). J Exp Biol 203:1887–1895
Kendall UD (1970) The anatomy of the tarsi of *Schistocerca gregaria* Forskål. Z Zellforsch 109:112–137
Kolega J (1991) Effects of mechanical tension on protrusive activity and microfilament and intermediate filament organization in an epidermal epithelium moving in culture. J Cell Biol 102:1400–1411
Lees AM, Hardie J (1988) The organs of adhesion in the aphid *Megoura viciae*. J Exp Biol 136:209–228
Mak AF (1986) Unconfined compression of hydrated viscoelastic tissues: a biphasic poroviscoelastic analysis. Biorheology 23:371–383
Maxwell MH (1978) Two rapid and simple methods used for the removal of resins from 1.0 μm thick epoxy sections. J Microsc 112:253–255
Meissner DH, Schwarz H (1990) Improved cryoprotection and freeze-substitution of embryonic qual retina a TEM study on ultrastructural preservation. J Electron Micr Techn 14:348–356
Neville AC (1975) Biology of the arthropod cuticle. Springer, Berlin Heidelberg New-York
Neville AC (1993) Biology of fibrous composites. Cambridge University Press, Cambridge
Persson BNJ (1998) Sliding friction. Physical principles and applications. Springer, Berlin Heidelberg New York
Petersen NO, McConnaughey WB, Elson EL (1982) Dependence of locally measured cellular deformation on position on the cell, temperature and cytochalasin B. Proc Natl Acad Sci USA 79:5327–5331
Radhakrishnan V (1998) Locomotion: dealing with friction. Proc Natl Acad Sci USA 95:5448–5455
Radmacher M, Fritz M, Kacher CM, Cleveland JP, Hansama PK (1996) Measuring the viscoelastic properties of human platelets with the atomic force microscope. Biophys J 70:556–567
Reynolds SE (1975) The mechanical properties of the abdominal cuticle of *Rhodnius* larvae. J Exp Biol 62:69–80
Roth LM, Willis ER (1952) Tarsal structure and climbing ability of cockroaches. J Exp Biol 119:483–517
Sanjeevi R (1982) A viscoelastic model for the mechanical properties of biological materials. J Biomech 15:107–109
Scherge M, Li X, Schaefer JA (1999) The effect of water on friction of MEMS. Tribol Lett 6:215–220
Schwarz H, Humbel BM (1988) Freeze-substitution and immunolabelling. Inst Phys Conf Ser 93:543–544
Slifer EH (1950) Vulnerable areas on the surface of the tarsus and pretarsus of the grasshopper (Acrididae, Orthoptera) with special reference to the arolium. Ann Entomol Soc Am 43:173–188
Snodgrass RE (1956) Anatomy of the honey bee. Comstock, New York
Spurr AR (1969) A low-viscosity epoxy resin embedding medium for electron microscopy. J Ultrastruct Res 26:31–43
Stork NE (1980) A scanning electron microscope study of tarsal adhesive setae in the Coleoptera. Zool J Linn Soc 68:173–306
Thoumine O, Ott A (1997) Time scale dependent viscoelastic and contractile regimes in fibroblasts probed by microplate manipulation. J Cell Sci 110:2109–2116
Vincent JFV (1975) Locust oviposition: stress softening of the extensible intersegmental membranes. Proc R Soc Lond Ser B 188:189–201
Vincent JFV (1981) Morphology and design of the extensible intersegmental membrane of the female migratory locust. Tissue Cell 13:18–31
Vincent JFV, Prentice JH (1973) Rheological properties of the extensible intersegmental membrane of the adult female locust. J Mat Sci 8:624–630
Vincent JFV, Wood SDE (1972) Mechanism of abdominal extension during oviposition in *Locusta*. Nature (Lond) 235:167–168
Walker G, Yule AB, Ratcliffe J (1985) The adhesive organ of the blowfly, *Calliphora vomitoria*: a functional approach (Diptera: Calliphoridae). J Zool Lond 205:297–307
Ziemann F, Rädler J, Sackmann E (1994) Local measurements of viscoelastic moduli of entangled actin networks using an oscillating magnetic bead micro-rheometer. Biophys J 66:2210–2216

J Comp Physiol A (2002) 187: 961–970
DOI 10.1007/s00359-001-0265-7

ORIGINAL PAPER

Contact behaviour of tenent setae in attachment pads of the blowfly *Calliphora vicina* (Diptera, Calliphoridae)

Senta Niederegger · Stanislav Gorb · Yuekan Jiao

Accepted: 31 October 2001 / Published online: 7 December 2001

Abstract To enable strong attachment forces between pad and substrata, a high proximity between contacting surfaces is required. One of the mechanisms, which can provide an intimate contact of solids, is a high flexibility of both materials. It has been previously presumed that setae of hairy attachment pads of insects are composed of flexible cuticle, and are able to replicate the surface profile. The aim of this work was to visualise the contact behaviour of the setae by freezing-substitution technique to understand setal mechanics while adhering to a smooth surface. This approach revealed considerable differences in the area of the setal tips between contacting and non-contacting pulvilli. Based on the assumption that setae behave like a spring pushed by the tip, a spring constant of 1.31 N m^{-1} was calculated from direct measurements of single setae by atomic force microscopy. In order to explain the relationship between the behaviour of the attachment setae at a microscale and leg movements, high-speed video recordings were made of walking flies. This data show that some proximal movement of the leg is present during contact formation with the substrate.

Keywords Cuticle · Adhesion · Pulvilli · Scanning electron microscopy · Atomic force microscopy

Introduction

The hairy attachment system of flies has been studied for more than 300 years. Numerous studies have been conducted to understand the anatomy and function of the system, as well as the design of the fly pretarsus. Nevertheless, there remains a lack of agreement on solutions concerning the attachment mechanism of the fly walking on smooth surfaces. Different hypotheses have been proposed to explain the mechanism of attachment. Theories of a sticky fluid (Dewitz 1884), microsuckers (Simmermacher 1884), and the action

S. Niederegger (✉) · S. Gorb · Y. Jiao
Biological Microtribology Group, Biochemistry Department, MPI of Developmental Biology, Spemannstr 35, 72076 Tübingen, Germany
e-mail: senta.niederegger@tuebingen.mpg.de

of electrostatic forces have been discussed (Gillett and Wigglesworth 1932). Based on experiments with beetles, these theories have been rejected and it has been concluded that cohesive forces, surface tension and molecular adhesion of the pad secretion are involved in the mechanism of attachment of the hairy system (Stork 1983b).

To enable strong attachment between pad material and diverse substrata, a high proximity between contacting surfaces is required. One of the mechanisms, which can provide an intimate contact of solids, is high flexibility of at least one of both materials. It has been previously presumed that setae are composed of flexible cuticle, and are able to replicate the surface profile (Bauchhenss 1979; Stork 1983b). If the single setae of the adhesive pads are adapted to deform under load and fit the microtexture of surfaces, contact with the maximum number of attachment points would be possible, independent of the substrate profile (Blackwell 1930; Bauchhenss and Renner 1977; Bauchhenss 1979; Stork 1980, 1983b; Röder 1984; Wigglesworth 1987). This hypothesis, however, has not been properly tested.

The aim of this work was to visualise the contact behaviour of the setae, to understand setal mechanics while adhering to a smooth surface. The general anatomy of the fly pretarsus and quantitative data on the arrangement of the tenent setae, studied in the scanning electron microscope (SEM), were an important set of data for the detection of differences between pulvilli contacting a substrate and pulvilli not in contact. The basic idea was to reveal differences in the area of the setal tips between contacting and non-contacting pulvilli in order to get an impression of the degree of material flexibility. Data obtained from direct measurements of the spring constant of a single seta by atomic force microscopy (AFM) was an additional approach to understanding setal mechanics. In order to explain the relationship between the behaviour of the attachment setae at a microscale and leg movements, high-speed video recordings were made of walking flies and their leg kinematics were analysed.

Materials and methods

Animals

Imagines of *Calliphora vicina* Robineau-Desvoidy (Diptera, Brachycera) were taken from laboratory colonies at MPI of Biological Cybernetics, Tübingen, and at the University of Würzburg. This species is a model for the study of hairy attachment pads and has been used in studies on the pad ultrastructure (Bauchhenss 1979) and in the measurements of attachment forces (Walker et al. 1985). In the present study, the pulvilli of male and female individuals of *C. vicina* were studied.

Light and electron microscopy

Semi-thin sections of the pulvillus were prepared from fixated pulvilli (2.5% glutaraldehyde in phosphate-buffered saline, pH 7.3) embedded in a low-viscosity resilin (Spurr 1969). They were dissected using a diamond knife, placed on round cover-slips, washed for 2 min in Maxwell solution (Maxwell 1978), washed in absolute acetone, transferred to absolute ethanol and critical point dried. After coating with gold-palladium(8 nm), the sections were viewed in the scanning electron microscope Hitachi S-800 at 20 kV. Another part of the sections were placed on glass microscope slides, stained with toluidine blue (80°C, 2 min), washed in running water, dried, mounted by polymerisation in Spurr resin at 70°C overnight, and observed under a Zeiss Axioplan light microscope.

Video recordings of leg movements

To obtain information about fly motion on the ceiling and wall, walking flies were recorded from one side by a high-speed camera KODAK Motion Corder Analyzer Model PS-220 at 250 frames s^{-1} (fps). Flies walked inside a Plexiglas tube with a Plexiglas platform running along the tube to hold the fly within focus of the camera. The flies walked horizontally both in upright and head down positions and vertically from bottom to top and from top to bottom. The rotation angle of the front leg and the maximum height of the leg pretarsus during the step cycle were measured from digital frames by the use of SigmaScan software separately for fore-, mid- and hindlegs.

Visualisation of setal tips in contact

Micrographs of pulvilli, which were in contact to a glass surface, were taken in order to visualise changes in shape of the contacting tenent setae. The tarsus was amputated from the freshly killed animal and immediately brought into contact with the glass surface of a small piece of thin cover-slip by a sliding movement in the proximal direction. To observe the contacting region of the pulvillus, the glass was mounted perpendicularly to a metal stub. No coating was applied to the samples to prevent artefacts in the contacting region of the setae. The samples were studied at reduced voltage of 5 kV in a SEM Hitachi S-800.

Drying experiment

The tarsi were cut off the body of a freshly-killed insect, and the pulvilli were immediately put in contact with a glass cover slip surface by a sliding movement in the proximal direction (Stork 1983b). The cover-slip was then put upside down onto small wax supports on microscope slides. The attachment points were observed at magnifications of × 400 and × 1000 under a light microscope (Zeiss, Axioplan) and pictures taken with a digital colour video camera (Sony Power HAD). The microscopic slides were stored in cardboard boxes at room temperature and humidity from 1 week up to 5 months. The legs of the fly were then removed from the glass cover-slips and mounted ventral side up onto metal stubs by the use of double-sided carbon tape, coated with carbon-platinum, and studied in the SEM. The area of the pulvilli, the number of setae and the setal tip areas were quantified from digital images by the use of SigmaScan software.

Freezing-substitution experiment

Legs were cut off the body of a freshly killed fly and immediately put into contact with a glass cover-slip surface with fine forceps by a sliding movement in the proximal direction. When contact was reached, liquid N_2 was poured over the leg. The frozen legs were put into absolute acetone at –196°C and fixed with a 1% solution of osmium tetroxide in absolute acetone at –80°C. After fixation the legs were washed in absolute acetone, gradually transferred to room temperature conditions (–80°C, –40°C, –20°C, +4°C, +20 °C), transferred to 100% ethanol, critical-point dried, mounted on metal stubs and, after sputter-coating with gold-palladium, viewed in the SEM Hitachi S-800 at 20 kV. The setal tip areas were measured using SigmaScan software. Only female individuals were used and the fore-, mid- and hindlegs were processed and analysed separately.

Some setae broke in the frozen pulvilli as the material became rigid at –196°C, but the low temperature conserved the shape of the setal tips in contact. After the substitution and drying procedure, setae no longer broke. In the drying experiment, setal shape in contact was preserved after removal from the glass. The slight discrepancy between data obtained from light microscope pictures and SEM micrographs of drying setae can be explained by the higher resolution of the electron microscope.

Atomic force microscopy

The force-distance curves were obtained with AFM (Nanoscope III, Digital Instruments, Calif., USA), by pressing the single seta with a 0.1 N m^{-1} cantilever (Park Scientific Instruments, Calif., USA). For details see Fig. 10C). The cantilever approached the seta (a-b), and jumped into contact with the seta (b–c), because of the secretion fluid on the setal tip (Walker et al. 1985). At point d, the cantilever returns to zero deflection. After point d, the seta is loaded. Information about deformation of the seta was obtained by subtracting the cantilever deflection from the cantilever displacement relative to point d. Setal deformation in the range d–e was obtained and plotted versus the force value applied to the cantilever. The hair is modelled as a spring, and the data of force versus contraction were fitted according to the equation $F = KD$, where F is applied force, D is contraction, and K the spring constant. For each of 12 measured setae, 3–4 force-distance curves were taken. Setal tip images were obtained in the height and deflection modus of the AFM.

Results

Anatomy of the pretarsus

The distal part of the terminal tarsomere is articulated with the pretarsus (Fig. 1A). The terminal tarsomere is a tube bearing the tendon of the claw flexor muscle (Weber 1930; Radnikow and Bassler 1991). The tendon is connected to the claws and attachment pads (pulvilli) through the unguitractor plate (Goel 1972). The unguitractor plate can be connected to the invagination of the tarsomere wall by cuticle outgrowths providing anchorage of the unguitractor plate, when the flexor muscle is contracted and claws are in contact with a support (Gorb 1996). A single claw consists of a stiff cuticle and runs like an arch from proximal to distal, approximately along the borderline of each pulvillus. The wider basal parts of the claws are connected through flexible cuticle to each other, to the dorso-distal part of the terminal tarsomere, and to the disto-lateral edge of the unguitractor plate. The length of the claw arch is about 350 μm. In the distal part of the unguitractor plate, an elongated cuticular process is located, called empodium, presumably functioning as a sensory organ.

Design of the pulvilli

Specialised attachment organs of the pretarsus are the pulvilli, which are connected to the unguitractor plate through bowl-like sclerites, the auxiliae. The pulvilli of *C. vicina* are oval shaped structures 250 μm long. They have proximally a height of about 120 μm and are flattened distally and laterally (Fig. 1E). The flatter regions forma lobe-like, transparent border surrounding the middle part. The dorsal side bears grooves that run disto-laterad. An important characteristic of the pulvillus is its material, composed exclusively of elastic types of cuticle (meso- and endocuticle). Dorsal longitudinal exocuticular strips and partly exocuticular auxiliae are the only stabilising structures, which, together with the relatively rigid basal part of the pulvillus, forma heel like stabilisation zone (Bauchhenss 1979).

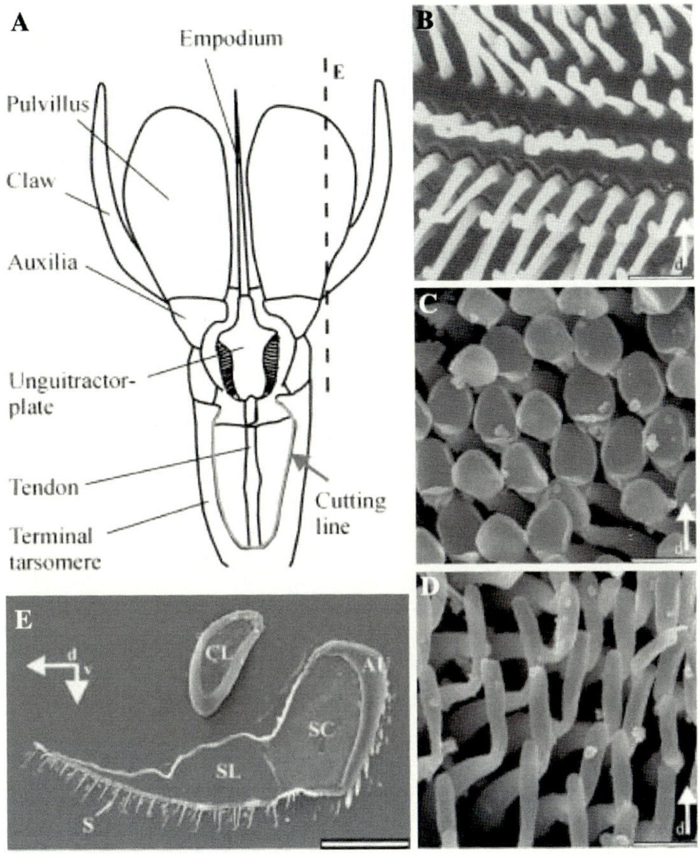

Fig. 1 A–E Anatomy of attachment pads. **A** Diagram of the pretarsus, ventral aspect. Cuticle of the terminal tarsomere is partly cut to expose the apodeme. *Broken line* indicates the region of the longitudinal section shown in **E**. **B–E** SEM-micrographs: **B** the setae of the pulvilli are arranged in alternating horizontal rows; **C** pulvilli frozen in contact, distal setae with endplate; **D** pulvilli frozen in contact, basal setae without endplate; **E** longitudinal semithin (1 μm) section of the pulvillus prepared according to Maxwell (1978). Scale bars: 6.5 μm (**B**), 2 μm (**C, D**), 50 μm (**E**). *AU* auxilia; *CL* claw; *S* setae; *SL* spongy layer; *SC* secretory cells; *v* ventral direction; *d* distal direction

The pulvillus bears no muscles. Secretory cells are located in the proximal part of the pulvillus, surrounded by the so-called spongy layer, where the secretion is deposited (Fig. 1E). The ventral side of each pulvillus is covered by 4,000–6,000 setae arranged in an alternating pattern of horizontal lines (Fig. 1B). The tenent setae are located in the ventral part of the pulvilli. Each seta is composed of a wide base, a long and narrow shaft and most of them have a bent spatula-like endplate. The setae are hollow and filled with the secretion (Gorb 1998). Two different types of setae are found on the pulvilli: the setae of the distal part of the pulvilli bear an endplate (distal type; Fig. 1C), whereas basal setae do not have a well-expressed endplate (proximal type; Fig. 1D). The density of the setae and the tip length remained more or less constant in the proximal and distal parts of the pulvillus (Fig. 2C, E). The tip area and tip width increased in the distal direction (Fig. 2D, F).

Sexual dimorphism in pulvilli area and setal design

Pulvillus area was the smallest in the forelegs and the largest in the hindlegs in both sexes. It was significantly larger in male individuals (forelegs: $P < 0.001$, $F = 80.61$; midlegs: $P < 0.001$, $F = 75.02$; hindlegs: $P < 0.001$, $F = 135.66$, one-way ANOVA; Fig. 2A, B). The difference in the foreleg and midleg was about 10,000 μm^2; in the hindleg 13,000 μm^2.

The setal density was higher in males than in females (Fig. 2E). The tip area of setae in the proximal part of the pulvilli was larger in female individuals whereas the tip area increased to a higher extent in males, which consequently had larger tip areas in the distal part of the pulvillus (Fig. 2F). This difference depended on the width of the tip, which was larger in females than in males (Fig. 2D). The length of the setal tip was almost equal in both sexes (Fig. 2C).

Leg movements

High-speed video recordings show that, during placement onto the substrate, pretarsi are pressed against the surface. Since the pulvillus moves along a circular trajectory, it is

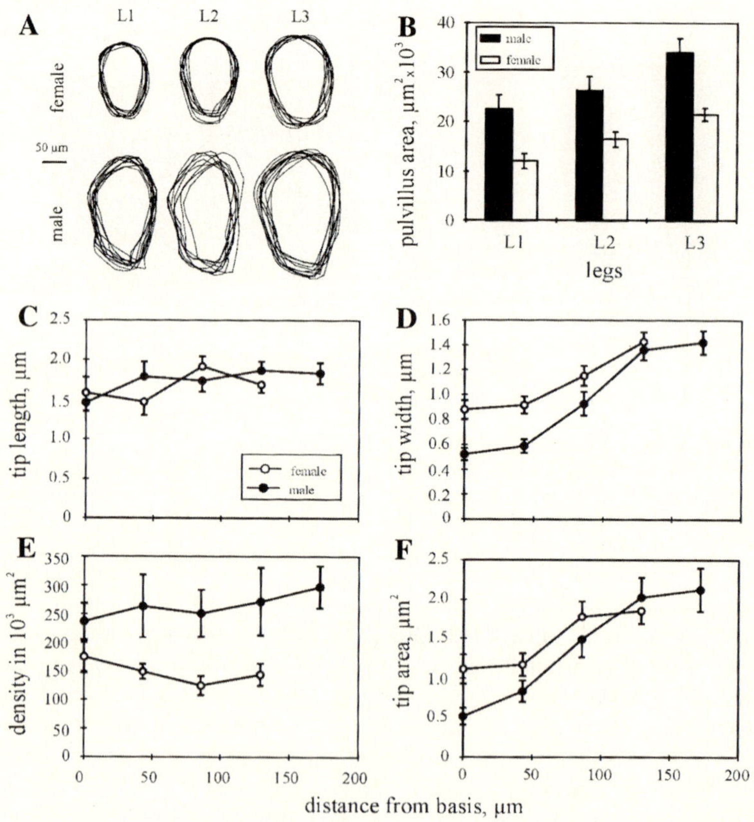

Fig. 2 **A–F** Sexual dimorphism in tenent setae. **A** Contours of male and female pulvilli arranged according to legs (L1– L3). **B** Pulvillus area of legs of males and females (diagram is based on measurements of areas shown in **A** (n = 16). **C–F** Different parameters of the setae versus the distance from the pulvillus basis; **C** length of the tips; **D** width of the tips; **E** Density of setae in 10^3 μm^2; F area of the tips

simultaneously pulled slightly in the proximal direction (Fig. 3). The angle of the leg rotation in the sagittal plane and the maximum pretarsus height in the step cycle, measured in the foreleg, did not vary significantly depending on the walking position of the fly (angle: $P = 0.095$, $F = 2.452$, one-way-ANOVA, $n = 60$; height: $P = 0.554$, $F = 0.598$, one-way-ANOVA, $n = 60$; Fig. 4A, B). The step height in the horizontal position was significantly higher in females than in males ($P < 0.001$, $t = 4.721$, t-test, $N = 4$ animals, $n = 5$ measurements/animal; Fig. 4A), but this difference was insignificant in the vertical and head-down positions (vertical: $P = 0.054$, $T = 79$, Rank Sumtest; head-down: $P = 0.600$, $t = -0.534$, t-test). No difference between sexes or positions was detected in the angle of the leg rotation ($P = 0.821$, $F = 0.437$, one-way ANOVA on pooled data, $n = 60$) (Fig. 4B). Females differed in step height in normal against vertical ($P = 0.043$, $t = 2.176$, t-test) and in vertical against head-down position ($P = 0.007$, $T = 141.000$, Mann-Whitney Rank Sumtest).

There were considerable differences in the step height between front, middle and hind pretarsi (Fig. 5). The highest steps occurred in the forelegs, and the lowest ones in the hindlegs (Fig. 5A). For pooled data for all positions, there were significant differences between fore- and midlegs ($P < 0.001$, $T = 1{,}172.000$, $n_1 = 28$, $n_2 = 29$), fore- and hindlegs ($P < 0.001$, $T = 358.000$, $n_1 = 26$, $n_2 = 28$, both Mann-Whitney Rank Sum test), and mid- and hindlegs ($P < 0.001$, $t = 5.232$, t-test; Fig. 5B). According to the walking position, there was a significant difference between vertical ($n = 14$) and head-down ($n = 9$) positions in the foreleg ($P < 0.001$, $t = 4.793$); between normal ($n = 5$) and head-down positions ($n = 10$) in the midleg ($P = 0.005$, $t = 3.398$); and between normal ($n = 5$) and head-down ($n = 7$) positions in the hindleg ($P = 0.009$, $t = 3.222$; all t-test).

Additionally, the time necessary for the placing of the pulvilli onto the surface was calculated from the high-speed recordings (Fig. 5C). The recordings were done with 250 pictures per second, which results in 4 ms per picture. The time was calculated from the

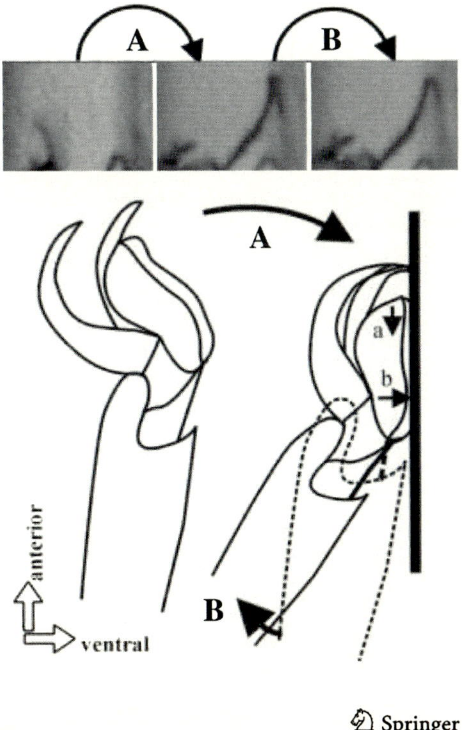

Fig. 3 Scheme of pretarsus placement of the foreleg on the surface of a vertical Plexiglas surface. It is based on the high-speed camera video recordings (250 frames s^{-1}; fps) shown in the upper part of the figure. The camera was oriented perpendicular to the animal's movement and parallel to the substrate. **A**: the pretarsus is put onto the surface at a certain angle and pulled slightly towards the body (*a*). **B**: the pretarsus is pressed to the surface (*b*) as the leg strives for an upright position. *anterior* indicates anterior direction of the fly body, *ventral* indicates ventral direction of the fly body

Fig. 4 **A, B** Foreleg movement in the step cycle. Data were obtained from males and females walking in different positions. **A** Maximal height of the foreleg pretarsus in the step cycle. **B** Maximal angle of the leg rotation in the sagittal plane

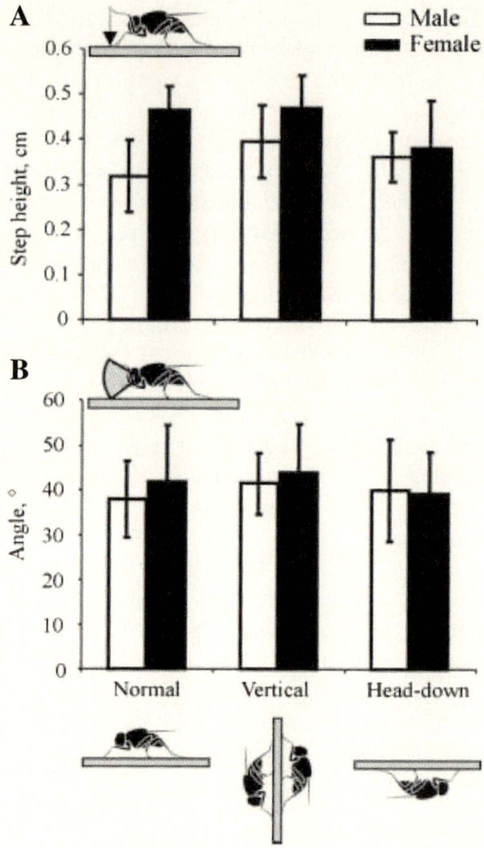

picture showing the pulvillus in the highest position to the picture showing the first contact. Pooling of all data revealed significant differences between fore-, mid- and hindlegs ($P < 0.001$, $H = 43.140$, $n_1 = 90$, $n_2 = 66$, $n_3 = 77$, ANOVA on ranks). Fore- and midlegs did not differ significantly ($P = 0.541$, $T = 5{,}352.000$, $n_1 = 90$, $n_2 = 66$, Mann-Whitney Rank Sum test), therefore only fore- and hindlegs were compared further. In normal ($P < 0.001$, $T = 898.000$, $n_1 = 26$, $n_2 = 27$), vertical ($P < 0.001$, $T = 811.000$, $n_1 = 30$, $n_2 = 23$) and head-down position ($P < 0.001$, $T = 1103.000$, $n_1 = 33$, $n_2 = 28$) fore- and hindlegs differed significantly.

These results show that acceleration distances available for foreleg pulvilli are greater than for hindleg pulvilli. In midleg pulvilli the time does not differ significantly from foreleg pulvilli, but as the distance is significantly larger in foreleg pulvilli, midleg pulvilli also are placed on the surface with a considerably lower acceleration rate.

Setal tip behaviour in contact

In the attached pretarsus, the pulvilli are slightly bent, and the claws are turned laterally (Fig. 6A). The setal shafts roughly keep their shapes and are sloped distally (Fig. 6B). It seems that they are not glued together (Fig. 6C). The end plates of the setae were always orientated distally (Fig. 6B, C). In the frozen contacting pulvillus, the setal tip area of the

Fig. 5 **A–C** Step height in the fly leg. **A** Single frames of the single step cycle of the fly walking in a horizontal position (high-speed video recordings, 250 fps). *Left frame* shows three legs (L1, R2, L3) in the stance phase; *right frame* shows all six legs in the stance phase.
B Maximal step height in fore-, mid- and hind legs ($N = 12$).
C Time required to bring the pulvillus from the highest position to the surface in fore-, mid- and hindlegs ($N = 12$)

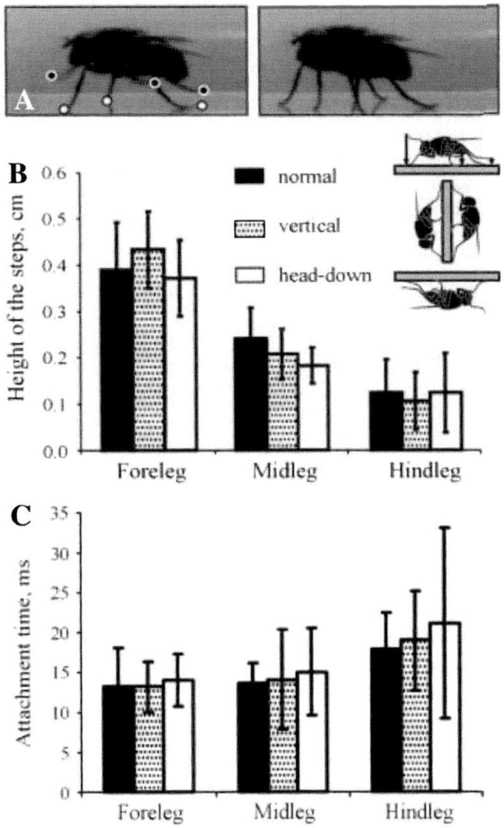

contacting part was significantly larger compared to the pulvillus, which was not in contact to the surface ($P < 0.001$, $F = 164.126$, one-way ANOVA, $N = 2$, $n = 100$; Fig. 7A–E). The difference between areas of contact in the distal setae was greater than in the proximal and middle setae (Fig. 7F). Area of setae in contact differed significantly from that not in contact ($P < 0.001$, basal setae: $t = 6.828$, middle setae: $t = 29.878$, distal setae: $t = 17.390$, t-test).

In the drying experiments, the pulvilli of the midleg, dried in contact with the glass surface for several weeks, were compared to air-dried pulvilli, which were not in contact. The setal tips of the pulvilli, which were in contact, can be clearly distinguished from tips, which were not in contact to the surface. This was verified by combining the light-microscope images taken of the fresh pulvilli in contact and SEM images obtained from same preparations dried out in contact to the glass (Fig. 8). Measurements based on light microscope images (Fig. 8B) demonstrated no significant changes in the tip area after removal of the pulvillus from the glass and SEM procedure ($P = 0.127$, $T = 2747$, Mann-Withey Rank Sumtest, $N = 2$, $n = 50$; Fig. 8C). However, areas of contacted and non-contacted tips differed significantly from one another ($P < 0.001$, $t = -39.694$, t-test, $N = 1800$; Fig. 8D).

In the disto-proximal direction, the area of the setal tips slightly increases from the distal to the middle part of the pulvillus, and rapidly decreases from the middle to the basal part. The largest setal tips were found in the middle and distal regions of the pulvillus,

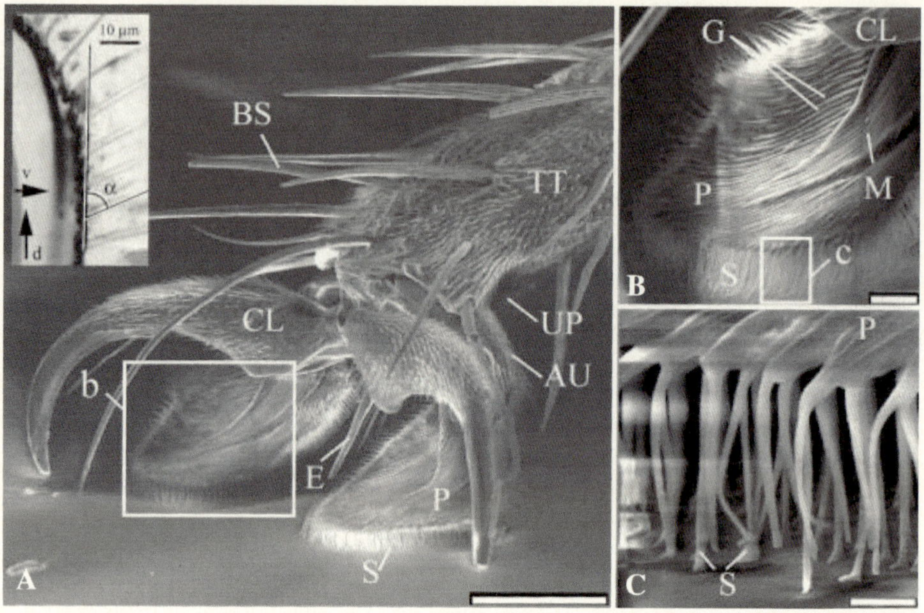

Fig. 6 **A–C** Scanning electron micrographs of the fly pulvilli attached to the glass cover-slip. **A** The whole pretarsus. *Inset*: light microscopic photograph of a semithin section of the pulvillus. *α* slope of a single seta; *v* ventral direction; d distal direction. **B** Right pulvillus. **C** Setae of the pulvillus. Scale bars: 50 µm (**A**), 20 µm (**B**), 5 µm (**C**). *AU* auxilia; *b* section shown in **B**; *BS* bristles; *c* section shown in C; *CL* claw; *E* empodium; *G* grooves on the dorsal surface of the pulvillus; *M* middle part of the puvillus; *P* pulvillus; *S* setae; *TT* terminal tarsomere; *UP* unguitractor plate

which are covered with distal-type setae ($P < 0.001$, $H = 191.619$, $N = 5$, $n = 50$, ANOVA on ranks; Fig. 9A, B). In the medio-lateral direction, the area of the setal tips significantly decreases from the middle to the edges of the pulvillus ($P < 0.001$, $F = 33.960$, $N = 4$, $n = 50$, one-way ANOVA; Fig. 9C). The total number of setae on the pulvillus amounts to 4,000–6,000. The contact area analysed under light and electron microscopes includes 1,300–2,000 setae corresponding to a portion of about one third of the total number of setae (Fig. 9D). This means that one-third of the setae is sufficient for pulvillus attachment to the glass surface.

Mechanics of a single seta

AFM images supported a hexagonal arrangement of single setae (Fig. 10A, B). After scanning of certain areas of the pulvillus, force-distance curves were recorded on a single seta (Fig. 10C). The seta waa pressed by the cantilever and cantilever deflection was measured. By subtracting this curve from the curve obtained on the hard sample (glass), the force-indentation dependence was obtained (Fig. 10D). However, it is not known exactly, whether the tip really pushed a seta down or penetrated its surface (the setal tip area is about 1 µm^2, and the cantilever tip is very sharp), or if both phenomena took place. Based on the assumption that setae behave like a spring pushed by the tip, the spring constants of setae were calculated. Spring constant of the single seta was 1.31 N m^{-1} (SD = 1.09, $N = 4$, $n = 12$).

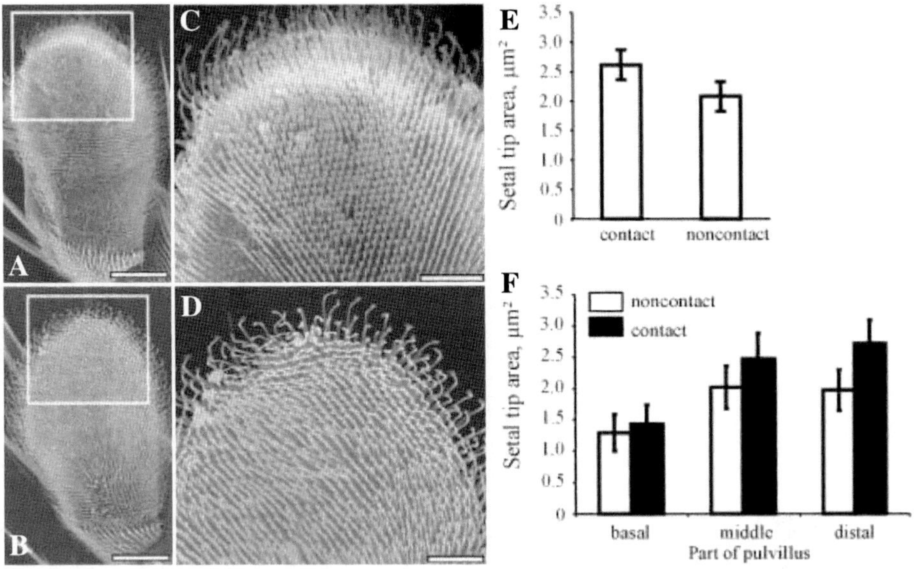

Fig. 7 **A–F** Results of shock- freezing experiments. **A–D** SEM micrographs of right pulvilli of the forelegs. **A, C** Pulvillus in contact with the surface. **B, D** Pulvillus not in contact. Scale bars: 50 μm (**A, B**); 20 μm (**C, D**). **E** Setal tip area of the distal part of the pulvilli. **F** Tip area of setae in contact and not in contact in basal, middle and distal regions of the pulvillus ($N = 3$, $n = 600$)

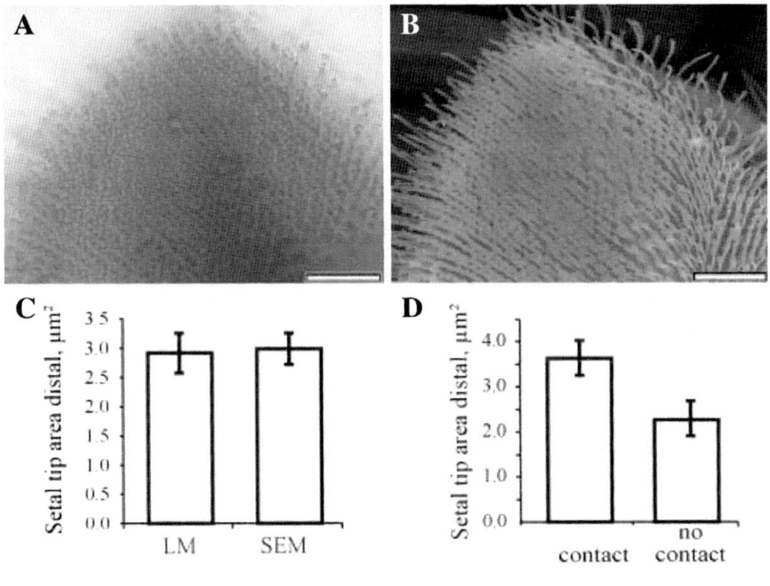

Fig. 8 **A–D** Results of drying experiments. **A** Light micro- scopic (LM) photographs of the fresh pulvillus manually attached to the glass surface. **B** SEM micrograph of the same pulvillus dried in contact. **C** Comparison of the setal tip area measured from light microscopic photographs and SEM micrographs. The data are not significantly different ($P = 0.127$, $T = 2747$, rank sum test, $N = 2$, $n = 50$). **D** Setal tip areas measured in middle legs, the pulvilli of which dried in contact to a glass surface, and those dried without contact. The difference is highly significant ($P < 0.001$, $t = 35.057$, t-test, $N = 4$, $n = 100$)

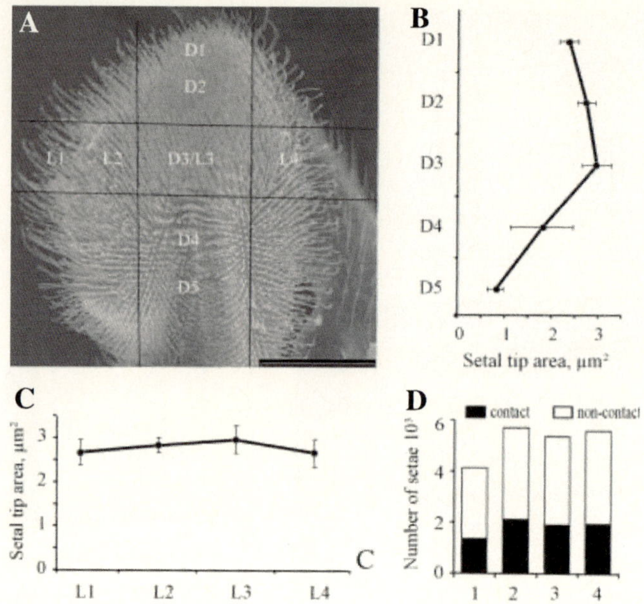

Fig. 9 **A–D** Results of drying experiments. **A** Typical SEM micrograph of a pulvillus dried in contact to a glass surface and used for measurements. **B** Gradient (*D1–D5*) of the setal tip area in the disto-proximal direction. **C** Gradient (*L1–L4*) of the setal tip area in the latero-medial direction. **D** Number of setae in contact and not in contact with the glass surface in four individual legs

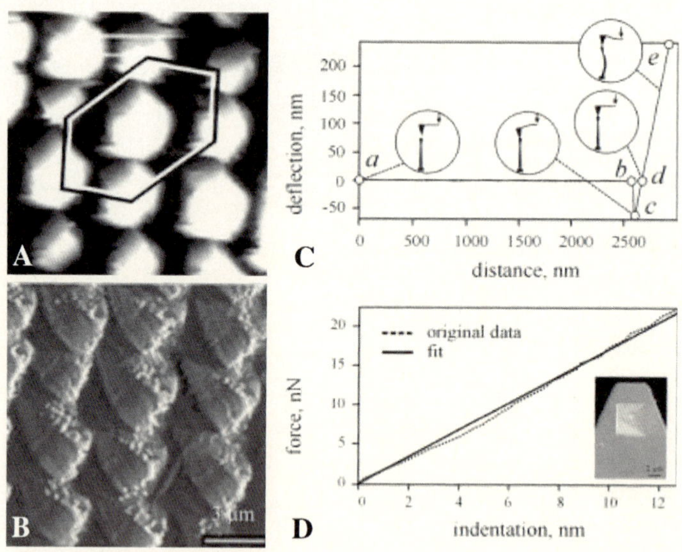

Fig. 10 **A–D** Atomic force microscopy (AFM) of the fly setae. **A** Height image of the fly setae. **B** Deflection image of the fly setae. **C** Deflection versus displacement obtained on a single seta. *Insets* show the cantilever position relative to the seta at each part of the curve. **D** Force-indentation curve used for spring constant determination of the single seta. Inset: cantilever tip (SEM micrograph). *a-b* approaching; *b-c* jump-into-contact; *c-d* returning to the zero position; *d-e* deflection under load in contact with the seta

Discussion

Material for intimate contact

Independent of the presumed basic attachment mechanism (capillary interactions, electrostatic interactions, viscous force, Van der Waals forces), a high proximity between contacting surfaces is required. High flexibility of the terminal elements of the pad material must be the most important property of hairy attachment systems. Pulvilli are composed of elastic types of cuticle, which is then also the case for material of setae. If the material is structured as outgrowths with high length-to-width aspect ratio, the flexibility of surface would by even higher. This is supported by the AFM data. Taking into account that the spring constant of a single seta is about 1 N m^{-1}, the total number of setae in 12 pulvilli about 60,000, and the fly weight about 3.953 mN (SD = 0.804, $N = 12$), it can be estimated to which extent setae can deformunder fly weight, if all setae of all legs are in contact. The calculation gives 0.064 μm. If the fly has only three legs in contact, which corresponds to the usual two-tripod walking pattern of most insects, expected deformation of single setae would be 0.128 μm. If only one-third of the setae are in contact, as in our experiments (Fig. 9D), the deformation would be even higher (0.384 μm).

The ability of insect pad setae to adhere to smooth surfaces has been previously shown for beetles (Stork 1983b). In the present study, it is quantitatively shown that the setal tips increase their areas in contact with the substrate. Although the pulvilli were not put into contact by the insect itself but manually, after amputation of the leg from the body, this approach provided some important information about the functional mechanism of setae. The hollow shafts of the setae appear to be stable, as the setae did not buckle under mechanical loads. However, the end parts of the setae seem to be composed of a rather flexible material, as the area of the tips increased by about 35% in contact. High flexibility of setal tips can be a mechanism of adaptation to the substrate nanotexture.

The setal tips in the basal part of the pulvillus are not round, and the area does not increase in the same range as in the middle and distal setae. However, the tip region of the proximal-type setae can bend, which result in an increase of contact area. A function of the proximal setae could be to ensure the stability of the pulvillus contact by supporting the proximal part of the pulvillus like a heel. The area of the setal tips increases in the distal part, thus providing large overall contact area under load. Such design results in stronger attachment forces. This assumption is supported by our experiments, in which, when a pulvillus of the cut off leg was brought into contact and left in this condition, only the distal part of the pulvillus attached to the glass surface.

Friction or adhesion?

Attachment force is composed of a variety of basic physical forces. Depending on structural, chemical, and mechanical properties of the surface, various forces can differently contribute to the overall attachment. Another important parameter is the direction to which an acting force is applied. Depending on the direction of the applied force, one divides forces resisting the acting force into friction and adhesion. One speaks about adhesion, when acting force is directed perpendicular to the surface and pulls two surfaces apart. Friction is the force resisting the sliding of two surfaces on each other (Bowden and Tabor 1986). Both types of force can differ in the composition of basic physical forces and may differently contribute to the overall attachment force depending on forces acting in contact.

Structural and experimental data of the present paper can give some insight into contribution of friction and adhesion to the overall attachment force of the fly pad.

Terminal spatula-like elements are known from other attachment pads, such as gecko hairs (Stork 1983a), spider (Foelix 1982), and beetle (Stork 1988) setae. Recent data on the single hair measurement in the gecko have revealed high forces during sliding (friction) and low forces in the direction perpendicular to the surface (adhesion) (Autumn et al. 2000). A short sliding is required to initiate hair attachment. Even in systems, in which the spatula-like ends are not orientated in one particular direction, one can expect anisotropic frictional properties. In the fly system, distal orientation of the end parts of the setae was found (Fig. 1B). Forces, acting on the pulvilli in the proximal direction during contact formation, will result in setal tip deformation in the distal direction, where setae are pre-oriented (Figs. 6, 8A, B). The aim of the proximal pull of the pulvillus observed in this study could be (1) spreading the spatula-like ends on the surface to optimise contact, and (2) release of the secretory fluid in the contact area. By a sliding movement, the fluid, which is essential for attachment, can be exuded underneath the endplate in the distal direction and also spread to optimise fluid film thickness between the seta and substrate. Probably, the secretion released in the distal part of the setae is not only needed for attachment as a coupling agent but also as a substance keeping the material of the tenent setae moist and flexible.

Structural and experimental data on the fly attachment system let us predict strong friction forces and small adhesion similar to the gecko system. Data obtained on the whole insect seems to support this hypothesis (Walker et al. 1985); however, it is difficult to judge from the measurements on the whole animal about direction of forces applied to the pulvillus and single setae. Presumably, pre-orientation of the spatula-like setal tips will result in anisotropy of frictional forces: higher forces are expected in the direction opposite to the spatula orientation (proximal). This is similar to the results obtained with the smooth pad attachment system of the grasshopper *Tettigonia viridissima*, in which distal orientation of stiff elements of the pad cuticle results in higher friction, when the pad slides in the proximal direction (Gorb and Scherge 2000).

Step height and pulvillus area

Pulvillus area increases (Fig. 2A) and the height of the steps decreases (Fig. 5B) from the forelegs to the hindlegs. In other words, the larger the pulvillus is, the lower is the acceleration distance of the leg prior to contact with the substrate. Assuming that a higher rising of the leg leads to the higher force impact during the pulvillus placement onto the surface and that the smaller pad needs a higher force to achieve the same attachment strength as a larger pad, there would be a trade-off between total area of the pulvillus and applied force required for the sufficient contact formation. This assumption is supported by the time calculation results (Fig. 5C). The contact forces of a single pretarsus are rather difficult to measure in a fly walking. However, such an attempt will be made in a future project by a force-sensitive platform using spring-FOS principle previously described (Gorb et al. 2000; Jiao et al. 2000).

Another explanation of different areas of pulvilli in different legs is a varied way of propulsive forces generation. Hind legs are used largely for pushing the insect forward (Crampton 1923). That means that pulvilli of hind legs are pushed in the distal direction, which is opposite to the optimal pre-orientation (distal direction) of spatula-like setal tips. Therefore, one can expect in the hindleg lower attachment forces, which are presumably compensated by the larger pulvillus area. The forelegs generate propulsive forces by pulling in an opposite direction to the hindlegs. Since pulling direction corresponds to the optimal preorientation of spatula-like setal tips, forelegs may require a lower pad area for sufficient attachment with the substrate. The midleg function is an intermediate in the generation of propulsive movements: they pull in the first half of the step cycle and push in the second one. Our results also demonstrated intermediate conditions of these legs in the step height and in the pulvillus area. How the pretarsus and pulvilli act during detachment will be demonstrated by future experiments.

It has been previously shown that flies *Musca domestica* change their gait pattern on the ceiling and use, in this situation, four contacting legs instead of three to increase overall contact area (Niederegger et al. 2001). The present study shows that in spite of different forces acting on legs in different walking positions, the height of steps of all three pairs of legs does not change in relation to the walking position.

Multiple function of insect attachment pads

Most attachment pads of insect legs were studied from the point of view of adaptations for locomotion. However, as the most biological system they may present multi-functional structures, which evolved in different lineages due to various sets of selective pressures. For example, beetles *Hemiphaerota cyanea* use attachment ability of their tarsal pads to defend themselves from predators' attacks (Eisner and Aneshansley 2000).

Another frequently occurring function of insect attachment pads is courtship and/or copulation. Sexual dimorphism in the morphology of attachment devices is usually correlated with such a function. The best-known example is a tarsal sucker, present only in male dytiscid beetles (Blunck 1912). A similar phenomenon is also known for attachment pad systems. The area of attachment pads in the males of chrysomelid and cerambycid beetles is usually larger than that of the females. Stork (1980) has demonstrated differences in attachment abilities of both sexes of the beetle *Chrysolina polita*. There are also dimorphisms in the structure and density of cuticle protuberances. Males of the Colorado potato beetle *Leptinotarsa decemlineata* (Coleoptera, Chrysomelidae), bear specialised setae on their tarsi for long-term fixation on the smooth female elytra. Such setae are absent in females (Pelletier and Smilowitz 1987). The specialised male setae may even prevent walking on smooth substrata because of too strong an adhesion.

Many other insects, which do not display obvious sexual dimorphism, may be different in terms of size and design of attachment organs in both sexes as in the case of *C. vicina*. This study shows that females, which are bigger and have a higher body weight, have smaller pulvilli and lower density of setae than males. The presence of sexual dimorphism in pads, which were thought to be a purely locomotory adaptation, suggests that these devices may represent multi-functional structures also used for other functions, such as holding on to a female while mating.

Acknowledgements Sincere thanks to Ch. Ziegler for access to the AFM, to H. Stadler and M. Mondon (all Universitat Tubingen, Germany) for introduction to the AFM controlling system. H. Schwarz and J. Berger (MPI für Entwicklungsbiologie, Tübingen, Germany) helped with SEM preparations. M. Heusel (MPI für biologische Kybernetik, Tübingen, Germany) and K. Scheller (Universität Würzburg, Germany) kindly supplied *Calliphora vicina*. W. Vötsch (MPI für Entwicklungsbiologie, Tübingen Germany) provided many useful suggestions about the use of video equipment. V. Kastner made linguistic corrections, which considerably improved an early draft of the manuscript. This work was supported by the Federal Ministry of Science of Germany (BMBF) grant BioFuture 0311851 to S.G.

References

Autumn K, Liang YA, Hsieh ST, Zesch W, Chan WP, Kenny TW, Fearing R, Full RJ (2000) Adhesive force of a single gecko foot-hair. Nature (Lond) 405:681–685

Bauchhenss E (1979) Die Pulvillen von *Calliphora erythrocephala* Meig. (Diptera, Brachycera) als Adhäsionsorgane. Zoomorphologie 93:99–123

Bauchhenss E, Renner M (1977) Pulvillus of *Calliphora erythrocephala* Meig. (Diptera; Calliphoridae). Int J Insect Morphol Embryol 6:225–227

Blackwell J (1930) Remarks on the pulvilli of insects. Trans Linn Soc Lond 16:487–492

Blunck H (1912) Beitrag zur Kenntnis der Morphologie und Physiologie der Haftscheiben von *Dytiscus marginalis* L. Z Wiss Zool 100:459–492

Bowden FP, Tabor D (1986) The friction and lubrication of solids. Clarendon Press, Oxford

Crampton GD (1923) Preliminary note on the terminology applied on the parts of an insect's leg. Can Entomol 55:126–132

Dewitz H (1884) Über die Fortbewegung der Tiere an senkrechten glatten Flächen vermittels eines Secretes. Archiv Ges Physiol 33:440–481

Eisner T, Aneshansley DJ (2000) Defense by foot adhesion in a beetle (*Hemisphaerota cyanea*). Proc Natl Acad Sci USA 97:6568–6573

Foelix R (1982) The biology of spiders. Harvard University Press, Massachusetts

Gillett JD, Wigglesworth VB (1932) The climbing organ of an insect, *Rhodnius prolixus* (Hemiptera, Reduviidae). Proc R Soc London Ser B 111:364–376

Goel SC (1972) Notes on the structure of the unguitractor plate in Heteroptera (Hemiptera). J Entomol 46:167–173

Gorb SN (1996) Design of insect unguitractor apparatus. J Morphol 230:219–230

Gorb SN (1998) The design of the fly adhesive pad: distal tenent setae are adapted to the delivery of an adhesive secretion. Proc R Soc Lond Ser B 265:747–752

Gorb SN, Scherge M (2000) Biological microtribology: anisotropy in frictional forces of orthopteran attachment pads reflects the ultrastructure of a highly deformable material. Proc R Soc Lond Ser B 267:1239–1244

Gorb SN, Jiao Y, Scherge M (2000) Ultrastructural architecture and mechanical properties of attachment pads in *Tettigonia viridissima* (Orthoptera Tettigoniidae). J Comp Physiol A 186:821–831

Jiao Y, Gorb SN, Scherge M (2000) Adhesion measured on the attachment pads of *Tettigonia viridissima* (Orthoptera, Insecta). J Exp Biol 203:1887–1895

Maxwell MH (1978) Two rapid and simple methods used for the removal of resins from 1.0 μmthick epoxy sections. J Microsc 112:253–255

Niederegger S, Gorb SN, Vötsch W (2001) Fly walking: a compromise between attachment and motion? Biona Report 15:327–330

Pelletier Y, Smilowitz Z (1987) Specialized tarsal hairs on adult male Colorado potato beetles, *Leptinotarsa decemlineata* (Say), hamper its locomotion on smooth surfaces. Can Entomol 119:1139–1142

Radnikow G, Bässler U (1991) Function of a muscle whose apodeme travels through a joint moved by other muscles: why the retractor unguis muscle in stick insects is tripartite and has no antagonist. J Exp Biol 157:87–99

Röder G (1984) Morphologische Untersuchungen an Prätarsen von Diptera and Mecoptera (Insecta), Dissertation, Nürnberg

Simmermacher G (1884) Haftapparate bei Wirbeltieren. Zool Gart 25:289–301

Spurr AR (1969) A low-viscosity epoxy resin embedding medium for electron microscopy. J Ultrastruct Res 26:31–43

Stork NE (1980) Experimental analysis of adhesion of *Chrysolina polita* (Chrysomelidae, Coleoptera) on a variety of surfaces. J Exp Biol 88:91–107

Stork NE (1983a) A comparison of the adhesive setae on the feet of lizards and arthropods. J Nat Hist 17:829–835

Stork NE (1983b) The adherence of beetle tarsal setae to glass. J Nat Hist 17:583–597

Stork NE (1988) Tarsal adhesive mechanisms in Coleoptera and their role in holding onto plant surfaces. In: Anon Proc 8th Int Congr Entomol. Vancouver, July 3rd–9th, 1988. Abstract and author index. Vancouver, pp 92–92

Walker G, Yule AB, Ratcliffe J (1985) The adhesive organ of the blowfly, *Calliphora vomitoria*: a functional approach (Diptera: Calliphoridae). J Zool Lond 205:297–307

Weber H (1930) Biologie der Hemipteren. Springer, Berlin Heidelberg New York

Wigglesworth VB (1987) How does a fly cling to the under surface of a glass sheet? J Exp Biol 129:363–367

Naturwissenschaften 85, 31–33 (1998)
DOI 10.1007/s00110050447

The Probable Purpose of the Mid-coxal Prong in Brachycera (Diptera)

Leonid Frantsevich · Stanislav Gorb

Received: 3 March 1997 / Accepted in revised form: 22 October 1997

The mid-coxal prong is a narrow out-growth from the distal sclerite of the coxa into the space between the coxa and the trochanter. It is present in members of most of families of flies (Brachycera). Its anterior face is hairy in large flies. The prong is the same in males and females and lacks any sensilla. It performs, presumably, some mechanical function. During flight the prong is clumped between the coxa and the trochanter, its anterior hairy face against a hairy field on the anterior coxal sclerite. The leg is put into its peculiar flight position by rotation of the trochanter. The trochanter presses the two hairy faces together. Morphology and biomechanics of the middle coxa suggest that the prong acts as a sort of a Velcro fastener, locking the middle leg in its flight position. However, excision of the prong impairs neither walking nor the ability to put the middle leg into a proper flight position and hold it against the strong head airstream.

The mid-coxal prong (MCP) is a tiny daggerlike process that protrudes from the posterior side of the coxa into the space between the coxa and the tro- chanter. It is present in all but a few families of the huge group of Muscomorpha, or Cyclorrhapha, a major subgroup of the brachycerous flies [1, 2]. The MCP has the same shape in males and females. Its covert position was among the reasons that entomologists neglected this organ. Its purpose was unknown. However, the sheer presence of the MCP in about 50,000 species of flies bears witness to its importance for functions of the wingbearing segment, the mesothorax.

In order to study the role of the MCP in *Eristalis tenax* L. (Syrphidae), *Calliphora vicina* Robineau-Desvoidy, *Lucilia caesar* L., and *L. sericata* Mg. (Calliphoridae) we studied the MCP on SEM microphotographs, dissected the subcoxo-coxal region, and photo-graphed and videotaped fly movements during walking, flight, and landing. Relevant events in videorecordings were analyzed frame by frame. Muscle potentials were picked up in tethered flies by a steel needle and amplified with conventional equipment.

The shape of the MCP is shown in Fig. 1. The MCP is soft and bare in a small species such as *Drosophila melanogaster* Mg. The MCP is sclero- tized in large flies. Its hind surface, facing the trochanter, is absolutely smooth. The trochanter in turn has a smooth groove on its anterodorsal face where the MCP glides during leg movements. The front side of the MCP in large species is covered with microtrichia. Viewed from the top the MCP resembles a

L. Frantsevich (✉) · S. Gorb
Schmalhausen-Institute of Zoology, Kiev 30, 252601 Ukraine,

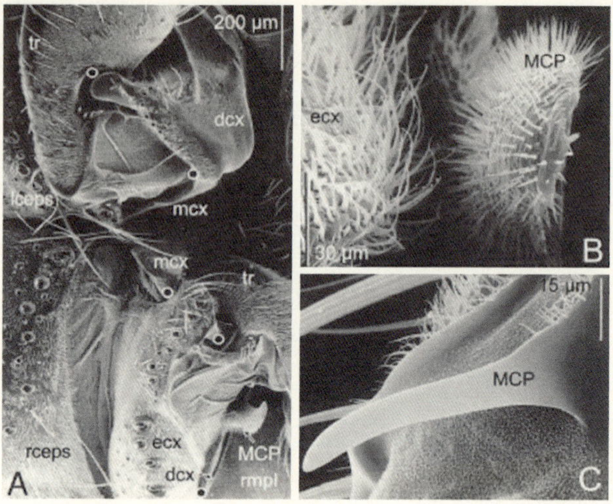

Fig. 1A–C Middle coxea and MCP in flies. A) Ventral view at the coxea of *Lucilia caesar* in pronating configurartion (*below*) and in supinating configuration (*above*). Craniad direction points to the *left side of the panel. Scale bar*, 200 mm. B) MCP in *Eristalis tenax*, as seen from the apex; the hairy anterior face of MCP against the hairy area of the eucoxite. C) MCP in *Drosophila melanogaster*. *ecx*, Eucoxite: *dcx*, distocoxite; *lceps*, left epistern: *MCP*, mid-coxal prong; *mcx*, mediocoxite; *rceps*, right epistern; *rmpl*, right meropleurite; *tr*, trochanter. *White circles filled black*, articulation points

hedgehog with a bare belly and spines on its back. (Fig. 1b). The same site which bears the MCP in muscomorph flies produces a pyramidal outgrowth of the articulation membrane in horse flies (Tabanidae); the front face of this out-growth is densely haired.

No sensilla were found on the MCP or at its base. No reflexes in the subcoxal muscles were obtained in preliminary electrophysiological observations upon tactile stimulation of the MCP. This process performs, presum-ably, some mechanical function.

The middle leg contributes to several modes of motion, including walking, grooming, grasping, flight start, flight posture, and the landing reflex. In all of these modes except the flight posture the trochanter and the femur (which act together) are more or less depressed, and the MCP slacks freely or slightly touches the trochanter. In 113 videorecorded runs of five specimens of *E. tenax* before and after removal of the MCP we found no significant difference in the gait, leg positioning relative to the body, or in parameters of the step (duration of the swing and stance phases and stride length versus velocity). The leg span increased by 5% after the operation. Excision of the MCP did not impair walking in *E. tenax*.

During flight the MCP is clumped be- tween the coxa and the trochanter. The position of the fly's legs during flight is rather peculiar (Fig. 2 a, b). The femora of the front and middle legs are elevated, tibiae are flexed, legs are pressed to the thorax. The tarsi point ahead and inward. In such a position the legs offer only slight aerodynamic resistance. The steering hind legs are elevated and extended and point backwards.

The middle femur points backwards and upwards. If the middle coxa were simply retracted, the femur raised, and the tibia flexed, the tarsus would point outwards instead of inwards (Fig. 2 c, d). The leg must also be ro- tated about the longitudinal axis of the femur in order to attain the flight position. Rotation is about 20° in *E. tenax*, 60 ° in *L. caesar*. Additional rotation is performed due to internal mobility of the middle coxa which consists of three separate sclerites [3–5], which we refer to below as the eu- coxite, the mediocoxite, and the distocoxite (Fig. 3 a).

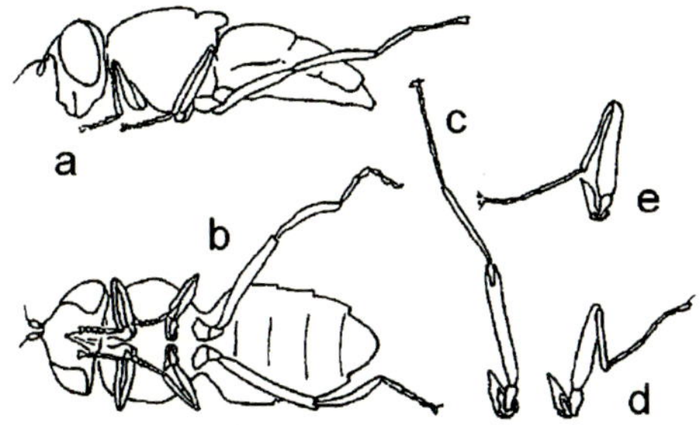

Fig. 2a–e Leg positions during flight in *Eristalis tenax* (from photographs). a) Side view. b) View from below, the fly yaws to the left, head turned, the left hind leg steering. c) d) e) Left middle leg configurations seen from below, as in b. c) Extended leg. d) The leg elevated, the tibia flexed, the femur pronated. e) The same, but the femur supinated

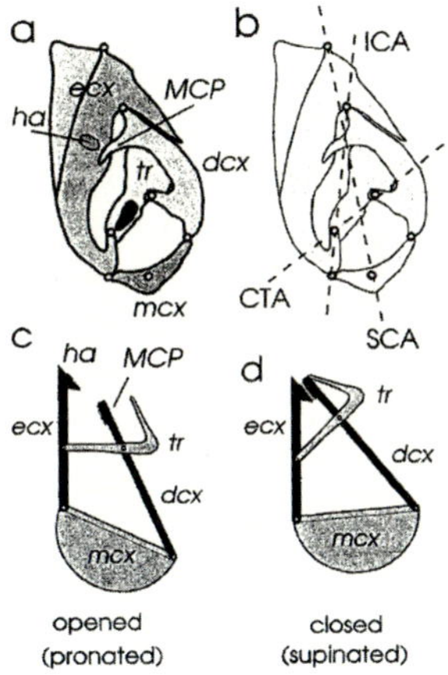

Fig. 3a–d Coxo-trochanteral mechanism in fies. a) Sclerites at the same projection as in Fig. 2cd. b) Equivalent kinematic scheme of the coxa. c) d) Flat model of the locking mechanism. Only the anterodorsal face of the trochanter is shown. d) In the supinated configuration hair brushes on the distocoxite (i.e., on the prong) and on the eucoxite are pressed together by the tochanter. Hollow circles, articulation points. Dashed line (c,d) shows position of the posterior elevator muscle. CTA, Coxo-trochanteral axis; ha, hairy area; ICA, intracoxal axis; SCA, subcoxo-coxal axis; other abbreviations see Fig. 1

The middle coxa attaches to the thorax at two fixed points: at the pleural ridge and at the evaginated peg on the furcasternum. The pleural ridge articulates with the eucoxite, the furcosternal peg is a pivot for the mediocoxite. Rotation of the coxa about this subcoxo-coxal axis (SCA) produces protraction and retraction of the leg. The trochanter articulates anteriorly with the eucoxite and posteriorly with the distocoxite. This pair of articulations defines the coxo-trochanteral axis (CTA). Rotation of the trochanter about CTA produces elevation or de-pression of the femur. Movements of the coxal sclerites relative to each other bring the distocoxite into the inferior or superior position, thus changing the direction of CTA. One may describe this movement as the rotation of the trochanter about the intracoxal axis (ICA). If the femur is depressed, this rotation adds to protraction-retraction. The elevated femur lies approximately parallel to the ICA, hence rotation about ICA in this configuration produces supination or pronation of the femur. Figure 1 a shows the difference between two configurations of the coxa: the exposed eucoxite in the pronated right leg and the exposed distocoxite in the supinated left leg. Three coxites and the trochanter form a closed kinematic chain of four segments with one degree of freedom (Fig. 3 b).

The supinated trochanter bends the MCP and presses it to the eucoxite. The part of the eucoxite against the hairy face of the MCP is covered with similar microtrichia in large flies. Mechanical interaction of the two hairy faces suggests that in this position they act as the two surfaces of a Velcro fastener. The coxa becomes strapped in the supinating configuration. When the trochanter presses on the MPC, the coxa is self-locked (Fig. 3 c, d). Similar friction clasps are known between the head and cervical sclerites in dragonflies (Odonata) [6, 7], the elytrae and tergites in beetles (Coleoptera) [8], and the wings and the thorax in sawflies (Hymenoptera) [9].

The proposed hypothesis explains the shape, position, and insensitivity of the MCP. It predicts that an individual without MCP will be awkward put- ting its legs in the flight position and holding the legs against the opposite airstream.

We experimented with tethered flights of *C. vicina* after one-side MCP removal. We selected 33 individuals which were able to acquire the proper flight posture in the head airstream of 3 m/s and flew steadily at higher wind velocity – up to 10–12 m/s. Over 300 flights were videorecorded. Several flies were operated bilaterally. No significant difference was observed between flight performance in intact and operated flies. Only four individuals showed remarkable instability of the operated leg. This small percentage of disabled individuals is safer to ascribe to damage of the op- erated leg than to damage of the MCP. Most of operated individuals had no difficulties in maintaining their perfect flight posture against the head airstream well above the physiological range of flight speed. There may exist some mechanisms (e.g., enhancement of the muscle activity) which compensate the lack of the presumed lock, but they are questionable.

The function of the MCP is still not proven. In any case, it is hardly believable that the structure, which is an obligatory character of so many families of Brachycera and has passed the route of evolution from a hairy evagination of the articulation membrane to the more elaborated sclerotized instrument, is of no significance to representatives of Brachycera.

Acknowledgements We thank K. Götz for valuable discussion and aid with the laminarized wind tunnel, and J. Berger for help with SEM microphotographs. This investigation was supported by the Max-Planck-Gesellschaft and by the International Science Foundation.

References

1. McAlpine JF et al (eds) (1981) Manual of Nearctic Diptera. Agriculture Canada, Monograph, no 27, Ottawa
2. Speight MCD (1987) External morphology of adult Syrphidae (Diptera). Tijdschr Entomol 130:141–175
3. Lowne BT (1890) The anatomy, physiology, morphology, and development of the blowfly (*Calliphora erythrocephala*). In: Porter RH (ed) A study in the comparative anatomy and morphology of insects, London
4. Crampton CG, Hasey WH (1915) The basal sclerites of the leg in insects. Zool Jahrb Anat 39:1–26
5. Bonhag PF (1949) The thoracic mechanism of the adult horsefly (Diptera, Tabanidae). Mem Cornell Univ Agric Exp Stn 285:3–39
6. Mittelstaedt HZ (1950) Physiologie des Gleichgewichtsinnes bei fliegenden Libellen. Z vergl Physiol 32:442–463
7. Gorb SN (1989) Functional morphology of arrester system in dragonflies. Vestn Zool N3:62–67
8. Hammond PM (1979) Wing-folding mechanisms of beetles, with special reference to investigations of adephagan phylogeny (Coleoptera). In: Ervin TL, Ball GE, Whitehead DR (eds) Carabid beetle, their evolution, natural history, and classification. Junk, The Hague, pp 113–180
9. Schrott A (1986) Vergleichende Morphologie und Ultrastruktur des Cenchrus-Dornenfeldapparates bei Pflanzenwespen (Insecta: Hymenoptera, Symphyta). Ber naturwissmed Ver Innsbruck 73:159–168

4. Attachment-related Behaviour

DOI 10.1007/978-1-4020-9033-2_23

Contact Mechanics at the Insect-Plant Interface: How Do Insects Stick and How Do Plants Prevent This?

Elena V. Gorb · Stanislav N. Gorb

Abstract Contacting surfaces in many insect attachment devices are subdivided into patterns of microor nanostructures with a high aspect ratio. This paper provides some explanations of the adhesion enhancement effect in such systems. However, attachment abilities of insects depend on the roughness and mechanical stability of the substrate features. Wax bloom of some plant surfaces causes a strong reduction of insect attachment. Four hypotheses of anti-adhesion mechanism of plant surfaces covered with epicuticular wax crystals are proposed, and two of them are experimentally supported.

Keywords Adhesion · Biotribology · Biomechanics · Functional surfaces

1 Contact Problem

Most of the one million insect species described so far are associated with plants. Such insects should be able to attach successfully to plant surfaces. Dealing with attachment, we have to consider a contact problem, in which two bodies are involved (Fig. 1). In our case, these are (1) insect attachment organs and (2) plant substrates. Both these parts may have various geometries, mechanical and chemical properties. Moreover, insects produce and deliver a secretory fluid into a contact zone [2, 10, 19, 23, 32]. The situation is usually even more complex, since many plant substrates bear specialised surface coverage [3, 9, 11, 20, 22, 24]. This paper summarises our previous structural and experimental studies on insect attachment and the effect of plant substrates on it.

F.M. Borodich (ed.), *IUTAM Symposium on Scaling in Solid Mechanics,* IUTAM Bookseries 10
DOI 10.1007/978-1-4020-9033-2_23

E. V. Gorb (✉)
Evolutionary Biomaterials Group, Department of Thin Films and Biological Systems
Max Planck Institute for Metals Research, Heisenbergstr. 3, 70569 Stuttgart, Germany
e-mail: o.gorb@mf.mpg.de

S. N. Gorb
, ,

Springer

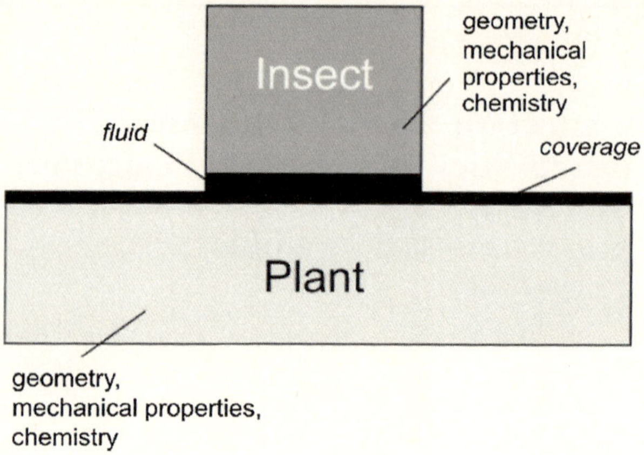

Fig. 1 Contact between an insect attachment organ and a plant surface

2 Hairy Attachment Devices

Most recent data on insect hairy attachment systems (Fig. 2) demonstrated their excellent adhesion and high reliability of contact [15, 16]. Interestingly, hairy systems appeared several times in insect evolution [17]. This fact may indicate that such surface structures must have an advantage for adhesion enhancement especially on real natural substrates, such as surfaces of plants. The physical background of the effect of adhesion enhancement was theoretically discussed in several recent publications [1, 6, 27, 28].

Comparison of a wide variety of animal groups revealed that the size of single contacting points gets smaller and their density increases as the body mass of the animal increases [30] (Fig. 3). This general trend is theoretically explained by applying the Johnson-Kendall-Roberts (JKR) theory, according to which splitting up the contact into finer subcontacts increases adhesion [1, 21]. There are a number of other design principles, extracted from studies on biological systems (Fig. 4). The effective elastic modulus of the fibre arrays is very small, which is of fundamental importance for adhesion on smooth and rough substrates [27]. It is predicted that an additional advantage of patterned surfaces is the reliability of contact on various surface profiles and the increased tolerance of defects at individual contacts. In a real situation, failure of some microcontacts because of dust

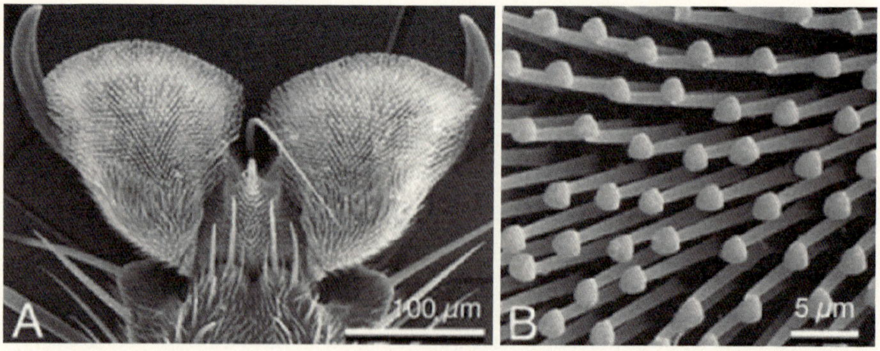

Fig. 2 Insect attachment system, viewed in a scanning electron microscope (SEM). Surface of adhesive pads (*pulvilli*) of the leg in the fly *Episyrphus balteatus* (**A**) is subdivided into setae with flattened tips (*spatulae*) (**B**)

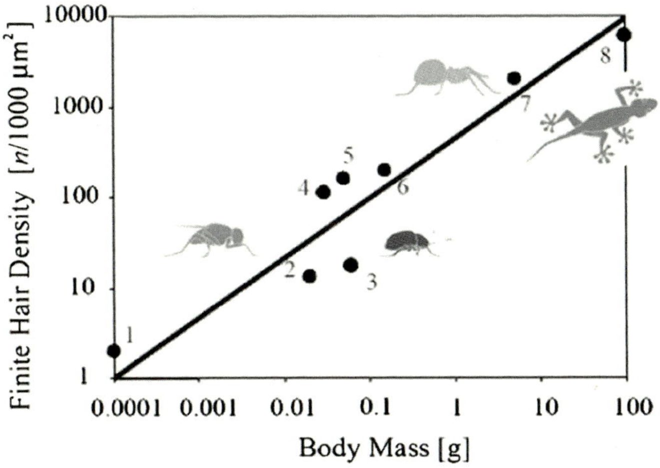

Fig. 3 Dependence of the finite hair density in the attachment pads on the body mass in hairy pad systems of representatives from diverse animal groups: 1, 2, 4, 5, flies; 3, beetle; 6, bug; 7, spider; 8, Gekkonid lizards. Adapted from [30]

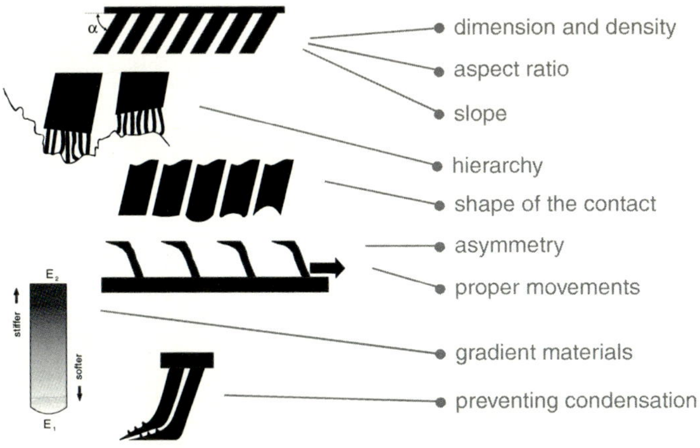

Fig. 4 Some functional principles, at which biological reversible adhesive systems operate. Their simultaneous implementation in one artificial system is hardly possible. However, depending on the requirement for a particular material or system, one principle or a combination of several of them can be implemented. From [7]

particles or mechanical damage of single hairs would minimally influence adhesion. In the case of a solitary contact, even slight damage of the contact due to the presence of contamination or surface irregularities will immediately lead to contact breakage, similar to the crack propagation in bulk material.

These theoretical considerations were proven experimentally. We have recently shown that structured surfaces have an increased tenacity (adhesion force per unit contact area) [18, 25, 31].

Why does the structured surface stick better? The effectiveness of the attachment system may be evaluated through energy. Contact energy is the difference between adhesion energy and elastic energy, stored in a contact. The structured sample has lower surface rigidity and, thus, higher flexibility. This increases the ability of the surface to adapt to substrate irregularities, in contrast to a flat sample. A flat sample is able to establish contact only at the tips of substrate irregularities and, therefore, generates rather low real contact area. Thus, one may formulate an ultimate requirement for an effective attachment system: *the system must reach maximal contact energy at the minimal elastic energy spent for formation of the contact area* [25].

There are two ways to increase adhesion energy of the contact. The first one is to increase the specific adhesion energy (SAE) at the constant contact area. The second one is to increase real contact area at the constant SAE. The first way has some limitations. SAE ranges from 50–70 (glass, silicon and GaAs) to 300–400 mJ/m^2 (metals). Additionally, a high SAE results in rather fast passivity of the surface even under normal conditions. Thus, under natural conditions in the presence of humidity, various gases and dust, the surface with high SAE will be contaminated very quickly and its SAE will drop rapidly, affecting a decrease of the surface stickiness.

Adhesion energy of contact may also be enhanced in a second way, namely by an increase of the real contact area between two surfaces. The relationship between real contact area and geometrical contact area, estimated by measuring of the resistance at the interface in metals, is about 1/10000 [5], and it depends mainly on the surface geometry. Surface patterning into structures with a high aspect ratio results in higher deformability of the surface, and therefore, in higher ability to build large areas of real contact especially with uneven substrates. A structured surface has a lower portion of elastic energy stored in the material. Such a decrease of the stored elastic energy of deformation results in an increase of the adhesion energy. This way does not have the disadvantages mentioned above for the first way. The area of real contact may vary widely, depending on the load applied, and is less dependent on contamination than a surface with high SAE. The presence of several hierarchical levels of structures in biological attachment systems may presumably increase the above effect.

A very small effective elastic modulus of the fibre array is of fundamental importance for adhesion on smooth and especially on rough substrates. During pulloff, the fibres may elongate many micrometers before the force in the fibre is high enough to break the bond to the substrate. Since the spring constant, associated with a long fibre, is very small, the displacement may be very large, leading to very large effective surface energy. Therefore, it was previously suggested that extension of "long bonds" may lead to an increase in the energy dissipation at the interface and by this to an increase in reliability of the adhesive contact [27, 28].

Another functional principle of patterned adhesion systems is the prevention of crack propagation. In a multicontact system, crack will be stopped at the level of a single contact element, and the new crack has to be formed at each subcontact during detachment. Formation of new cracks affects energy dissipation, which prevents contact breakage. The behaviour of crack propagation on the patterned adhesive films leads to a significant enhancement of fracture energy [6].

An additional functional principle is related to the spatula-shaped tips of the setae, which are responsible for proper contact formation with the substrate due to the low bending stiffness of the plates with or without a minimum of normal load [29]. While sliding over the surface, thin plate-like spatulae may easily make contact with the surface by adapting to the surface profile and replicating surface irregularities of a certain length scale.

3 Anti-Adhesive Plant Surfaces

It is well known that insects attach well to macroscopically rough surfaces and use their claws to interlock with surface irregularities [4, 8]. To attach to smooth and microrough substrates, insects apply adhesive pads [15]. Since plant surfaces range from rather smooth to those covered with trichomes or wax crystals, insect-plant interactions may often rely on insect attachment ability to the particular plant surface. In the course of the last five years, we carried out numerous experiments in order to understand insect-plant interactions at the level of the contacting surfaces. The studies were mainly performed with insects having hairy adhesive pads, the surfaces of which are densely covered with microscopic hairs.

Plant surfaces have a wide range of textures and microsculptures. They may be smooth or structured, i.e. covered with various hairs (trichomes) or bear microscopic crystals of epicuticular waxes [3, 20] (Fig. 5). To study the role of different structures of a plant surface in insect attachment, various plant surfaces were screened [11].

Attachment ability of the beetle *Chrysolina fastuosa* was tested on 99 surfaces of 83 plant species, belonging to 45 families. Among these surfaces were smooth, dry hairy, felt-like, moist hairy and waxy ones. Insects attached successfully to smooth, dry hairy and felt-like substrates, but they could not attach properly to surfaces covered with wax crystals. To explain anti-adhesive properties of substrates, structured with crystalline waxes, four hypotheses were proposed (Fig. 6):

A. **Roughness-hypothesis.** Wax crystals cause microroughness, which considerably decreases the real contact area between the substrate and setal tips of insect adhesive pads.
B. **Contamination-hypothesis.** Wax crystals are easily detachable structures that contaminate pads.
C. **Wax-dissolving-hypothesis.** Insect pad secretion may dissolve wax crystals. This would result in the appearance of a thick layer of fluid, making the substrate slippery.
D. **Fluid-absorption-hypothesis.** Structured wax coverage may absorb the fluid from the setal surface.

The influence of surface roughness on insect attachment has been tested experimentally in several studies [12, 15, 26]. Centrifugal experiments on the attachment ability of insects on artificial substrates with varying surface roughness showed that the fly *Musca domestica*

Fig. 5 Diversity of plant surfaces

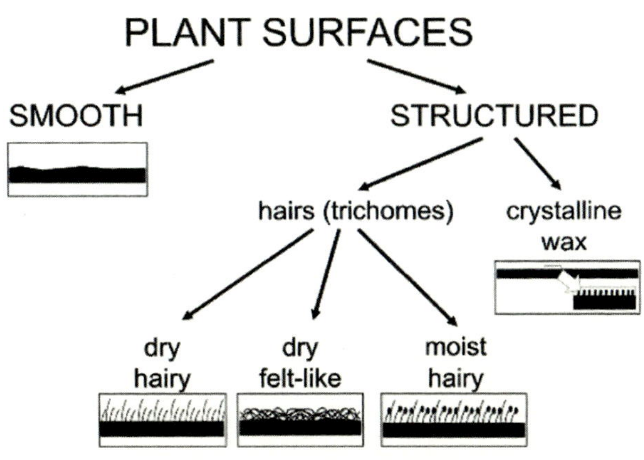

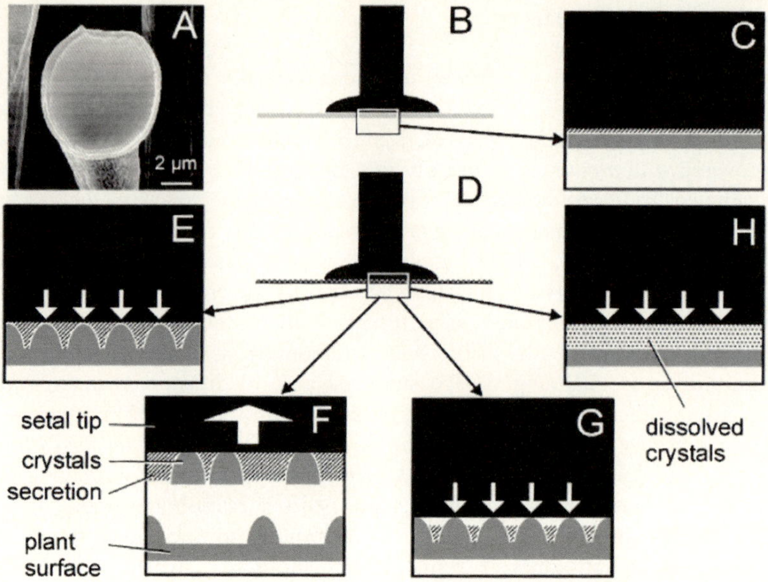

Fig. 6 Hypotheses of anti-adhesive mechanisms of a waxy plant surface. (**A**). Tenent seta of the adhesive pad of the *Chrysolina fastuosa* beetle. (**B**), (**C**). Setal contact with a smooth surface. (**D**)–(**H**). Setal contact with a waxy surface. (**E**). Roughness-hypothesis. (**F**). Contamination-hypothesis. (**G**). Fluid-absorption-hypothesis. (**H**). Wax-dissolving-hypothesis. After [11]

and the beetle *Gastrophysa viridula* generated much higher forces on either smooth or rough surfaces with a particle size, exceeding 3.0 µm (Fig. 7)

This effect may be explained by spatula-like terminal elements of setae that are able to generate sufficient contact with large surface irregularities. The worst attachment was observed on substrates with a roughness of 0.3 – 1.0 µm, corresponding to the size of plant wax crystals. Because of the small size of wax crystals, the area of real contact between these substrates and the tips of insect spatulae is very small. Since the attachment force depends on the area of real contact, insects are not able to attach successfully to surfaces with such a microroughness.

Since many wax crystals are easily erodable structures, they may contaminate adhesive pads, and this may reduce proper functioning of the pads. We proved the contamination-hypothesis experimentally for several insects and a series of plant species. Our data on the contamination of *Chrysolina fastuosa* pads by wax crystals of 12 waxy plant surfaces [13] showed that the plants differ essentially in their contaminating effects on insect pads (Fig. 8). Differences were found in both the nature of the contamination, i.e. the structure of the contaminating material and the presence of recognisable crystals, and the degree of contamination (the portion of setal tip surface, covered with contaminating particles, and the portion of setae, covered with the wax). These differences were hypothesised to be caused by various micromorphologies of waxes in the plant species studied. Analysing the relationship between the contaminating ability and geometrical parameters of wax crystals, we found that the contamination is related to both the largest dimension and the largest aspect ratio of crystals.

Pad contamination was also observed in our study on the waxy surface of the slippery zone in trapping organs (pitchers) of the carnivorous plant *Nepenthes alata* [14]. The wax

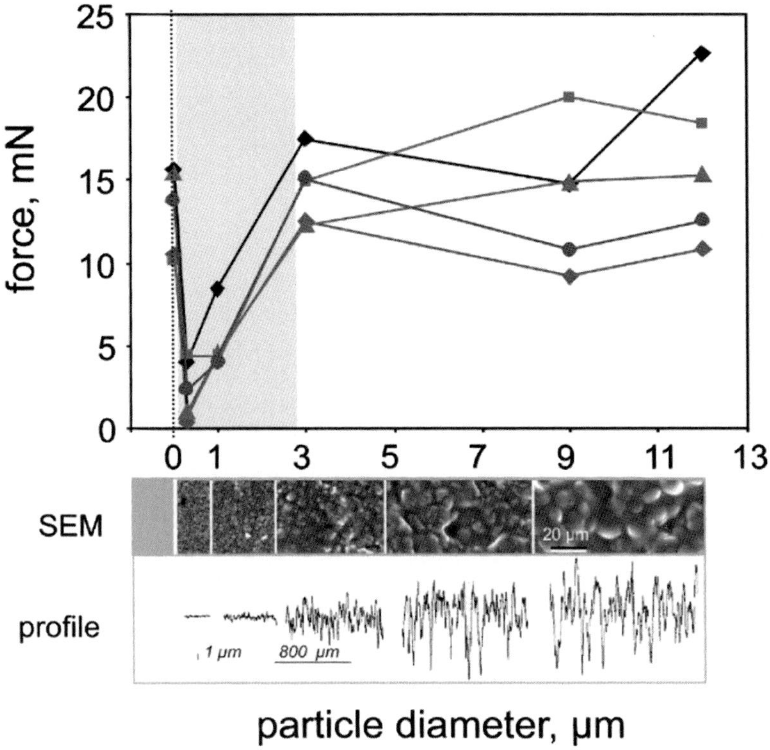

Fig. 7 Attachment forces of the beetles *Gastrophysa viridula* on substrates with different roughness. After [26]

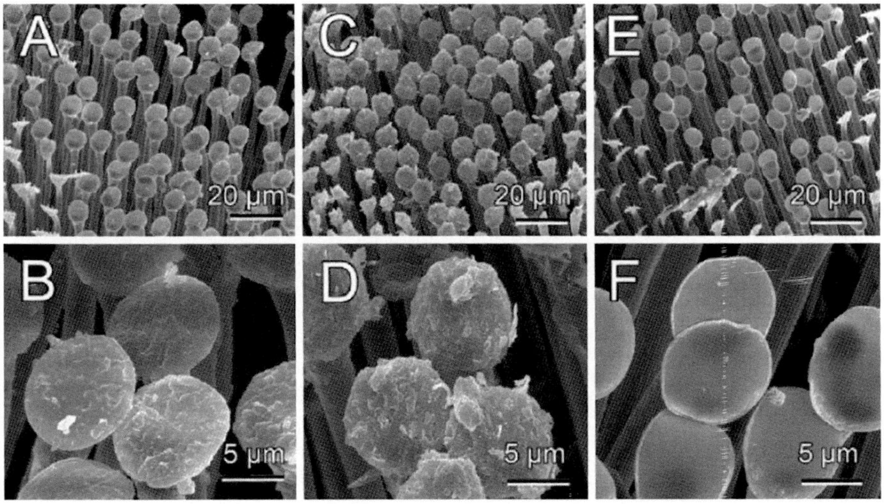

Fig. 8 SEM micrographs of the pad surface in the third tarsomere of the beetle *Chrysolina fastuosa* after walking on waxy plant surfaces. (**A**), (**B**). Leaf of *Chenopodium album*. (**C**), (**D**). Young stem of *Acer negundo*. (**E**), (**F**). Leaf of *Berberis vulgaris*. After [13]

coverage consists of two clearly distinguished superimposed layers, differing in their structure, chemical composition and mechanical properties. Laboratory experiments with tethered beetles *Adalia bipunctata* showed that both wax layers essentially reduce the attachment force of insects, however, in different ways. Crystals of the upper wax layer contaminate the insects' adhesive pads, whereas the lower wax layer leads to the reduction of the real contact area of the insects' feet with the pitcher surface.

4 Conclusions

Contacting surfaces in many insect attachment devices are subdivided into patterns of microor nanostructures with a high aspect ratio. This paper provides some explanations of the adhesion enhancement effect in such systems. The attachment abilities of insects depend on the roughness and mechanical stability of the substrate features. Wax bloom found on the surface of some plant surfaces is an adaptation against insect attachment. Four hypotheses of anti-adhesion mechanism of plant surfaces covered with epicuticular wax crystals are proposed, and two of them are experimentally supported.

Acknowledgement This work was supported by the Federal Ministry of Education, Science and Technology, Germany (project InspiRat01RI0633C) and by the Landesstiftung Baden-Württemberg (project Plants as Source for Biomimetics of Anti-Adhesive Surfaces). Travel grant to the IUTAM Symposium on Scaling in Solid Mechanics (25–29 June 2007, Cardiff, UK) was supported by IUTAM.

References

1. Arzt E, Gorb S, Spolenak R. "From micro to nano contacts in biological attachment devices", *Proc. Nat. Acad. Sci. USA*, **100**, pp. 10603–10606, 2003
2. Attygalle AB, Aneshansley DJ, Meinwald J, Eisner T. "Defence by foot adhesion in a chrysomelid beetle (*Hemisphaerota cyanea*): characterization of the adhesive oil", *Zoology*, **103**, pp. 1–6, 2000
3. Barthlott W, Neinhuis C, Cutler D, Ditsch F, Meusel I, Theisen I, Wilhelmi H. "Classification and terminology of plant epicuticular waxes", *Bot. J. Linn. Soc.*, **126**, pp. 237–260, 1998
4. Betz O. "Performance and adaptive value of tarsal morphology in rove beetles of the genus Stenus (Coleoptera, Staphylinidae)", J. Exp. Biol., **205**, pp. 1097–1113, 2002
5. Bowden FP. "Adhäsion und Reibung", *Endeavour*, **16**, pp. 5–18, 1957
6. Chung JY, Chaudhury MK. "Roles of discontinuities in bio-inspired adhesive pads", *J. R. Soc. Interface*, **2**, pp. 55–61, 2005
7. Creton C, Gorb S. "Sticky feet: from animals to materials", *MRS Bull.*, **32**, pp. 466–468, 2007
8. Dai Z, Gorb S, Schwarz U. "Roughness-dependent friction force of the tarsal claw system in the beetle *Pachnoda marginata* (Coleoptera, Scarabaeidae)", *J. Exp. Biol.*, **205**, pp. 2479–2488, 2002
9. Duffey SS. "Plant glandular trichomes: their partial role in defence against insects", *Insects and the plant surface*, Juniper BE, Southwood TRE. (eds.), London, Edward Arnold, pp. 151–172, 1986
10. Federle W, Riehle M, Curtis ASG, Full RJ. "An integrative study of insect adhesion: mechanics and wet adhesion of pretarsal pads in ants", *Integr. Comp. Biol.*, **42**, pp. 1100–1106, 2002
11. Gorb E, Gorb S. "Attachment ability of the beetle *Chrysolina fastuosa* on various plant surfaces", *Entomol. Exp. Applicata*, **105**, pp. 13–28, 2002
12. Gorb E, Gorb S. "Combination of the surface profile and chemistry reduces the attachment of the beetle *Gastrophysa viridula* on the *Rumex obtusifolius* leaf surface", *Proceedings of the 5th Plant Biomechanics Conference*, Salme'n L. (ed.), Stockholm, Sweden, **Vol. 2**, Stockholm, pp. 537–542, 2006
13. Gorb E, Gorb S. "Do plant waxes make insect attachment structures dirty? Experimental evidence for the contamination hypothesis", *Ecology and biomechanics: A mechanical approach to the ecology of animals and plants*, Herrel A, Speck T, Rowe N. (eds.), Boca Raton, pp. 147–162, 2006
14. Gorb E, Haas K, Henrich A, Enders S, Barbakadze N, Gorb S. "Composite structure of the crystalline epicuticular wax layer of the slippery zone in the pitchers of the carnivorous plant *Nepenthes alata* and its effect on insect attachment", *J. Exp. Biol.*, **208**, pp. 4651–4662, 2005

15. Gorb S. *Attachment devices of insect cuticle*, Dordrecht, Boston, London, Kluwer Academic Publishers, 2001
16. Gorb SN "Uncovering insect stickiness: structure and properties of hairy attachment devices", *Am. Entomol.*, **51**, pp. 31–35, 2005
17. Gorb SN, Beutel RG. "Evolution of locomotory attachment pads of hexapods", *Naturwissenschaften*, **88**, pp. 530–534, 2001
18. Gorb S, Varenberg M, Peressadko A, Tuma J. "Biomimetic mushroom-shaped fibrillar adhesive microstructure", *J. R. Soc. Interface*, **4**, pp. 271–275, 2007
19. Ishii S. "Adhesion of a leaf feeding ladybird *Epilachna vigintioctomaculata* (Coleoptera: Coccinellidae) on a vertically smooth surface", *Appl. Ent. Zool.*, **22**, pp. 222–228, 1987
20. Jeffree CE. "The cuticle, epicuticular waxes and trichomes of plants, with references to their structure, functions and evolution", *Insects and the plant surface*, Juniper BE, Southwood TRE. (eds.), London, Edward Arnold, pp. 23–64, 1986
21. Johnson KL, Kendall K, Roberts AD. "Surface energy and the contact of elastic solids", *Proc. R. Soc. Lond. A*, **324**, pp. 301–313, 1971
22. Juniper BE. "Waxes on plant surfaces and their interactions with insects", *Waxes: chemistry, molecular biology and functions*, Hamilton RJ. (ed.), West Ferry, Dundee, Oily, pp. 157–174, 1995
23. Kosaki A, Yamaoka R. "Chemical composition of footprints and cuticula lipids of three species of lady beetles", *Jpn. J. Appl. Entomol. Zool.*, **40**, pp. 47–53, 1996
24. Levin DA. "The role of trichomes in plant defence", *Q. Rev. Biol.*, **48**, pp. 3–15, 1973
25. Peressadko A, Gorb SN. "When less is more: experimental evidence for tenacity enhancement by division of contact area", *J. Adhesion*, **80,** 247–261, 2004
26. Peressadko A, Gorb S. "Surface profile and friction force generated by insects", *Proceedings of the 1st International Conference Bionik,* Boblan I, Bannasch R. (eds.), Hannover, Germany, pp. 257–263, 2004
27. Persson BNJ. "On the mechanism of adhesion in biological systems", *J. Chem. Phys.*, **118**, 7614–7621, 2003
28. Smith BL, Schäffer TE, Viani M, Thompson JB, Frederick N, Kindt J, Belcher A, Stucky GD, Morse DE, Hansma PK. "Molecular mechanistic origin of the toughness of natural adhesives, fibers and composites", *Nature,* **399**, pp. 761–763, 1999
29. Persson BNJ, Gorb S. "The effect of surface roughness on the adhesion of elastic plates with application to biological systems", *J. Chem. Phys.*, **119**, pp. 11437–11444, 2003
30. Scherge M, Gorb SN. "*Biological microand nanotribology*", Berlin, Springer, 2001
31. Varenberg M, Peressadko A, Gorb S, Arzt E. "Effect of real contact geometry on adhesion"*Appl. Phys. Lett.*, **89**, N 121905, 2006
32. Vötsch W, Nicholson G, Müller R, Stierhof Y-D, Gorb S, Schwarz U. "Chemical composition of the attachment pad secretion of the locust *Locusta migratoria*", *Insect Biochem. Mol. Biol.*, **32**, pp. 1605–1613, 2002

Arthropod-Plant Interactions
DOI 10.1007/s11829-007-9007-2

ORIGINAL PAPER

Insect walking techniques on thin stems

Dmytro Gladun · Stanislav N. Gorb

Received: 30 January 2007 / Accepted: 27 April 2007

Abstract The attachment ability of insects on surfaces are associated not only with the micro- and nanostructure of the adhering part of an attachment device, but also with the global scale kinematics responsible for contact formation and release. In the present study, the locomotory techniques of several representatives of insects from four different orders (Orthoptera, Heteroptera, Coleoptera, and Hymenoptera), possessing different types of attachment structures, are described. The study is based on video recordings of insects walking on a flat surface and on cylindrical rods of various thickness, imitating plant stems. Attachment devices of tarsi and pretarsi were visualized using Scanning Electron Microscopy. The results show a different manner in the use of adhesive structures on substrates with various curvatures. Insects bearing attachment pads on proximal tarsomeres usually touch flat and curved substrates using all tarsomeres, whereas insects with their attachment devices on the distal tarsomeres usually walk on flat surfaces using the distal tarsomeres of the overextended tarsus. On substrates, with diameters comparable to or larger than the tarsus length, insects walk above the stem by clasping the stem with the bent tarsi. On thin stems, insects clasp the stem between their tarsi and hang under the stem. Thus, on thin and thick rods, forces applied to attachment organs act in opposite directions. There are two methods of leg positioning for walking on a rough flat substrate. In the first case, the tarsus is straightened and the rough substrate is gripped between the claws and the proximal complex of attachment devices (tarsal euplantulae, fossulae spongiosa, and terminal spurs of tibiae). In the second case the tibia does not touch the substrate; the insect is supported only by distal tarsomeres. The tarsus is in an overextended condition. On rods, with diameters comparable to or larger than the tarsus length, insects walk by clasping the

Handling Editor: Heikki Hokkanen

D. Gladun · S. N. Gorb
Department of Insect Ethology and Sociobiology, Schmalhausen Institute of Zoology, Chmielnicki str., 15, 01601 Kiev, Ukraine
e-mail: gladun.dmytro@gmail.com; gladun@bigmir.net

S. N. Gorb (✉)
Evolutionary Biomaterials Group, Department of Thin Films and Biosystems, Max-Planck-Institute for Metals Research, Heisenbergstr. 3, 70569 Stuttgart, Germany
e-mail: s.gorb@mf.mpg.de

stem with the bent tarsi. This posture is characteristic for the majority of insects independent of the tarsal position they normally use while walking on a plane. If the rod's diameter is smaller than the tarsus length, walking insects usually clutch it between contralateral tarsi. Using such a posture they are supported by interlocking or by strong friction, generated by attachment devices of the proximal tarsomeres, and do not use attachment devices of the pretarsus. Contact with the substrate is reinforced due to the coordinated contralateral clutch using all supporting legs. It is concluded that the use of different types of attachment structures correlates with locomotory techniques.

Keywords Adhesive organs · Attachment · Insecta · Leg · Locomotion · Tarsus · Walking

Introduction

The most simple mode of terrestrial locomotion in insects is walking on horizontal substrates such as soil, thick horizontal tree branches or flat leaves. Holding onto the substrate is achieved by using terminal parts of the leg (tarsus and tibia). While walking on a plane substrate, there are two methods of leg positioning. In the first method, all tarsomeres touch the substrate. Such leg positioning is typical for some representatives of Orthoptera (Imms 1951, Figure 254; Kazakova 1988). The insect either compresses a rough substrate between terminal spurs of the tibia and claws, or adheres to a smooth substrate with the aid of specialized attachment devices such as arolia, euplantulae, pulvilli, and hairy coverage of the ventral surfaces of tarsomeres (Kazakova 1988; Gorb 1998, 2001; Gorb and Scherge 2000). In this case, the tarsus is straightened. In the second method, the tibia does not touch the substrate and the animal is supported only by distal tarsomeres. This method has been described for tiger beetles (Cicindelidae) (Nachtigall 1996), fast running cockroaches (Frazier et al. 1999), and flies (Niederegger et al. 2002; Niederegger and Gorb 2003). Rough substrate is compressed between distal spurs of the terminal tarsomere, proximally pointed setae of the ventral side of the tarsus, and claws. Attachment to a smooth substrate is achieved by specialized attachment pads. The tarsus is in an overextended position.

Flexion of the tarsus in insects is accomplished by the *retractor unguis* muscle, originating in the tibia and femur (Snodgrass 1956). When the muscle is contracted, the tendon is drawn proximally. This action results in the ventral bending of the tarsal chain, flexion of claws, and in hymenopterans and several other groups, in ventral overturn of the arolium. To attach to the rough plane, even a slight flexion of claws is sufficient to get a grip on the surface. The substrate is compressed between claws of the legs, forming a tripod while the insect is walking. The vectors of lateral forces are directed along each leg to the body, thus contributing to this grip (Full et al. 1995).

On the smooth substrate, claws slip over the substrate and drift apart, and the arolium spreads and presses to the substrate (Snodgrass 1956; Federle et al. 2001; Frantsevich and Gorb 2002). Similar behavior of the tarsal chain is described for flies (Niederegger and Gorb 2003). However, the anatomical range of the tarsus flexion possible is much greater than what is necessary for walking on a flat substrate. In the stick insect *Carausius morosus* (Phasmodea: Lonchodidae), completely protracted tendons of the claw retractor muscle result in pretarsus deflection relative to the straightened position at an angle of 270° (Walther 1969). In the hornet, *Vespa crabro* (Hymenoptera: Vespidae), the pretarsus can be bent to an angle of 107° (Frantsevich and Gorb 2002). In a natural environment, the extreme bending ability of the tarsus is used by insects to walk on thin stems and rods of plants (Ermolenko 1972; Kazakova 1988; Hassenstein and Hustert 1999).

Leg kinematics during locomotion on a narrow substrate has been previously studied in beetles, bugs, and stick insects (Frantsevich et al. 1995, 1996; Frantsevich and Cruse 1997, 2005). However, these works do not consider the role of attachment devices in contact formation with the ground. In the present study, locomotory techniques of several representatives of insects from different orders, possessing different types of attachment structures on their legs, are described. The study is mainly based on the video recordings of insects walking on a flat substrate and on cylindrical substrates of various thicknesses. Plant stems were imitated using cylindrical rods. Special attention was paid to methods of the tarsus placement on the substrate. The present study is important for understanding how the attachment structures are used on various substrates. It is known that the attachment abilities of insects are associated not only with the type of micro- and nanostructures of an attachment device, but also with global scale kinematics responsible for contact formation and release. Previous studies have revealed that the manner of foot placement on the surface is an important factor contributing to contact formation and release in such animals as tree frogs (Hanna and Barnes 1990), insects (Gorb 2001; Niederegger and Gorb 2003), and geckoes (Russell 1975; Gao et al. 2004).

Materials and methods

The list of species studied is given below. Orthoptera: *Metrioptera bicolor* (Philippi 1830) (Tettigoniidae); Heteroptera: *Himacerus apterus* (Fabricius 1798) (Nabidae), *Pyrrhocoris apterus* Linnaeus, 1758 (Pyrrhocoridae); Coleoptera: *Coccinella septempunctata* Linnaeus, 1758 (Coccinnellidae); Hymenoptera: *Arge melanochroa* (Gmelin 1790) (Tenthredinidae), *Coelichneumon sugillatorius* (Linnaeus 1758) (Ichneumonidae), *Mellinus arvensis* (Linnaeus 1758) (Sphecidae), *Eumenes coarctatus* (Linnaeus 1758) (Vespidae), *Formica polyctena* Förster, 1850 (Formicidae), *Apis mellifera* Linnaeus, 1758 (Apidae).

All species were collected in the field in the suburbs of Kiev (Ukraine). In some insects (*A. mellifera, C. sugillatorius, M. arvensis*, and *E. coarctatus*), the wings were ablated prior to experiments. Insects, walking on different substrates, were videotaped using a camcorder Panasonic NV-A3. Three different geometries of substrates were used: (1) flat, (2) thin rods thinner than the tarsus length of the insect, (3) thick rods, with a diameter surpassing tarsus length. The following substrates were used: steel wire (diameter = 0.4 mm), straw (diameter = 1 mm), polished wooden rods (diameter = 2 and 3 mm). In experiments, the rods were oriented horizontally and vertically. On horizontal rods, the insect's locomotion on the upper and lower surface of the rod were videotaped. In some cases, the camera was mounted on a binocular microscope MBS-9, in order to zoom some details of the foot placement on the substrate.

We videotaped at the least ten passages of each insect on each type of substrate. For fast-running insects up to 30 passages were videotaped for obtaining of sufficient set of clear frames. Only even passages were analyzed (without turns, falls, etc.). Several steps were observed in the field of view during each passage. In Table 2, we registered all observed types of tarsus placement on each substrate (usage of the different positions was not analyzed). Video recordings were digitized and further processed using ATI Video In 6.3, ATI Video Editor 6.2, AVIedit, and Adobe Photoshop 6.0 software.

For microscopic observations of leg attachment structures, air-dried tarsi were mounted on metal stubs, sputter-coated with a 6 nm thick layer of gold-palladium, and viewed in a field emission scanning electron microscope (SEM) (Hitachi S-4800) at 3 kV. In the present study, tarsomeres are counted from the distal one (T1) to the proximal ones.

Results

Attachment devices of insects studied

Orthoptera

Similar to other representatives of Ensifera, each tarsus of the bush cricket *Metrioptera bicolor* bears four tarsomeres (Fig. 1D). Second, third, and fourth tarsal segments are equipped with euplantulae on their ventral sides (Fig. 1A, C, D). Euplantulae are soft pads adapted to attachment to smooth substrates (Jiao et al. 2000). The second tarsomere is laterally extended and bears a heart-shaped euplantula. Fore and middle tibiae of *M. bicolor* do not bear terminal spurs (Fig. 1D). Hind legs bear one pair of terminal tibial spurs and one pair of tarsal spurs (Fig. 1A, B). The cuticle of tettigoniid euplantulae is composed of relatively stiff filaments branching in the vicinity of the surface, which is

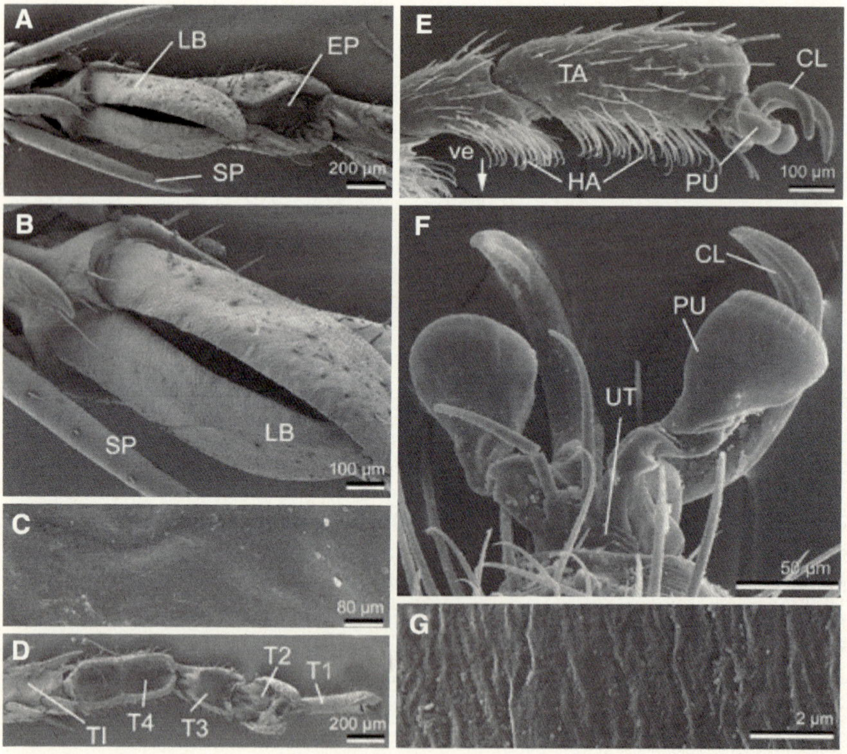

Fig. 1 SEM micrographs of insect tarsi. (**A–D**) *Metrioptera bicolor* (Orthoptera, Tettigoniidae), ventral aspect of the tarsus. (**A**, **B**) Hind leg. (**C**) Surface of the euplantula. (**D**) Foreleg. (**E–G**) *Pyrrhocoris apterus* (Heteroptera, Pyrrhocoridae), tarsus. (**E**) Lateral aspect. (**F**) Pretarsus, ventral aspect. (**G**) Ventral surface of the pulvillus. *CL* claw, *EP* euplantula, *HA* hairs, *LB* lobe-like spurs of the basal tarsomere, *PU* pulvillus, *SP* tibial spur, *TA* tarsus, *TI* tibia, *T1–T4* tarsomeres, *UT* unguitractor, *ve* ventral direction

covered with thin epicuticle (Gorb et al. 2000; Gorb 2001). The pretarsus is equipped with simple claws and lacks structures specialized for attachment on smooth substrates.

Heteroptera

Tarsi of the bug *Pyrrhocoris apterus* contain three segments. Tarsomeres of the fore- and midlegs are dorsally and laterally covered with rather long stiff setae (Fig. 1E). The ventral side of tarsomeres is covered with sharp curved setae, presumably softer than the dorsal and lateral setae (Fig. 1E). The pretarsus bears simple claws and slightly corrugated pulvilli (Fig. 1F, G) that belong to the smooth type of attachment devices (Gorb 2001). The pulvilli are able to generate rather high attachment forces on smooth substrates, as it has been previously experimentally demonstrated for the bug *Coreus marginatus* (Gorb and Gorb 2004).

In the bug *Himacerus apterus*, the distal ends of fore and mid-tibiae bear oval fields of small bent setae (*fossulae spongiosa*) (Fig. 2D). These organs were previously studied in representatives of reduviids (Gillett and Wigglesworth 1932; Weirauch 2005). The edge of the field is covered with longer and slightly flattened setae sloped latero-proximally (Fig. 2D, E). The ends of hind tibiae are covered with long, thick, stiff setae (Fig. 2B). All tarsi consist of three tarsomeres and are covered with sparsely distributed, stiff setae, denser on the ventral side of the tarsomeres (Fig. 2A–C). The pretarsus is free of adhesive pads and equipped only with a pair of simple claws (Fig. 2A).

Coleoptera

Tibiae of mid and hind legs of a ladybird beetle *Coccinella septempunctata* bear a pair of small terminal spurs. Two proximal tarsomeres have hairy coverage (Fig. 3A, B), composing an organ adapted for attachment to relatively smooth substrates. There are two types

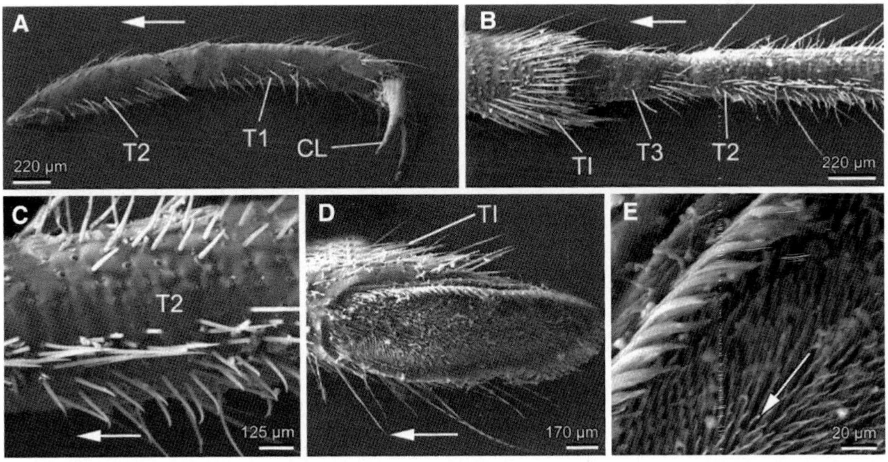

Fig. 2 SEM micrographs of the *Himacerus apterus* (Heteroptera, Nabidae) tarsi. (**A**) Foreleg, first and second tarsomeres, lateral aspect. (**B**) Hind leg, distal end of the tibia, second and third tarsomeres, ventral aspect. (**C**) Setae on the second tarsomere of the hind leg. (**D**) *Fossula spongiosa* at the distal end tibia of the fore leg (tarsus is ablated). (**E**) Hairs of the *fossula spongiosa*. *White arrow*, proximal direction; *CL* claw, *T1–T3* tarsomeres, *TI* tibia

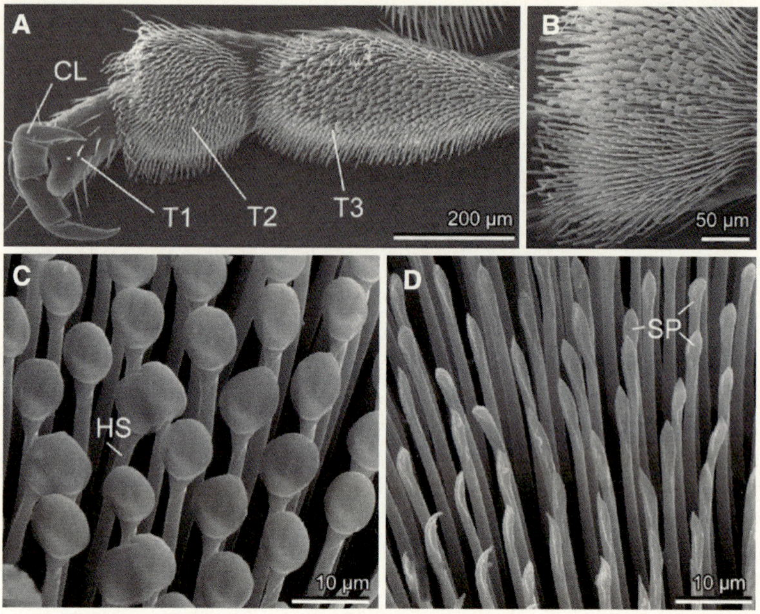

Fig. 3 SEM micrographs of the *Coccinella septempunctata* (Coleoptera, Coccinnellidae) tarsi. (**A**) Ventral aspect of the foreleg tarsus. (**B**) Second tarsomere of the foreleg. (**C**) Mushroom-like setae (hairs) of the second tarsomere. (**D**) Spatula-like compressed setae (hairs) of the second tarsomere. *CL* claw, *HS* hair shaft, *SP* spatula, *T1–T3* tarsomeres

of tenent setae: one with mushroom-like tips (Fig. 3B, C) and one with spatula-like compressed tips (Fig. 3B, D). The pretarsus is free of adhesive pads and equipped only with simple claws (Fig. 3A).

Hymenoptera

Each tibia of the sawfly *Arge melanochroa* bears two terminal spurs (Fig. 4A). All tarsi are composed of five segments regularly covered with fine setae. The second and third tarsomeres also bear several slightly longer, stiff setae on the distal edge. Four main tarsal segments are provided with plantulae presumably originating from transformed spurs (Fig. 4A, C). Pretarsus is equipped with simple claws and arolium (Fig. 4A, B). The arolium surface contains microfolds running perpendicular to the longitudinal axis of the tarsus (Fig. 4D).

The parasitic wasp *Coelichneumon sugillatorius* has foreleg tibia having one large terminal spur (Fig. 4E). Tibiae of mid and hind legs have two terminal spurs each. The tarsus is composed of five segments ventrally covered with fine setae (Fig. 4E, F). Setae are strongly compressed and slightly curved proximally (Fig. 4H). There is a gradient in the distribution of setae: setae are much stiffer at the distal part of each tarsomere. The pretarsus is equipped with simple claws and an arolium (Fig. 4G).

The tibiae of the forelegs of the sphecid wasp, *Mellinus arvensis*, have one terminal spur, whereas mid and hind tibiae bear two spurs each (Fig. 5A). Tarsi of *M. arvensis* are composed of five segments each. The ventral side of the tarsomeres is covered with fine hairs (acanthae) (Fig. 5B, C). Similar to *C. sugillatorius*, distal margins of the ventral side

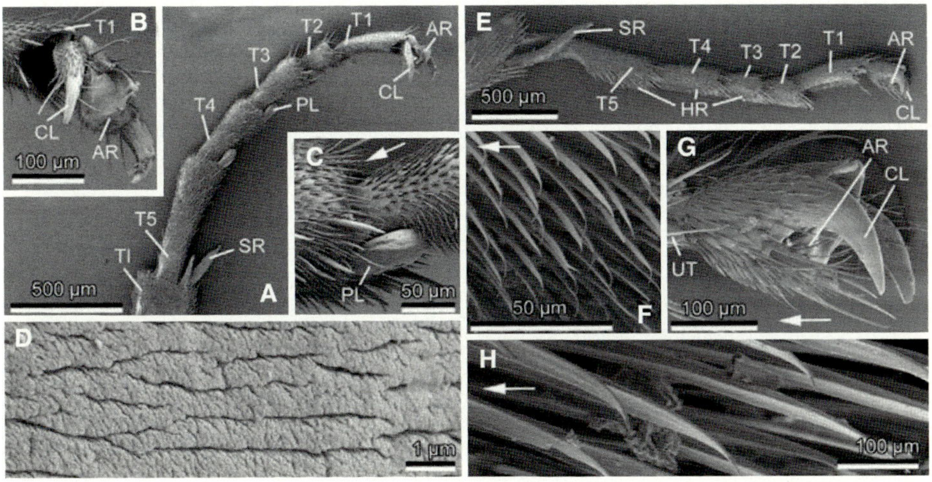

Fig. 4 SEM micrographs of hymenopteran tarsi. (**A–D**) *Arge melanochroa* (Hymenoptera, Tenthredinidae). (**A**) Lateral aspect of the tarsus. (**B**) Pretarsus with the arolium and claws. (**C**) Plantulae originating from transformed spines. (**D**) Surface of the arolium. (**E–H**) *Coelichneumon sugillatorius* (Hymenoptera, Ichneumonidae). (**E**) Lateral aspect of the foreleg tarsus. (**G**) Pretarsus of the foreleg. (**F**, **H**) Hairy coverage of the ventral surface of the hindleg and midleg, respectively. *White arrow*, proximal direction; *AR* arolium, *CL* claw, *HR* hairy coverage of the ventral surface of tarsomeres, *PL* plantulae, *SR* tibial spur, *T1–T5* tarsomeres, *TI* tibia, *UT* unguitractor

of tarsomeres are covered with stiff setae and also bear spines (Fig. 5A, B). Four tarsal segments (T2–T5) are provided with plantulae presumably originating from transformed spurs (Fig. 5A–C). The pretarsus bears arolium and simple claws (Fig. 5A, B).

Tibiae of fore- and midlegs of the wasp, *Eumenes coarctatus*, have one terminal spur. Hind legs bear two terminal spurs on each tibia (Fig. 5D). Tarsi are composed of five tarsomeres covered with setae (acanthae) (Fig. 5F) similar to those in *C. sugillatorius* and *M. arvensis*. All tarsomeres bear spines on their distal margins (Fig. 5E, F). The pretarsus has an arolium and bifurcated claws (Fig. 5D, E).

Tibiae of fore- and mid legs of the honeybee, *Apis mellifera*, have one terminal spur. Hind legs have no spurs on tibiae. Tarsi are composed of five segments each, ventro-distal surfaces of which are covered with stiff setae (Fig. 6A, C). Basal tarsomere T5 is larger than T1–T4 (Fig. 6A). An arolium and bifurcated claws are present (Fig. 6A, B).

In *Formica polyctena* ants, all tibiae contain one terminal spur. Each tarsus contains five segments (Fig. 6D). The ventral surface of each tarsomere is covered with stiff setae (Fig. 6D, E). Additionally, tarsomeres bear spines on the ventral side and on distal margins. The pretarsus has an arolium and simple claws (Fig. 6E, F). The surface of the arolium is strongly corrugated at the microscale (Fig. 6G).

Walking techniques and leg positioning on different substrates

In the bush cricket, *Metrioptera bicolor*, the tarsus is straight, and all tarsomeres (with the exception of the first one) and claws contact the flat substrate (Figs. 7A, 17A). On a thin, vertical cylindrical rod, the insect walks by supporting itself with the distal end of the tibiae and proximal tarsomeres (Figs. 7B, C, 17F). In such a posture, the tarsus is bent, embracing

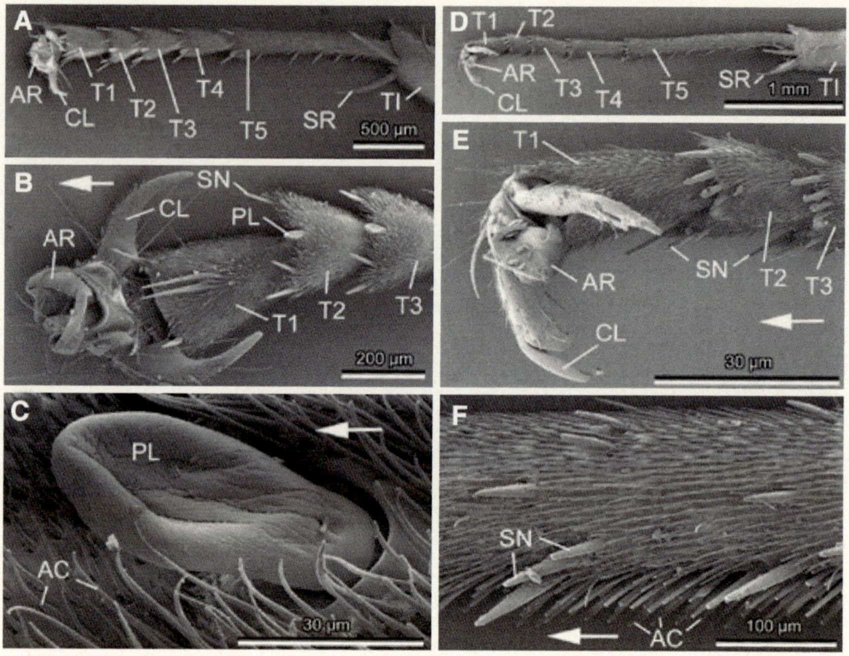

Fig. 5 SEM micrographs of hymenopteran tarsi. (**A**–**C**) *Mellinus arvensis* (Hymenoptera, Sphecidae). (**A**) Ventro-lateral aspect of the hindleg tarsus. (**B**) Foreleg pretarsus with the arolium and claws. (**C**) Plantulae originating from transformed spines. (**D**–**F**) *Eumenes coarctatus* (Hymenoptera, Vespidae). (**D**) Ventro-lateral aspect of the hindleg tarsus. (**E**) Pretarsus of the hind leg with the arolium and claws. (**F**) Hairy coverage of the ventral surface of the hind leg. *White arrow*, distal direction; *AR* arolium, *CL* claw, *AC* hairy coverage (acanthae) of the ventral surface of tarsomeres, *PL* plantulae, *SN* spines, *SR* tibial spur, *T1–T5* tarsomeres, *TI* tibia

the rod (Fig. 7C). However, the pretarsus and large euplantula do not take part in the insect's attachment to such a substrate (Fig. 7B).

When the substrate diameter is approximately equal to or larger than the length of the tarsus, the tarsus bends ventrally and all its tarsomeres (sometimes with the exception of the first) and claws attach to the rod with their ventral sides (Figs. 7D–F, 17D, E). In such a position, attachment devices of the three proximal tarsomeres participate in contact formation. Thus, tarsal compliance results in the same area of contact on plane substrates and on those with rather large radii. The bush cricket readily walked down a vertical rod, but less so upwards. The bush cricket was not able to walk on thin horizontal rods (diameter = 0.4–3.0 mm) (Table 1).

In the ladybird, *Coccinella septempunctata*, (Fig. 8A–C), positions of the tarsi on different substrates are similar to those in *M. bicolor*. *C. septempunctata* walks perfectly well on both horizontal and vertical rods in all directions (Table 1).

The bug *Himacerus apterus* walks on a plane substrate leaning on distal ends of tibiae and claws. A rough flat substrate is gripped between claws and the distal end of the tibia (Figs. 9A, 17B), while the tarsal chain remains straight. On rods, the bug uses the same leg positions as *M. bicolor* and *C. septempunctata* (Fig. 9B–F). *H. apterus*, however, has difficulties walking down vertical rods (Table 1).

Another bug species, *Pyrrhocoris apterus*, walks on a flat substrate leaning on the first and second tarsomeres. The tarsus is overextended (Figs. 10A, 17C). On thick rods, tarsi of

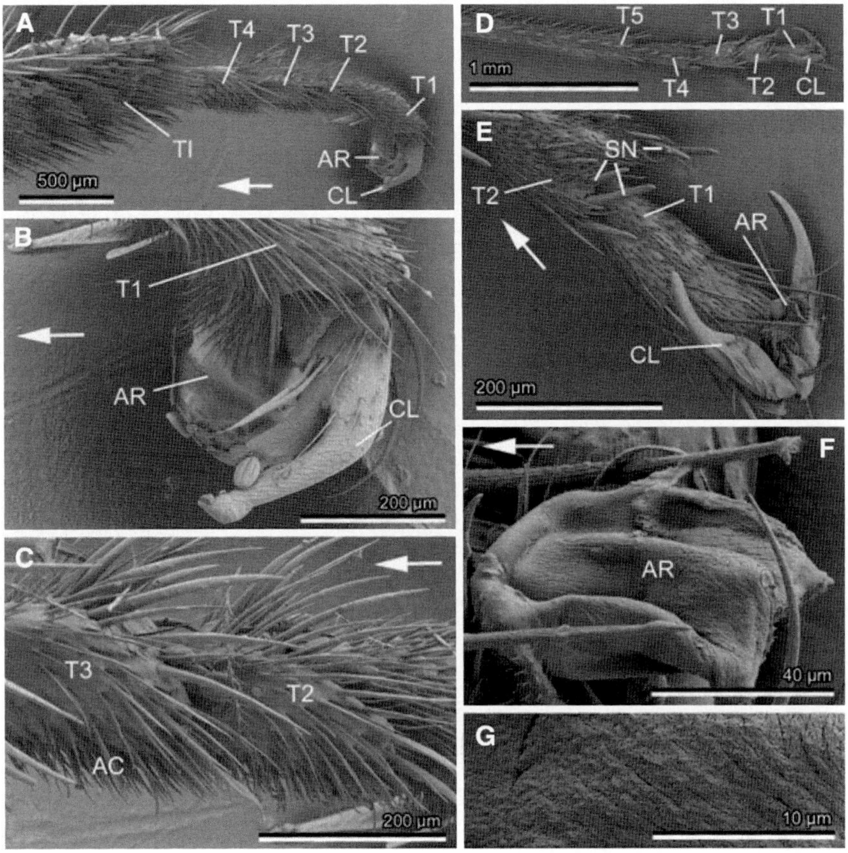

Fig. 6 SEM micrographs of hymenopteran tarsi. (**A–C**) *Apis mellifera* (Hymenoptera, Apidae), ventro-lateral aspect of the tarsus. (**A**) Tarsus. (**B**) Pretarsus. (**C**) Area of the joint between the second and third tarsomeres. (**D–G**) *Formica polyctena* (Hymenoptera, Formicidae). (**D**) Ventral aspect of the second leg tarsus. (**E**) Ventro-lateral aspect of the pretarsus with the terminal tarsomere. (**F**) Ventral aspect of the arolium in the spread position. (**G**) Surface of the arolium. *White arrow*, proximal direction; *AR* arolium, *CL* claw, *AC* hairy coverage (acanthae) of the ventral surface of tarsomeres, *SN* spines, *T1–T5* tarsomeres, *TI* tibia

the fore- and hind legs may be positioned on the substrate similar to that on the flat substrate. Tarsi of the mid legs bend ventrally and clutch the substrate with the ventral surface (Fig. 17D). There is no correlation between substrate orientation (horizontal or vertical), position of the insect relative to the substrate (above or below), direction of walking (up or down) or the grip technique of the insect (Fig. 10B–H). On thin rods, *P. apterus* bends fore and hind legs such that the pretarsus, first and in some cases second tarsomere contact the substrate (Fig. 17G). Mid legs clutch a thin rod between basal tarsomeres of the ventrally bent tarsi (Figs. 10I, J, 17f). The bug is not able to walk down a thin rod (Table 1).

Walking technique of the sawfly *Arge melanochroa* on a flat substrate is similar to that of *P. apterus*. Only the pretarsus and distal edge of the second tarsomere contact the substrate. The tarsus is overextended (Fig. 17C). Its tarsomeres are bent in an arc, convex ventrad (Fig. 11A). A thin rod is clutched between basal tarsomeres of the ventrally bent

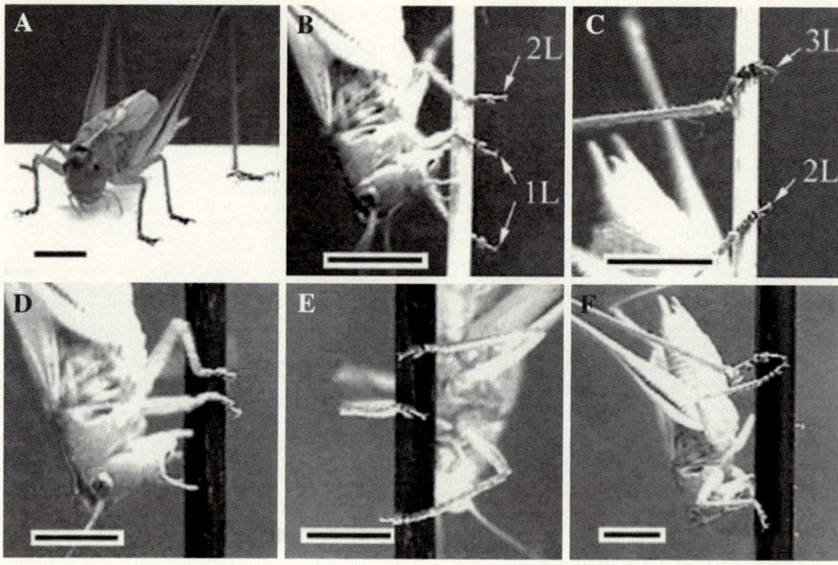

Fig. 7 Grasshopper *Metrioptera bicolor* (Orthoptera, Tettigoniidae) walking on the flat substrate (**A**) and on rods of different diameters: $d = 1$ mm (**B**, **C**), 2 mm (**D**, **E**), and 3 mm (**F**). *Scale bars*, 4 mm

Table 1 Insect walking abilities on thin vertical and horizontal rods of various diameters

Species	*d* = 3 mm				*d* = 2 mm				*d* = 1 mm				*d* = 0.4 mm			
	↑	↓	O̲	/	↑	↓	O̲	/	↑	↓	O̲	/	↑	↓	O̲	/
Metrioptera bicolor	±	+	−	+	±	+	−	+	±	+	−	±	−	−	−	±
Himacerus apterus	+	±	+	+	+	±	+	+	+	±	±	+	+	−	±	+
Pyrrhocoris apterus	+	+	+	+	+	+	+	+	+	+	−	+	+	−	−	+
Coccinella septempunctata	+	+	+	+	+	+	+	+	+	+	+	+	+	+	−	+
Arge melanochroa	+	+	+	+	+	+	+	+	+	+	±	+	+	±	−	+
Coelichneumon sugillatorius	+	−	+	−	+	−	+	−	+	−	−	+	+	−	−	±
Mellinus arvensis	+	−	+	−	+	−	+	−	+	−	±	±	−	−	−	−
Eumenes coarctatus	+	−	±	−	+	−	−	±	+	−	−	±	−	−	−	−
Formica polyctena	+	±	+	+	+	±	+	+	+	±	±	+	+	−	−	±
Apis mellifera	+	±	±	+	+	±	−	+	+	−	−	+	±	−	−	±

↑, Up the vertical rod; ↓, down the vertical rod; O̲, above the horizontal rod; /, below the horizontal rod; +, insect walked well; ±, insect walked hesitatingly or unwillingly; −, insect walked poorly or do not walk

tarsi and the pretarsus (claws and arolium) does not touch the substrate (Figs. 11B, 17F). On relatively thick rods, *A. melanochroa* presses all tarsomeres of the ventrally bent tarsi to the substrate (Figs. 11C, 17D). The sawfly walked well on and under horizontal rods and ascended vertical rods. This species has only refused to descend thin vertical rods (diameter = 0.4 mm) or walk on horizontal rods of the same diameter (Table 1).

The parasitic wasp, *Coelichneumon sugillatorius*, walks on a flat substrate and thin rods in a similar manner as the sawfly does (Figs. 12A–C, 17C, F). The wasp easily ascended the vertical rod. On the horizontal rod, a floundering walk was observed. On thick

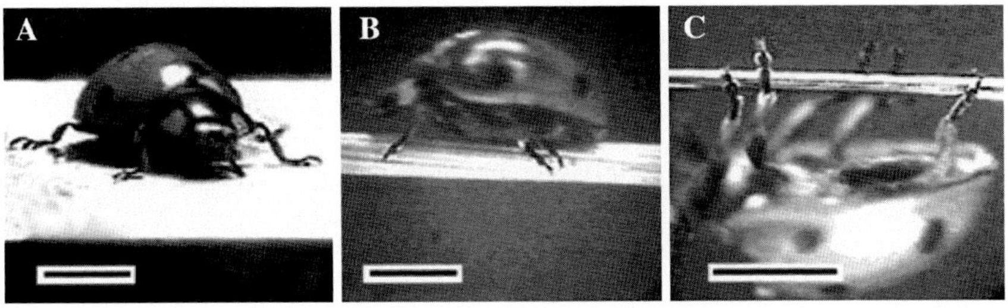

Fig. 8 Beetle *Coccinella septempunctata* (Coleoptera, Coccinnellidae) walking on the flat substrate (**A**) and on rods with diameters of $d = 1$ mm (**B**) and 0.4 mm (**C**). *Scale bars*, 2 mm

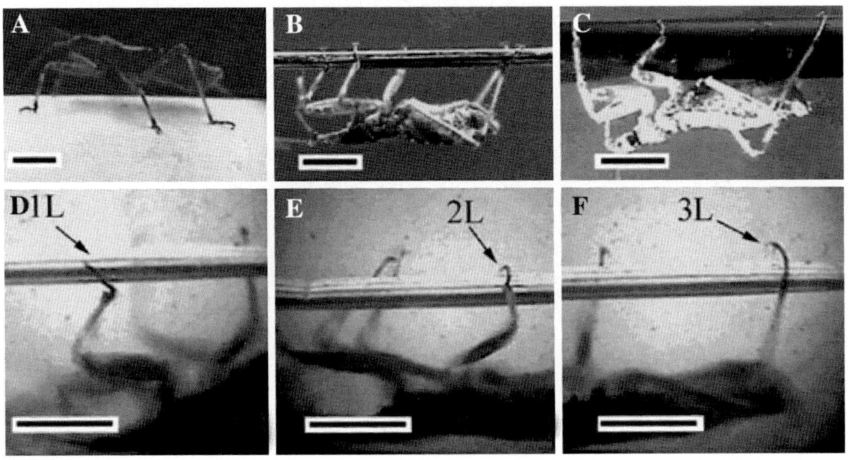

Fig. 9 Bug *Himacerus apterus* (Heteroptera, Nabidae) walking on a flat surface (**A**) and on rods of different diameters: $d = 1$ mm (**B**, **D**, **E**, **F**) and 3 mm (**C**). *Scale bars*, 3 mm

horizontal stems (diameter = 2 and 3 mm), the insect was able to walk above. On thin rods (diameter = 0.4 mm), the insect walked only in an upside down position hanging below the rod (Table 1). On thick rods, positions of the tarsi of fore and hind legs in *C. sugillatorius* were similar to that on the flat surface. However, in some runs, tarsi were bent and all tarsomeres were pressed down to the substrate (Fig. 17C, D). Middle tarsi were always bent and completely pressed to the rod.

Mellinus arvensis (Sphecidae) walks on a flat surface on distal tarsomeres of overextended tarsi (Figs. 13A, 17C). On a thin horizontal rod, it walked poorly: fore and hind legs rely either on distal tarsomeres (similar to that on the flat substrate) (Fig. 13B, C) or on the center of the overextended or slightly bent tarsus (Fig. 17I). In this latter case, the distal tarsomeres do not touch the substrate. This method was most often observed in the mid legs. When *M. arvensis* walks upward on a thin rod, the positioning of tarsi is different in different pairs of legs. Fore tarsi contact the substrate with the distal tarsomeres and pretarsus (Fig. 17J), whereas mid and hind tarsi contact the substrate with basal tarsomeres or even with the distal edge of the tibia (Fig. 17H). On thick rods, fore and hind tarsi function similar to the situation on a flat substrate; the middle tarsus, however, is pressed to the rod with all its tarsomeres (Fig. 17D). On a horizontal rod, *M. arvensis* tried to walk above the rod. This species is not able to walk down a thin rod (Table 1).

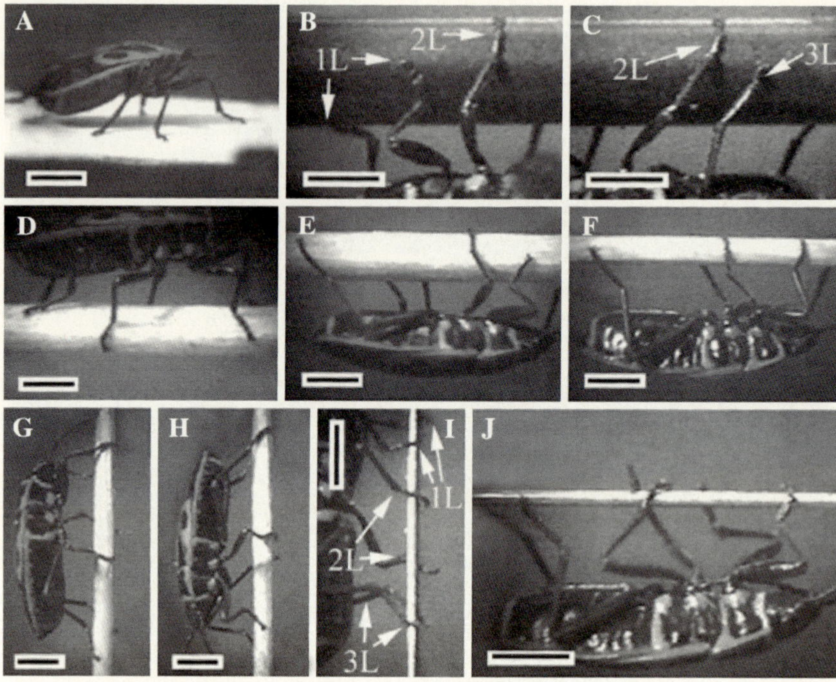

Fig. 10 Bug *Pyrrhocoris apterus* (Heteroptera, Pyrrhocoridae) walking on a flat substrate (**A**) and on rods with different diameters: $d = 3$ mm (**B**, **C**), 2 mm (**D**, **E**), 1 mm (**F**–**H**), and 0.4 mm (**I**, **J**). *Scale bars*, 2 mm

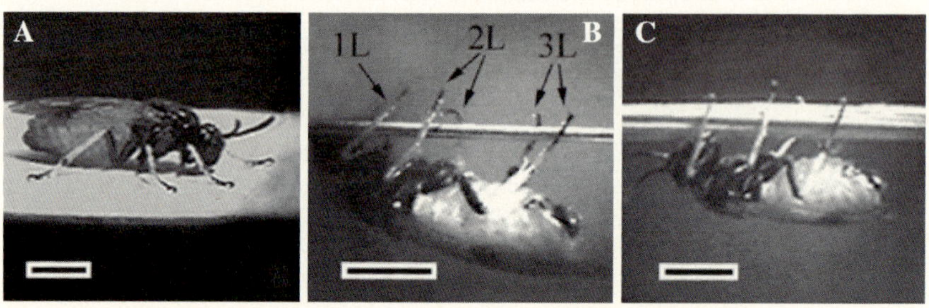

Fig. 11 Sawfly *Arge melanochroa* (Hymenoptera, Tenthredinidae) walking on a flat substrate (**A**) and on rods with diameters $d = 0.4$ mm (**B**) and 1 mm (**C**). *Scale bars*, 2 mm

Methods of leg positioning in the wasp, *Eumenes coarctatus*, is basically similar to *M. arvensis* in all experimental situations (Fig. 14A–C). However, *E. coarctatus* walks poorly on horizontal rods and is not able to walk down vertical rods (Table 1).

In the ant, *Formica polyctena*, walking on a flat substrate, legs are supported by terminal tarsomeres of overextended tarsi (Figs. 15A, B, 17C). On thin rods, these tarsomeres do not contact the substrate (Fig. 15C), and the bent tarsus is pressed to the substrate either at the level of the proximal tarsomeres or at the distal edge of the tibia (Fig. 17F, I). However, while walking up a thin rod, the fore tarsi often clutch the substrate with the claws

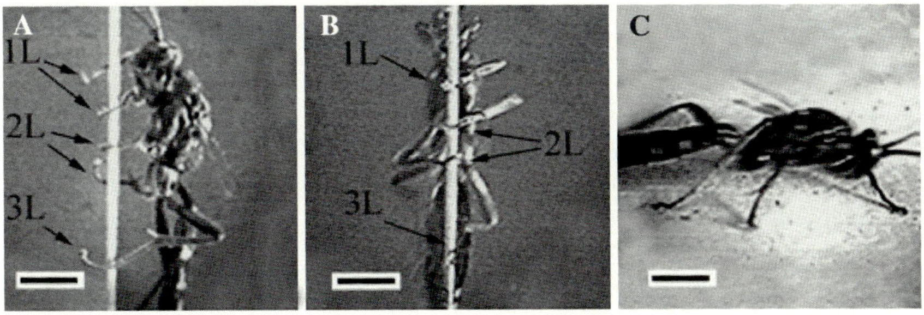

Fig. 12 Parasitic wasp *Coelichneumon sugillatorius* (Hymenoptera, Ichneumonidae) walking on a rod $d = 0.4$ mm (**A**, **B**) and on a flat substrate (**C**). *Scale bars*, 2 mm

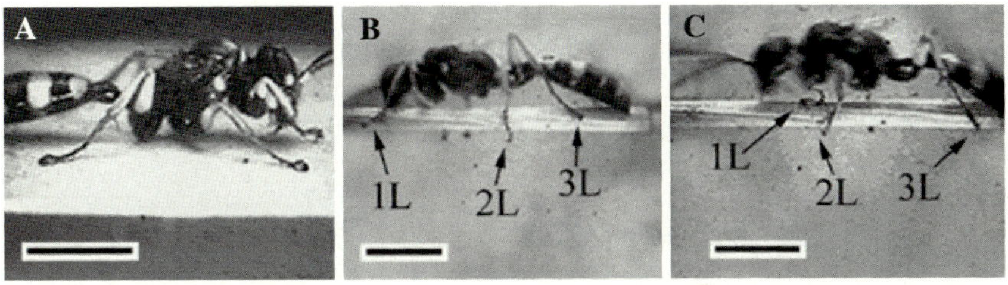

Fig. 13 Wasp *Mellinus arvensis* (Hymenoptera, Sphecidae) walking on a flat surface (**A**) and on a rod $d = 1$ mm (**B**, **C**). *Scale bars*, 3 mm

(Fig. 17G). On thick rods, the ant presses all tarsomeres of the bent tarsi to the substrate (Fig. 15D, F, 17D). Fore and hind tarsi are often used on thick substrates in a way similar to that on a flat substrate (Fig. 15E). *F. polictena* unwillingly walks down thin vertical rods (Table 1).

The honeybee, *Apis mellifera*, mainly uses pretarsus and 1–4 tarsomeres for attachment to substrates (Fig. 16A–D). On a plane substrate, the bee is supported by pretarsi and distal tarsomeres of an overextended tarsus (Fig. 16A, 17C). While walking on a thick rod, the fore and mid tarsi are used in the same way as on the plane. For the mid legs, the following position is typical: the tarsomeres 1–4 are bent and pressed to the substrate but they remain straightened relative to the tibia and basal tarsomere (Fig. 17K). In this position, basal tarsomeres contact the substrate with their distal edges (Fig. 16B). On thin rods, the tarsus is bent (Fig. 17J) and either the pretarsus and tarsomeres 1–4 (Figs. 16D, 17J) or the pretarsus and the first tarsomere contact the substrate (Figs. 16C, 17G). The latter position was observed when the insect was walking on a very thin rod (diameter = 0.4 mm). While walking upward on a thin vertical rod, the bee usually clutches the substrate using pretarsi of the fore legs (Fig. 17G) and leans on basal enlarged tarsomeres of the hind legs (Fig. 17J). The bee walks poorly down vertical rods (Table 1).

Methods of leg positioning during walking on flat substrate and rods are summarized in Table 2 and Fig. 17.

Difficulty of some insects to attach to the thin rods may be explained by their relatively big size, therefore they walk on the such substrates hesitatingly or fall from them.

It is possible, that the preference to walk up/down the vertical rods or above/below the horizontal rods correlates with other behavioral adaptations of each species.

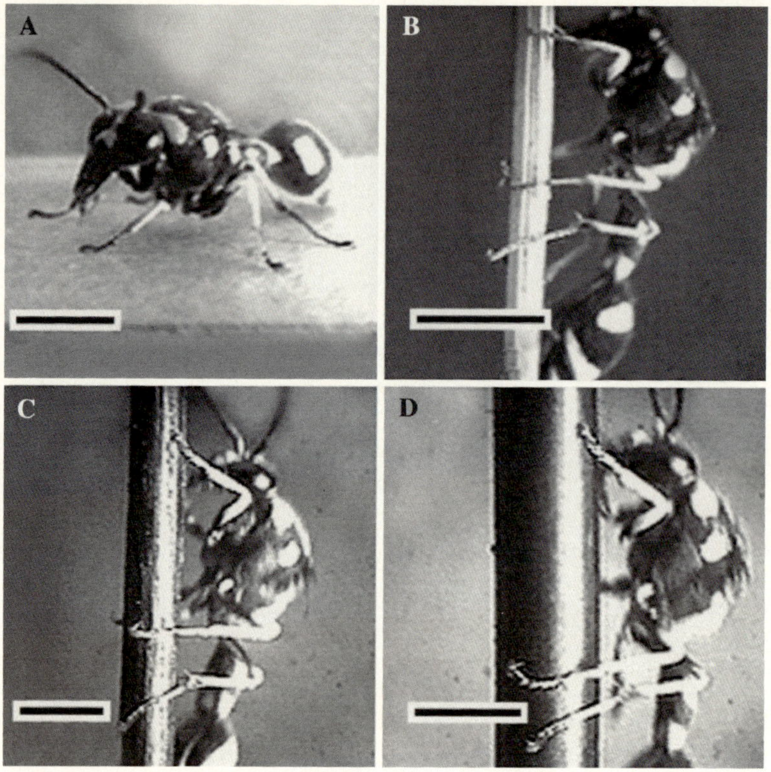

Fig. 14 Wasp *Eumenes coarctatus* (Hymenoptera, Vespidae) walking on a flat substrate (**A**) and on rods with different diameters: $d = 1$ mm (**B**), 2 mm (**C**), and 3 mm (**D**). *Scale bars*, 3 mm

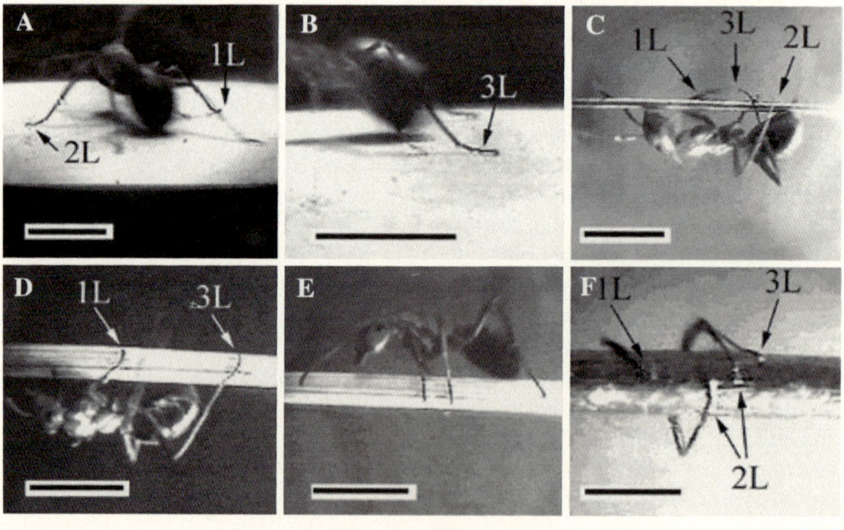

Fig. 15 Ant *Formica polyctena* (Hymenoptera, Formicidae) walking on a flat substrate (**A**, **B**) and on rods with different diameters: $d = 0.4$ mm (**C**), 1 mm (**D**, **E**), and 3 mm (**F**). *Scale bars*, 3 mm

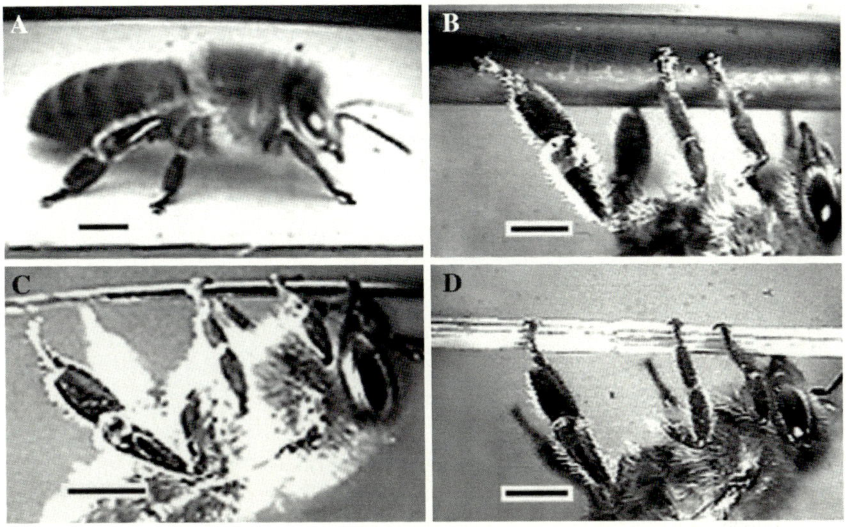

Fig. 16 Honeybee *Apis mellifera* (Hymenoptera, Apidae) walking on a flat surface (**A**) and on rods with different diameters: 3 mm (**B**), 0.4 mm (**C**), and 1 mm (**D**). *Scale bars*, 2 mm

Discussion

We have previously demonstrated that some insects such as Diptera and Hymenoptera are capable of strong bending and overextension of their tarsi depending on the behavioral situation and type of substrate used (Niederegger and Gorb 2003; Frantsevich and Gorb 2004). This paper describes various methods of contact formation with substrates of different geometries by insects from different lineages. Undoubtedly, the ability for strong bending and overextension of the tarsus in insects is an adaptation to a life in vegetation, to walking on substrates having various geometries. Also, numerous specialized attachment devices for walking on an enormous diversity of plant substrates has appeared in the evolution of insects (Beutel and Gorb 2001, 2006; Gorb and Beutel 2001).

Some attachment devices are preferably used on smooth substrates (arolium, pulvilli, and plantulae) and others on rough ones (claws, spurs, spines, and specialized setae). In most cases, an insect relies on several different types of attachment organs, using them in concert, depending on the substrate type. The shape of the substrate and its roughness are of great importance for the use of particular leg structures, in order to hold onto the substrate. On very rough substrates, claws are applied in order to generate strong traction forces (Dai et al. 2002). In addition, many insects use the spurs on the distal end of the tibia (*E. coarctatus*, *M. arvensis*, and *A. melanochroa*) or plantulae on the ventral side of tarsomeres ('Symphyta', Schulmeister 2003, some Pompilidae and Apidae, *M. arvensis*) for fixation while walking on thin stems, but do not use them while walking on a plane surface. Previous authors have reported the use of tibial spurs by cockroaches in interlocking with the substrate as well as for support on porous or mesh-like substrates. At the critical range of substrate roughness (0.3–3 μm), both claws and adhesive pads fail in generating strong contact forces (Gorb 2001; Peressadko and Gorb 2004b).

Curved rods, corresponding to plant stems, are probably the most challenging substrates for insects to walk on and, according to our observations, on these curved rods insects mostly use attachment devices of different types simultaneously. That is why leg and tarsus

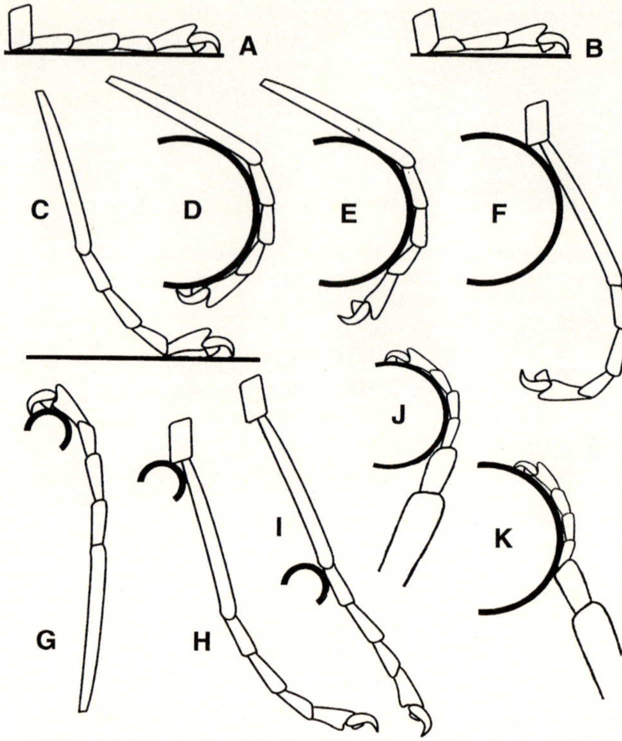

Fig. 17 Methods of leg positioning while walking on a flat substrate and on rods (also see Table 2). (**A**) Tarsus is straight, distal part of the tibia, pretarsus and all tarsomeres except the first one contact the substrate. (**B**) Tarsus is straight, distal part of the tibia, pretarsus and sometimes the basal tarsomere contact the substrate. (**C**) Tarsus is overextended, only the pretarsus, first tarsomere or distal part of the second tarsomere contact the substrate. (**D**) Tarsus is bent, all tarsomeres (sometimes also the distal part of the tibia) and claws contact the substrate. (**E**) Tarsus is bent, all tarsomeres except the distal one contact the substrate. (**F**) Tarsus is bent, only proximal tarsomeres (sometimes also the distal part of the tibia) contact the substrate. (**G**) Tarsus is bent and supported by the pretarsus and the distal tarsomere. (**H**) Tarsus is overextended, proximal tarsomeres (sometimes also the distal part of the tibia) contact the substrate. (**I**) Tarsus is straight or slightly bent; it contacts the substrate with mid tarsomeres. (**J**) Tarsus is bent, all tarsomeres except the basal one contact the substrate. (**K**) The chain of tarsomeres, from the first to fourth tarsomere, is bent and pressed to the substrate, but straightens relative to the tibia and basal tarsomere

positioning on the substrate is of great importance. On their ventral side, insect tarsomeres may bear organs specialized for generating strong adhesion on relatively smooth substrates. These organs may be smooth and soft like euplantulae in *M. bicolor* and other orthopterans or hairy, such as fields of specialized setae in *C. septempunctata* and other beetles from the families of Coccinellidae, Buprestidae, Cerambycidae, Chrysomelidae, etc. Plantulae, spines, and stiff setae are often located on the distal edge of tarsomeres. On the tibia, such devices are spurs, in some cases furnished with adhesive pads, as in some representatives of Cimbicidae (Ermolenko 1972), with stiff setae (*P. apterus*), and fields of specialized adhesive setae (*fossulae spongiosa* on the fore and mid legs of *H. apterus*). While walking on a very rough flat substrate, interlocking is often realized by clutching of the substrate features between claws and attachment devices that are located at the distal end of the tibia or/and tarsomeres. This method of attachment is also known in aphids (Dixon et al. 1990; Kennedy 1986) and rove beetles (Betz 2002).

Table 2 Methods of leg positioning during walking on a flat substrate and on rods

Species	Leg	Plane	$d = 3$ mm	$d = 2$ mm	$d = 1$ mm	$d = 0.4$ mm
Metrioptera bicolor	1	A	D	D, E	F	
	2	A	D	D, E	F	
	3	A	D	D, E	F	
Himacerus apterus	1	B	D	D	D, F	F
	2	B	D	D	F	F
	3	B	D	D	F	F
Pyrrhocoris apterus	1	C	C, D	C, D	C, D	G
	2	C	D	D	D	F
	3	C	C, D	C, D	C, D	G
Coccinella septempunctata	1	A	D, E	D, E	D, E	F
	2	A	D, E	D, E	D, E	F
	3	A	D, E	D, E	D, E	F
Arge melanochroa	1	C	D	D	D	F
	2	C	D	D	D	F
	3	C	D	D	D	F
Coelichneumon sugillatorius	1	C	C, D	C, D	F	F
	2	C	D	D	F	F
	3	C	C, D	C, D	F	F
Mellinus arvensis	1	C	C	C	C, I, J	
	2	C	D	D	H, I	
	3	C	C	C	C, H, I	
Eumenes coarctatus	1	C	C	C	J	
	2	C	D	D	J	
	3	C	C, H, I	C, H, I	C, H, I	
Formica polyctena	1	C	C, D	C, D	C, D	F, G, I
	2	C	D	D	D	F, I
	3	C	C, D	C, D	C, D	F, I
Apis mellifera	1	C	C	C, K	J	G
	2	C	K	K	J	G
	3	C	C	C, K	J	G

Letters marking the position of tarsi correspond to those given in Fig. 17. A. Tarsus is straight, distal part of the tibia, pretarsus and all tarsomeres except the first one contact the substrate. B. Tarsus is straight, distal part of the tibia, pretarsus and sometimes the basal tarsomere contact the substrate. C. Tarsus is overextended, only the pretarsus, first tarsomere or distal part of the second tarsomere contact the substrate. D. Tarsus is bent, all tarsomeres (sometimes also the distal part of the tibia) and claws contact the substrate. E. Tarsus is bent, all tarsomeres, except the distal one, contact the substrate. F. Tarsus is bent, only proximal tarsomeres (sometimes also the distal part of the tibia) contact the substrate. G. Tarsus is bent and supported by the pretarsus and the distal tarsomere. H. Tarsus is overextended, proximal tarsomeres (sometimes also the distal part of the tibia) contact the substrate. I. Tarsus is straight or slightly bent; it contacts the substrate with mid tarsomeres. J. Tarsus is bent, all tarsomeres, except the basal one, contact the substrate. K. Chain of tarsomeres from the first to the fourth tarsomere is bent and pressed to the substrate, but straighten, relative to the tibia and basal tarsomere

In many cases, the tarsomeres differ not only in their size and shape, but also in the presence of particular attachment devices. In *M. bicolor*, all tarsomeres with the exception of the first one are equipped with euplantulae on their ventral side. The first tarsomere does not bear any attachment device and serves as a support for pretarsal structures (in this case, only for claws). While walking on a flat substrate, the first tarsomere does not contact the surface, whereas a rough substrate is gripped between claws and the proximal complex of attachment devices (tarsal euplantulae and terminal spurs of tibiae). The same functional principle also applies to the beetle *C. septempunctata* and to the bug *H. apterus*. In the beetle, the proximal complex of attachment devices is represented by fields of specialized setae on the ventral side of tarsomeres and terminal spurs on tibiae. In the bug *H. apterus*, tarsomeres lack any specialized attachment devices. However, such devices (*fossulae spongiosa*) are located on the distal end of the tibia. In this case, tarsomeres do not contact the substrate, which is gripped between the pretarsus (claws) and distal end of the tibia. It is, however, not known whether *fossulae spongiosa* may function efficiently on a flat plane surface.

It is also important to mention that *M. bicolor*, *C. septempunctata*, and *H. apterus* do not have any adhesive organs in the distal area of their tarsi, and, therefore rely on the attachment devices of proximal tarsomeres (*M. bicolor* and *C. septempunctata*) or tibia (*H. apterus*) while attaching to smooth substrates. Adhesive pads (euplantulae) of *M. bicolor* belong to the smooth type. It has been previously shown that euplantulae, due to their material structure (Slifer 1950; Kendall 1970) and liquid secretion (Vötsch et al. 2002), are attachment devices with particular adhesive and frictional properties (Gorb et al. 2000; Jiao et al. 2000; Gorb and Scherge 2000; Perez Goodwyn et al. 2006). Also it has been previously demonstrated that old cockroaches, having smooth euplantulae and arolia as adhesive organs, have difficulty climbing a smooth incline of 45°, because their legs slip much more extensively compared to younger animals. Morphological data supports the hypothesis that slipping may be caused by degeneration of tarsal adhesive organs (Ridgel et al. 2003).

Adhesive organs, where contact is subdivided into many single contacts (hairy systems), are presumably more effective on rough substrates than smooth systems. In an adhesion experiment with structured and smooth surfaces, both made of the same polymer, structured polymer surfaces have demonstrated stronger adhesion on uneven glass substrates compared to smooth polymer surfaces (Peressadko and Gorb 2004a). Different shapes of adhesive setae tips are possible adaptations to different roughness of potential substrates (Persson and Gorb 2003). It is possible that the mushroom-like setae of *C. septempunctata* are responsible for strong attachment on smooth substrates, whereas spatula-like setae are for rough ones.

All other insects investigated here (*P. apterus*, *A. melanochroa*, *C. sugillatorius*, *M. arvensis*, *E. coarctatus*, *F. polyctena*, and *A. mellifera*) put their tarsus down in an overextended state while walking on a plane. A rough plane is gripped only between claws and the distal edge of the second and third tarsomeres. However, attachment devices are normally present on all tarsomeres, and sometimes on tibiae. They are located on the ventral side close to the distal edge of each tarsomere. Attachment devices on the tibiae and proximal tarsomeres do not contact the substrate while walking on a plane and are used by these insects while walking on stems only. In such a situation, the proximal tarsomeres, bearing soft smooth plantulae (*A. melanochroa* and *M. arvensis*) and stiff cuticle protuberances (spurs, spikes, and setae), work in combination with claws.

Walking on stems, where the diameter is equal or slightly larger than the tarsus length, the majority of insects studied clasp the substrate with a bowed tarsus. In addition, terminal spurs of tibiae often contact the stem. The mid tarsi are always completely pressed to the stem. This behavior does not depend on the position of mid tarsi on the plane (overextended or not). However, fore and hind tarsi are used in the same way, as when walking on the plane. This behavior may be explained by orientation of mid tarsi relative to the body. While walking, mid tarsi are oriented approximately perpendicular to the longitudinal body axis and to the stem, whereas fore and hind tarsi are oriented either parallel to the body axis or at a sharp angle to it.

If the diameter of the stem is smaller than the tarsus length, walking insects usually clutch it between their contralateral tarsi. In such a position, they are supported by interlocking or strong friction generated by proximal tarsomeres and do not use attachment devices of the pretarsus. Contact with the substrate is reinforced by the coordinated contralateral clutch using all supporting legs. This posture is characteristic for the majority of insects, independent of the tarsal position they normally use while walking on a plane. Only *A. mellifera*, and presumably some other species of bees, bear the modified basal tarsomere, which does not touch the thin rods.

It was shown that, while walking up thin stems, some Hymenoptera studied (*M. arvensis*, *E. coarctatus*, and *F. polyctena*) use a very special technique. The rod is gripped between the proximal tarsomeres of the mid tarsi; the fore tarsi clutch the stem with claws, and the hind tarsi rest on the terminal spurs of the tibiae.

We may conclude that the ability of insects to considerably bend their tarsi is an adaptation to life in vegetation, enabling them to walk on plant stems greatly varying in diameter, shape and roughness, as well as on the upper and lower surfaces of leaves having diverse surface profiles (Gorb and Gorb 2002). The results of the present study clearly demonstrate that the choice of attachment devices used depends on geometry and degree of substrate roughness as well as on spatial orientation of substrate and the direction of the insect's movement. For insects living on plants, two basic strategies of contact formation with substrate may be assumed. The first strategy applies to insects with relatively short tarsi consisting of few segments, the ventral surface of which is entirely covered with adhesive structures (*M. bicolor* and *C. septempunctata*). These devices contact the surface during walking on both a plane and on rods. The second strategy applies to insects with long tarsi consisting of many segments and capable of significant bending (*A. melanochroa*, *C. sugillatorius*, *M. arvensis*, *E. coarctatus*, and *F. polyctena*). The attachment devices are located on the distal edge of tarsomeres and on the pretarsus. Usually, they contact the substrate while walking on stems, when the bent tarsus clasps the stem. Although the attachment devices do not entirely cover the ventral surface of tarsomeres, a significant flexibility of the tarsus provides good contact to the substrate and results in the generation of strong friction forces contributing to a firm hold by the insect to the stem.

The present study shows that knowledge of the morphology and forces generated by different attachment devices of insects should be considered in the broader behavioral context. Also, the use of attachment devices of different pairs of legs must be reconsidered depending on the substrate and on the manner of locomotion. During the past few years, groups of engineers have been working on the development of robots able to climb on various surfaces (see for example Ritzmann et al. 2004; Daltorio et al. 2005a, b). The main challenge is to combine different attachment mechanisms thus enabling such devices to attach to, and walk on unpredictable surface curvature and roughness. The data presented here may inspire engineers to design mechanical devices with gripping strategies similar to those found in insects.

Acknowledgments This work was supported by the grant 05.07/00194 of SFFR (State Fund of Fundamental Research, Ukraine) and by the Federal Ministry of Education, Science and Technology, Germany (project InspiRat 01RI0633C). The authors thank V. M. Ermolenko and P. V. Puchkov for the identification of insects used in this study. Both authors greatly appreciate the valuable criticism and suggestions of

L. I. Frantsevich (Schmalhausen Institute of Zoology, Kiev, Ukraine) as well as linguistic corrections of V. Kastner (Tübingen, Germany).

References

Betz O (2002) Performance and adaptive value of tarsal morphology in rove beetles of the genus *Stenus* (Coleoptera, Staphylinidae). J Exp Biol 205:1097–1113

Beutel RG, Gorb SN (2001) Ultrastructure of attachment specializations of hexapods (Arthropoda): evolutionary patterns inferred from a revised ordinal phylogeny. J Zool Syst Evol Res 39:177–207

Beutel RG, Gorb SN (2006) A revised interpretation of the evolution of attachment structures in Hexapoda with special emphasis on Mantophasmatodea. Arthropod Syst Phylogeny 64:3–25

Dai Z, Gorb SN, Schwarz U (2002) Roughness-dependent friction force of the tarsal claw system in the beetle *Pachnoda marginata* (Coleoptera, Scarabeidae). J Exp Biol 205:2479–2488

Daltorio KA, Horchler AD, Gorb S, Ritzmann RE, Quinn RD (2005a) A small wall-walking robot with compliant, adhesive feet. In: Proceedings of the 2005 IEEE/RSJ international conference on intelligent Robots and systems (IROS '05), August 2–6, 2005, Edmonton, Canada, pp 4018–4023

Daltorio KA, Gorb S, Peressadko A, Horchler AD, Ritzmann RE, Quinn RD (2005b) A robot that climbs walls using micro-structured polymer feet. In: Proceedings of the international conference on climbing and walking robots (CLAWAR), September 13–15, 2005, London, UK, pp 131–138

Dixon AFG, Croghan PC, Gowing RP (1990) The mechanism by which aphids adhere to smooth surfaces. J Exp Biol 152:243–253

Ermolenko VM (1972) Horntails and saw-flies. Part 2 Tenthredinoidea. Fauna of the Ukraine. Band 10. Naukova Dumka Publishers, Kiev, p 204

Federle W, Brainerd EL, McMahon TA, Hölldobler B (2001) Biomechanics of the movable pretarsal adhesive organ in ant and bees. Proc Natl Acad Sci USA 98:6215–6220

Frantsevich LI, Cruse H (1997) The stick insect, *Obrimus asperrimus* (Phasmida, Bacillidae) walking on different substrates. J Insect Physiol 43:447–455

Frantsevich LI, Cruse H (2005) Leg coordination during turning on an extremely narrow substrate in a bug, *Mesocerus marginatus* (Heteroptera, Coreidae). J Insect Physiol 51:1092–1104

Frantsevich L, Gorb S (2002) Arcus as a tensegrity structure in the arolium of wasps (Hymenoptera: Vespidae). Zoology 105:225–237

Frantsevich L, Gorb S (2004) Structure and mechanics of the tarsal chain in the hornet, *Vespa crabro* (Hymenoptera: Vespidae): implications on the attachment mechanism. Arthropod Struct Dev 33:77–89

Frantsevich LI, Mokrushov PA, Shumakova ID, Gladun DV (1995) Kinematics of insect walking on narrow ground. Vestn Zoologii 4:41–48

Frantsevich LI, Mokrushov PA, Shumakova ID, Gorb SN (1996) Insect rope-walkers: Kinematics of walking on thin rods in a bug, *Graphosoma italicum* (Heteroptera, Pentatomidae). J Zool 238:713–724

Frazier SF, Larsen GS, Neff D, Quimby L, Carney M, DiCaprio RA, Zill SN (1999) Elasticity and movements of the cockroach tarsus in walking. J Comp Physiol A 185:157–172

Full RJ, Jamauchi A, Jindrich DL (1995) Maximum single leg force production: cockroach righting on photoelastic gelatin. J Exp Biol 198:2441–2452

Gao H, Wang X, Yao H, Gorb S, Arzt E (2004) Mechanics of hierarchical adhesion structures of geckos. Mech Mater 37:275–285

Gillett JD, Wigglesworth VB (1932) The climbing organ of an insect, *Rhodnius prolixus* (Hemiptera, Reduviidae). Proc R Soc Lond B 111:364–376

Gorb EV, Gorb SN (2002) Attachment ability of the beetle *Chrysolina fastuosa* on various plant surfaces. Entomol Exp Appl 105:13–28

Gorb SN (1998) The design of the fly adhesive pad: distal tenent setae are adapted to the delivery of an adhesive secretion. Proc R Soc Lond B 265:747–752

Gorb SN (2001) Attachment devices of insect cuticle. Kluwer, Dordrecht, Boston, London, p 305

Gorb SN, Beutel RG (2001) Evolution of locomotory attachment pads of hexapods. Naturwissenschaften 88:530–534

Gorb SN, Gorb EV (2004) Ontogenesis of the attachment ability in the bug *Coreus marginatus* (Heteroptera, Insecta). J Exp Biol 207:2917–2924

Gorb S, Jiao Y, Scherge M (2000) Ultrastructural architecture and mechanical properties of attachment pads in *Tettigonia viridissima* (Orthoptera Tettigoniidae). J Comp Physiol A 186:821–831

Gorb S, Scherge M (2000) Biological microtribology: anisotropy in frictional forces of orthopteran attachment pads reflects the ultrastructure of a highly deformable material. Proc R Soc Lond B 267:1239–1244

Hanna G, Barnes W-JP (1990) Adhesion and detachment of the toe pads of tree frogs. J Exp Biol 155:103–125

Hassenstein B, Hustert R (1999) Hiding responces of locusts to approaching objects. J Exp Biol 202:1701–1710

Imms AD (1951) A general textbook of entomology, including the anatomy, physiology, development and classification of insects. Methuen, London, p 727

Jiao Y, Gorb SN, Scherge M (2000) Adhesion measured on the attachment pads of *Tettigonia viridissima* (Orthoptera, Insecta). J Exp Biol 203:1887–1895

Kazakova IG (1988) Morphological adaptations of Acrididae to the use of grasses as substrates for feeding, migration, and oviposition. In: Zolotarenko GS (ed) Landscape ecology of insects. Nauka Publishers, Novosibirsk, Russia, pp 26–34

Kendall UD (1970) The anatomy of the tarsi of *Schistocerca gregaria* Forskål. Z Zellforsch Mikrosk Anat 109:112–137

Kennedy CE (1986) Attachment may be a basis for specialization in oak aphids. Ecol entomol 11:291–300

Nachtigall W (1996) Locomotory behaviour in a population of the tiger beetle species *Cicindella hybrida* on a small, hot, sandy area (Coleoptera: Cicindellidae). Entomol Gen 20:241–248

Niederegger S, Gorb SN, Jiao Y (2002) Contact behavior of tenent setae in attachment pads of the blowfly *Calliphora vicina* (Diptera, Calliphoridae). J Comp Physiol A 187:961–970

Niederegger S, Gorb S (2003) Tarsal movements in flies during leg attachment and detachment on a smoothe substrate. J Insect Physiol 49:611–620

Peressadko AG, Gorb SN (2004a) When less is more: experimental evidence for tenacity enhancement by division of contact area. J Adhes 80:1–15

Peressadko AG, Gorb SN (2004b) Surface profile and friction force generated by insects. In: Boblan I, Bannasch R (eds) First international industrial conference Bionik, April 22–23, 2004, vol 15. Fortschritt-Berichte VDI, Hannover, Germany, pp 257–263

Perez Goodwyn P, Peressadko A, Kastner V, Schwarz H, Gorb S (2006) Material structure, stiffness and adhesion: why attachment pads of the grasshopper (*Tettigonia viridissima*) adhere more strongly than those of the locust (*Locusta migratoria*) (Insecta: Orthoptera). J Comp Physiol A 192:1233–1243

Persson BNJ, Gorb S (2003) The effect of surface roughness on the adhesion of elastic plates with application to biological systems. J Chem Phys 119:11437–11444

Ridgel AL, Ritzmann RE, Schaefer PL (2003) Effects of aging on behavior and leg kinematics during locomotion in two species of cockroach. J Exp Biol 206:4453–4465

Ritzmann RE, Gorb S, Quinn RD (2004) Arthropod locomotion systems: from biological materials and systems to robotics. Arthropod Struct Dev 33:183–185

Russell AP (1975) A contribution to the functional analysis of the foot of the Tokay, *Gekko gecko* (Reptilia, Gekkonidae). J Zool Lond 176:437–476

Schulmeister S (2003) Morphology and evolution of the tarsal plantulae in Hymenoptera (Insecta), focussing on the basal lineages. Zool Scr 32:153–172

Slifer EH (1950) Vulnerable areas on the surface of the tarsus and pretarsus of the grasshopper (Acrididae, Orthoptera) with special reference to the arolium. Ann Entomol Soc Am 43:173–188

Snodgrass RE (1956) Anatomy of the honeybee. Comstock Publication Association, Ithaca, p 334

Vötsch W, Nicholson G, Müller R, Stierhof Y-D, Gorb S, Schwarz U (2002) Chemical composition of the attachment pad secretion of the locust *Locusta migratoria*. Insect Biochem Mol Biol 32:1605–1613

Walther C (1969) Zum Verhalten des Krallenbeugersystem bei der Stabheuschrecke *Carausius morosus* Br. Z Vergleichende Physiol 62:421–460

Weirauch C (2005) Pretarsal structures in Reduviidae (Heteroptera, Insecta). Acta Zool 86:91–110

Arthropod Plant Interactions
DOI 10.1007/s11829-007-9021-4

ORIGINAL PAPER

Plant surface–bug interactions: *Dicyphus errans* stalking along trichomes

Dagmar Voigt · Elena Gorb · Stanislav Gorb

Received: 11 July 2007 / Accepted: 17 October 2007

Abstract In contrast to many arthropods whose locomotion on plant surfaces is impeded by trichomes, the omnivorous mirid bug *Dicyphus errans* Wolff (Heteroptera, Miridae, Bryocorinae) lives on pubescent plants and preys on a variety of phytophagous arthropods. Morphological (slim body, long slender legs, elongated curved claws) and behavioural (locomotion) adaptations to hairy plant substrates result in higher predation effectiveness and fecundity, as well as a shorter developmental cycle of the bug compared to insects on plants without trichomes. To understand the bug–plant interactions from the biomechanical point of view, the bug's attachment system and the leaf surfaces of various plant species were analysed. Bug attachment ability was estimated in an inversion experiment on the adaxial and abaxial sides of leaves in 40 plant species. Furthermore, bug traction forces on the abaxial leaf side of 14 plant species were measured. Morphometrical variables of trichomes and the adhesive properties of plant surfaces were estimated. The bugs' traction force ranged from 0.07 mN on *Brassica oleracea* (Brassicaceae) to 1.21 mN on *Plectranthus ambiguus* (Lamiaceae) and *Solanum melongena* (Solanaceae). Bugs performed considerably better on hairy surfaces where a significant positive correlation between the force and both the trichome length and diameter was found. The trichome density and aspect ratio did not influence the force. Adhesion properties of plant surfaces covered with trichomes may also significantly impede the traction force. Based on the results obtained, it is concluded that hairy plants provide a more suitable environment for *D. errans* than either smooth ones or those covered with wax crystals. Hairy plant surfaces are predicted to support stronger attachment and therefore more reliable locomotion of the bug.

Handling editor: Heikki Hokkanen

D. Voigt (✉) · E. Gorb · S. Gorb
Evolutionary Biomaterials Group, Department of Thin-Films and Biological Systems,
Max-Planck Institute for Metals Research, Heisenbergstraße 03, 70569 Stuttgart, Germany
e-mail: voigt@mf.mpg.de

E. Gorb
e-mail: o.gorb@mf.mpg.de

S. Gorb
e-mail: s.gorb@mf.mpg.de

Keywords Adhesion · Attachment · Biomechanics · Bryocorinae · Heteroptera · Insect–plant interactions · Locomotion · Miridae · Traction force · Trichomes

Introduction

Many arthropods have the ability to attach to plant surfaces. Thus, they have access to widely expanded, complex and greatly diverse habitats provided by numerous plants. More than two-thirds of the insects in Central Europe are associated with plants at least during part of their life (Zwölfer 2003). Plants play a crucial role as interactive components in the interplay between entomophagous arthropods and herbivores (Cortesero et al. 2000). Many different kinds of plant–arthropod interactions have been intensively studied, e.g. pollination and phytophagy (e.g. Bernays and Chapman 1994; Jolivet 1998; Schoonhoven et al. 2005). Numerous adaptations that evolved in both partners (insects and plants) have been reported (Southwood 1973; Becerra 2003; Grimaldi and Engel 2004).

Structural features of the plant surface have been mainly considered in the context of direct plant defense and mechanical resistance against phytophagous insects (e.g. Chapman 1977; Juniper and Jeffree 1983; Begon et al. 2006). Peeters (2002) supposed that composition of an herbivore assemblage is more strongly correlated with leaf structural traits than with chemical leaf constituents. Both epicuticular waxes and plant hairs (trichomes) are well known plant structures which act as physical barriers for phytophages causing antibiotic and antixenotic reactions in insects (Smith 2005). Plant surfaces covered with wax crystals are slippery and inaccessible substrates for many arthropod species (Stork 1980b, 1986; Jeffree 1986; Juniper 1995; Eigenbrode 1996; Brennan and Weinbaum 2001; Gorb and Gorb 2002, 2006b; Müller 2006). The impact of plant trichomes on herbivores was previously reported for representatives of several arthropod families (in particular Chrysomelidae, Aphididae, Aleurodidae, Cicadellidae, Sphingidae, Noctuidae, Tetranychidae), based mainly on the presence of trichomes (Lee et al. 1986; Haddad and Hicks 2000), their density (e.g. Fordyce and Agrawal 2001; Kowalewski and Robinson 1978; Zvereva et al. 1998; Rudgers et al. 2004), shape (Andres and Connor 2003), length (Hoxie et al. 1975; Hoffmann and McEvoy 1985; Dalin et al. 2004), and glandular function (e.g. Duffey 1986). Hook-shaped trichomes on *Phaseolus* sp. and *Passiflora* sp., for example, can impale small insects (McKinney 1938; Johnson 1953; Gilbert 1971; Pillemer and Tingey 1978; Gepp 1977). Glandular trichomes, contaminating insects with their sticky exudates, have been reported for *Solanum* spp. (e.g. Gibson 1971; Gibson and Picket 1983; Gregory et al. 1986; Lapointe and Tingey 1986; Avé et al. 1987; Malakar and Tingey 2000), *Lycopersicon* spp. (e.g. McKinney 1938; Johnson 1956; Goffreda et al. 1989; Musetti and Neal 1997; Leite et al. 1999; Ecole et al. 2001; Kennedy 2003; Simmons et al. 2003, 2004), *Nicotiana* spp. (Thurston and Webster 1962; Thurston et al. 1966; Thurston 1970; Goundoudaki et al. 2003), and *Datura* spp. (van Dam and Hare 1998; Hare and Elle 2002; Hare 2005; Hare and Smith 2005). Plant pubescence may affect population dynamics (e.g. abundance, mobility, oviposition, development, mortality), body weight, behaviour, host-plant choice and acceptance, as well as the performance and fitness of phytophagous arthropods (e.g. Ringlund and Everson 1968; Gibson 1971; Hoxie et al. 1975; Benedict et al. 1983; Roberts and Foster 1983; Zvereva et al. 1998; Malakar and Tingey 2000; Valverde et al. 2001; Ranger and Hower 2002; Andres and Connor 2003).

However, smaller and/or specialized (mostly long-legged) species having an ability to attach and live on waxy and hairy plant surfaces, sometimes benefit from predator/

parasitoid-free space, because natural enemies and their influence on regulation of phytophagous arthropods may be negatively affected by plant phenotypic characteristics (Kennedy 1986; Kareiva and Sahakian 1990; Coll 1998; Gassman and Hare 2005).

Such predators as representatives of Coccinellidae, Chrysopidae, and Anthocoridae may be hampered in their mobility on thick wax blooms (Shah 1982; Grevstad and Klepetka 1992; Eigenbrode 1996, 2004; Eigenbrode et al. 1996). For pubescent surfaces, Shanower et al. (1996) found that the longer and denser the trichomes are, the less are the number of pest insects that occur. Moreover, searching and consumption times become longer, and natural enemies are more frequently chemically repelled. The walking speed, searching efficiency, and encounter rate of Coccinellid larvae have been reported to be reduced on hairy plant substrates (Shah 1980, 1982; Belcher and Thurston 1982; Obrycki and Tauber 1984; Carter et al. 1984; Grevstad and Klepetka 1992). The ladybird beetles themselves and their integument may also be harmed particularly by hook-shaped trichomes (Putman 1955; Walters 1974; Dixon 1986; Quilici and Iperti 1986; Obrycki 1986; Eisner et al. 1998). Some species of Chrysopidae are hindered by a highly dense pubescence (Arzet 1972; Elsey 1974; Dos Santos et al. 2003; Simmons and Gurr 2004, 2005). Predacious thrips have been observed to be impaled by hooked trichomes (Sengonca and Gerlach 1984). Parasitoids are frequently deterred and hurt by the complex pubescence of plant surfaces (Rabb and Bradley 1968; Keller 1987; Andow and Prokrym 1990; Kashyap et al. 1991; Jervis and Kidd 1996; Gingras and Boivin 2002). True bug larvae of the family Anthocoridae have been reported to be entangled by hooked and glandular trichomes (Evans 1976; Coll et al. 1997).

Although contact phenomena and mechanical interactions between insect tarsi and plant surfaces are very important in insect–plant interactions, they have been relatively rarely studied in detail. The effect of different characteristics of the plant surface on attachment and locomotion of insects has been reported formerly in Chrysomelidae: *Phaedon chochleariae* Fab. (Stork 1980b), *Chrysolina fastuosa* Scop. (Gorb and Gorb 2002, 2006b), *Gastrophysa viridula* de Geer (Gorb and Gorb 2006a), Coccinellidae: *Adalia bipunctata* L. (Gorb et al. 2005), Calliphoridae: *Calliphora vicina* R.-D. (Gaume et al. 2004; Gorb et al. 2004), Aphididae: *Megoura viciae* Buckt. (Lees and Hardie 1988), several Formicidae (Federle et al. 1997, 2000), and Pyrrhocoridae: *Pyrrhocoris apterus* L. (Gorb et al. 2004). However, pubescent plant surfaces were considered only by Gorb and Gorb (2002) for *C. fastuosa* attachment ability on various plant surfaces, and by Betz (2002), estimating climbing performance of several *Stenus* spp. (Staphylinidae) on the adaxial side of leaves in *Glyceria maxima* (C. Hartm.) Holmb. and *Phragmites communis* Trin. (Poaceae).

Some predatory insect species, such as the palaearctic mirid bug *Dicyphus errans* Wolff (Heteroptera, Miridae, Bryocorinae), prefer hairy plant surfaces as a substrate for living, and for prey capture. Our previous bionomical studies have demonstrated a higher fecundity and predation efficiency, as well as a shorter developmental cycle of the bug on hairy plant surfaces compared to smooth ones (Voigt 2004, 2005). This generalized mirid bug lives omnivorously on various host plants and preys on a wide range of small arthropods. Plants play a crucial role for this bug species as protected habitat, hunting ground, oviposition site as well as a source of water and nutrition. Similar to other representatives of the tribus Dicyphini among the subfamily Brycorinae (Miridae, Heteroptera), *D. errans* seems to be adapted to live on and to adhere to hairy and glandular hairy plant surfaces because of morphological and behavioural adjustments. The dicyphine bugs stalk rather quickly along plant trichomes, using their long and slender legs (Kullenberg 1946; Wheeler 2001; Voigt 2005). Using video-recordings, Southwood (1986) described several modi of locomotion of *D. errans* on a glandular hairy *Ononis* sp. surface.

Springer

The bug's body is held away from the plant surface, whereas the long tibiae are oriented almost vertically and the claw is at an angle parallel to the plant surface. The hind tibiae are especially long, so that the hind femur is situated above body height. Moving one leg, the femur is raised and the tibia moved almost vertically, without contacting any trichome. The mirid bugs have a small contact area with the plant surface and adhere to trichomes only with tarsal tips, the claws, mostly without encountering the lamina between trichomes. One part of the paired claw, supported by adhesive organs (pseudopulvilli and parempodia), may grip precisely one single trichome. These claws are called "*Dicyphus*-like" with large "pseudoarolia" (pseudopulvilli) (Seidenstücker 1967) and narrow "shepherd's crook" (Southwood 1986), and are thought to enable the bug's attachment to trichomes. Wigglesworth (1959) and Jordan (1962) have assumed that adhesive organs enable the dicyphine bugs to attach not only to plants, but to smooth surfaces as well. Although the function of pulvilli and pseudopulvilli in trichome-stalking Heteroptera is not really understood, Wagner (1955) has supposed that true bugs with long, free pseudopulvilli are adapted to live on plant surfaces covered with sticky trichomes.

The goal of this study was to analyse the influence of different plant surfaces, in particular pubescent ones, on the attachment of the predatory mirid bug *D. errans* and the role of biomechanics in the bug–plant relationship.

Three main questions were asked: (1) Does the generalist *D. errans* demonstrate equal attachment ability on different plant surfaces? (2) How do different types of plant surfaces affect the locomotion and attachment force of *D. errans*? (3) Which insect–plant interactions, related to the attachment ability, can be quantitatively analysed, considering the bug tarsal morphology, trichome geometry, and plant surface adhesive properties?

The pretarsal morphology of *D. errans* was microscopically analysed. The adaxial and abaxial leaf surfaces of 40 plant species belonging to 25 families were screened according to the mirid bug attachment strength in an inversion experiment. Furthermore, 14 of these species were used in traction force tests. Leaf surfaces were described and classified into several types. Adhesive properties, trichome length and diameter of the different leaf surfaces used in the traction force test were measured and the density of hairs estimated.

Materials and methods

Insects

Adults of *Dicyphus errans* were taken from a stock culture, reared on the host plant *Plectranthus ambiguus* Bolus and Codd (Lamiaceae) with several arthropod prey species at a temperature of 22 ± 5°C/20 ± 3°C day/night, a relative humidity between 60 and 70%, and a 16 h photoperiod.

Detailed information about the attachment system was obtained from light, fluorescence, and scanning electron microscopy. Fresh tarsi of adult bugs were cut off with a razor blade, mounted on glass slides in polyvenylalcohol (Moviol), covered with glass cover-slips, and observed in a fluorescent microscope Zeiss Axioplan, equipped with HBO 103 mercury vapor lamp and XBO 75 xenon short-arc lamp (Carl Zeiss MicroImaging GmbH, Jena, Germany) as well as an integrated digital video-camera AxioCam MRc (AxioVision GmbH, München-Hallbergmoos, Germany). Fluorescence microscopy was used to localize the elastic protein resilin in the pretarsus. This protein has an auto-fluorescence with an emission maximum at 420 nm (Andersen and Weis-Fogh 1964). Superimposed digital images, obtained at green (excitation 512–546 nm, emission 600–640 nm), red (excitation

710–775 nm, emission 810–890 nm), and UV wavelengths (excitation 340–380 nm, emission 425 nm) were analysed to assess the presence of resilin (Gorb 1999).

Scanning electron microscopy studies were carried out using a cryo-SEM Hitachi S-4800 (Hitachi High-Technologies Corp., Tokyo, Japan) equipped with a Gatan ALTO 2500 cryo-preparation system (Gatan Inc., Abingdon, UK). Fresh samples of mirid bug legs were mounted on metal holders, frozen in the preparation chamber at −140°C, sputter-coated with gold–palladium (3 nm) and examined in a frozen state in the cryo-SEM at 3 kV and −120°C. From digital images of 10 randomly selected pretarsi of adult specimens, the inner length of the claw, diameter of a circle fitting the concavity of the claw, and length of the setiform parempodia were estimated, using Sigma Scan Pro 5 (SPSS Inc.) software. To observe, in detail, how the claw interlocks with a single plant trichome, live insects attached to pubescent plant surfaces were prepared, following the same cryo-SEM procedure as described above.

Plant surfaces

The screening of plant surfaces in the inversion experiment included 40 plant species and cultivars with different surface types, belonging to 25 plant families (30 herbs, 6 lianas, 2 shrubs, 1 tree). Single leaves, located between the top and the middle of the plants, were collected at the Botanical Garden of Dresden University of Technology (Dresden, Germany). To define the type of plant surface, fresh leaves were studied using a binocular microscope Leica MZ 12.5 with integrated digital video-camera Leica ICA (Leica GmbH, Wetzlar, Germany). For detailed surface description, fresh plant samples were cut off with a razor blade, mounted on metal holders, and examined in cryo-SEM as described above for insects.

The abaxial leaf surfaces of 14 plant species (10 families), belonging to different surface types, were tested in the traction force experiment with females of *D. errans*. In hairy surfaces, the trichome length, density per 1 mm^2, and diameter (in the middle part of the trichome) were estimated from digital images and SEM-micrographs using Sigma Scan Pro 5 (SPSS Inc.) software. The values were averaged for all trichome types occurring on one leaf surface, except for peltate and clavate ones, which obviously did not allow the bug to attach according to preliminary observations. The aspect ratio of trichomes was calculated using the estimated mean values of the trichome length and diameter. To assess the trichome density, only the visible trichome bases in randomly selected areas were considered.

Adhesion force measurements of plant surfaces

For 14 abaxial leaf surfaces used in traction force experiment, adhesive properties were estimated in the adhesion force test. The measurements were carried out using a force transducer (10 g capacity, Biopac Systems Ltd., Santa Barbara, CA, USA), attached to a motorised micromanipulator DC3314R with controller MS314 (World Precision Instruments Inc., Sarasota, FL, USA). Fresh plant samples (2 cm^2) were cut out with a razor blade and mounted, with the abaxial side up, on a horizontal plexiglass stage using double-sided tape (Fig. 1). A polished sapphire ball (500 μm diameter corresponding to the total claw length as supposed contact area between mirid bug tarsi and plant surface if all claws are interlocked with trichomes, Good Fellow Cambridge Limited, Huntingdon, UK) was glued to an insect pin (diameter 0.25 mm, original Karlsbader, Germany) with superglue (ergo 5300, Kisling Deutschland GmbH, Schwäbisch Hall, Germany). The sapphire ball was used because it is regularly shaped, smooth, and chemically neutral which gives the possibility to compare adhesive properties of different surfaces. The pin was firmly attached to the force transducer. The sapphire ball was moved continuously with a velocity 20 μm/s, brought in contact with the leaf surface, pre-loaded to maximum 20 μN force, and then withdrawn while the pull-off force was measured. Force-time curves were used to estimate the maximum pull-of (adhesion) force. For each plant surface, 20 measurements on different sites, and 280 single measurements in total were carried out (N = 14 plant species, n = 20 measurements per plant species). Kruskal–Wallis one way ANOVA on ranks followed by all pairwise multiple comparision procedures (Tukey test) was used to evaluate differences in the adhesion force values between plant surfaces (software SigmaStat 3.1.1®, Systat Software, Inc., Richmond, California, USA). The relationship between traction force values and plant surface adhesion was analysed by linear regression method.

Inversion experiment

A bug specimen was placed on the plant surface (adaxial or abaxial leaf side), and the surface then inverted (180°). As shown in Fig. 2, the attachment ability was categorised as follows: the specimen (0) fell immediately after inversion, (1) fell after inversion and slight shaking at a frequency (f) of 2 Hz and an amplitude (A, up and down shaking movements) of 50 mm, (2) fell after inversion and strong shaking (f = 4 Hz, A = 100 mm), and (3) did not fall at all. Shaking was conducted considering the plant movement by wind and other environmental influences to which the bugs may be exposed in nature. In order to allow the specimen to rest between different plant surface trials, the bugs were allowed to walk over dry filter paper for 3 minutes. For each leaf surface, 5 females and 5 males were tested. The sequence of plant species used in the experiment was varied randomly. A total of 800 individual inversions (N = 80, n = 10 adult bugs) were carried out at 23 ± 2°C and

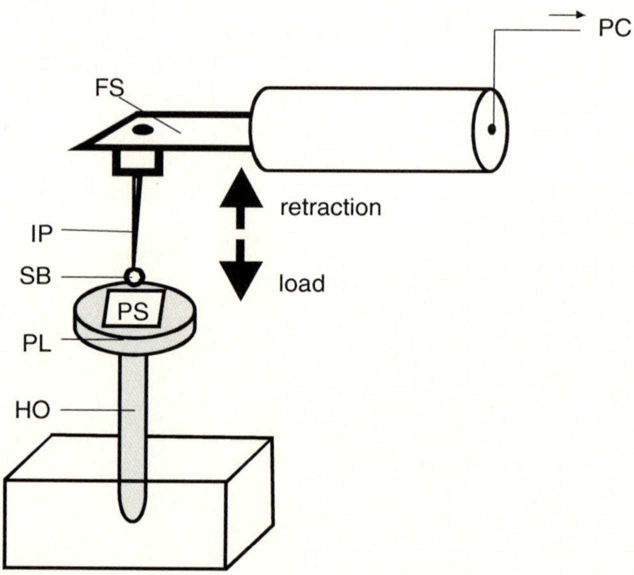

Fig. 1 Experimental setup for estimation of adhesive properties in plant surfaces. The freshly cut plant sample (PS) was attached to a horizontal platform (PL) on a holder (HO). A sapphire ball (SB) was glued to an insect pin (IP), adhered to a force sensor (FS). The sensor with ball was moved with a motorized micromanipulator until contact between the sapphire ball and plant surface was formed at a load of 0.2 mN, and then pulled back. The sensor signal was recorded and processed further in the computer (PC)

55 ± 3% relative humidity. Data were analysed by χ^2 analysis of contingency tables (software SigmaStat 3.1.1®, Systat Software Inc., Richmond, California, USA).

Traction force experiment

Traction force measurements were carried out according to Gorb et al. (2004) using a force transducer (as described above for adhesion experiments). Insects were anaesthetised with CO_2 and attached to a 10 cm long human hair with a small molten wax droplet and their forewings glued together. Bugs were weighed using an analytical balance AG 204 Delta Range (Mettler Toledo GmbH, Greifensee, Switzerland). Following anaesthesia, insects were allowed to recover for one hour, after which the free end of the hair was connected to the force sensor. Fresh plant samples with an area of 5–25 cm^2 were cut out with a razor blade from the middle part of leaves (6-leaf-stage plants) and attached to a horizontal glass plate using double-sided tape. The traction force was measured in *D. errans* females, walking distally on the abaxial leaf surface and on a glass surface, used as a reference substrate before and after walking on the plant surface. Force–time curves were used to estimate the maximal traction force produced by bugs within 30 s of the experiment on the different test substrates. For each plant surface, five individual females were tested.

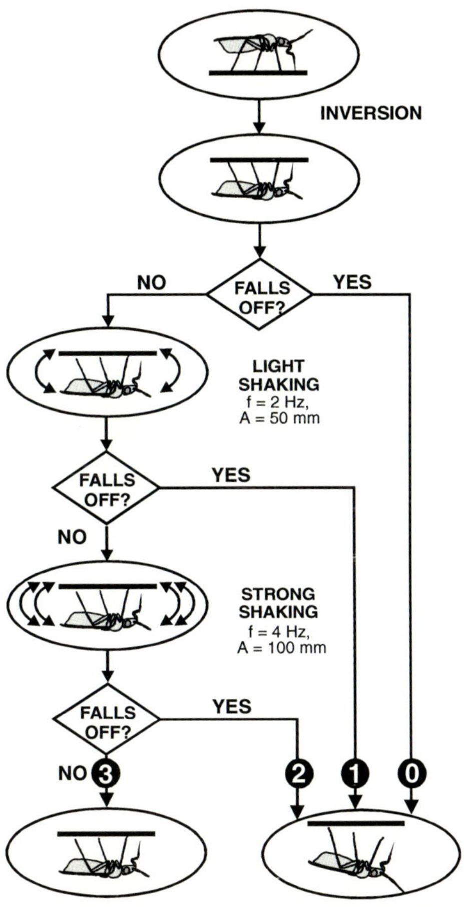

Fig. 2 Scheme of the inversion experiment, carried out on both abaxial and adaxial leaf surfaces of 40 plant species to estimate the attachment ability of *Dicyphus errans*. Four categories of the attachment strength were assigned: (0) fell immediately after inversion, (1) fell after inversion and light shaking (f = 2 Hz, A = 50 mm), (2) fell after inversion and strong shaking (f = 4 Hz, A = 100 mm), and (3) did not fall at all

Totally, 70 individual tests with 70 different specimens were carried out (N = 14 plant species, n = 5 individual tests). One way ANOVA followed by all pair-wise multiple comparision procedures (Tukey test) was applied to find out differences in force between plant surfaces (software SigmaStat 3.1.1®, Systat Software, Inc., Richmond, California, USA). The relationship between absolute force values and pubescent plant surface parameters was analysed using a linear regression method. Paired *t*-test was used to calculate differences in the traction force on glass before and after plant surface trials. A posteriori, traction force measurements on glass allowed an estimation of possible tarsus contamination during walking on plant surfaces. To visualize contamination of the bug's attachment system, feet of randomly selected specimens were cut off with a razor blade immediately after testing on the plant, without allowing the specimen to groom. These were mounted on holders by means of conductive carbon double-sided adhesive tape, air-dried, sputter-coated with gold-palladium (10 nm) and observed in the SEM at 3 kV and room temperature. The degree of tarsus contamination was estimated from SEM-micrographs as proportion of contaminated area to the entire pretarsus area, averaged for 5 randomly selected pretarsi for each tested plant surface.

Results

The attachment system of *Dicyphus errans*

The inner length of the claw (n = 10), without considering the basal lobe-like structure, was 74.0 ± 10.9 μm (claw tooth length: 2.5 ± 0.7 μm). The parempodia were 38.7 ± 5.6 μm long. The average diameter of an ideal circle fitting the concavity of the claw was 16.6 ± 0.9 μm.

The fluorescence microscopy images (Fig. 3A, B) showed the presence of resilin, an elastic protein, in the material of pseudopulvilli and membraneous areas surrounding the unguitractor plate, especially in their transitional region to the claw base. As shown in the cryo-SEM micrographs (Fig. 3C–E), the pretarsus of *D. errans* consists of an unguitractor plate bearing paired, narrow, elongated claws (ungues) with a basal lobe-like structure and a tapered tooth at the tip, two approximately equilateral trapezium-shaped rounded pseudopulvilli as well as two thin setiform parempodia. Dorsally, a stronger seta arises from the third tarsomer. The claws are strongly curved close to their basal lobe-like thickening and slightly curved only distally, while the shaft is almost straight. The claw surface has longitudinal groove-like striae, and is somewhat chiseled, polygonal in the cross section. The ventral side of the pseudopulvilli is smooth (Fig. 3D), whereas the dorsal side is structured with nearly parallel-oriented grooves, 0.5 μm apart, ending at about 1 μm to the distal margin of the pseudopulvilli (Fig. 3C).

According to ocular and cryo-SEM observations of bug locomotion on *Nicotiana sylvestris* (Fig. 3F), a single claw of the pair present on the tarsus, interlocks with a single trichome, using the border between the basal lobe-like structure and the ventral surface of the claw shaft that apparently leaves an imprint in the trichome wall (Fig. 3F, arrowhead). The use of the pseudopulvilli and parempodia in claw interlocking with trichomes was not observed.

Plant surface characterisation

A great variety of plant surfaces studied requires systematic description (Appendix) and classification following the parameters defined in Fig. 4. A plant surface may be generally

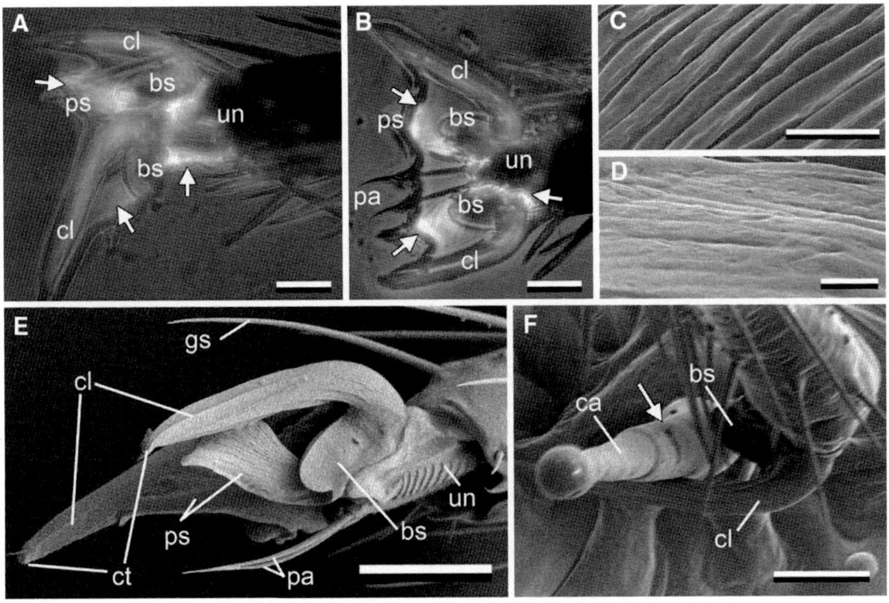

Fig. 3 Attachment system of *Dicyphus errans*. The pretarsus bears an unguitractor plate (*un*), paired curved claws (*cl*) each with a claw tooth (*ct*), and basal lobe-like structure (*bs*), pseudopulvilli (*ps*), setiform parempodia (*pa*), stronger, so called guard setae (*gs*). (**A**, **B**) Shiny blue areas in fluorescence light microscopy images demonstrate the presence of resilin in the pseudopulvilli and the membranous areas surrounding unguitractor plate (arrows); dorso-lateral view (**A**), dorsal view (**B**). (**C**, **D**) Cryo-SEM micrographs showing the surface structure of the dorsal (**C**) and ventral sides (**D**) of the pseudopulvillus. (**E**) The whole pretarsus. (**F**) The claw interlocking with a glandular capitate trichome (*ca*) of *Nicotiana sylvestris;* arrow points to a transverse imprint of the trichome stem, possibly caused by bug claw attachment. Scale bars: **A**, **B**, **E**, **F** = 25 μm; **C**, **D** = 2 μm

smooth, waxy or hairy, however, combinations of these features frequently occur. Various structures can even be found on the same plant organ (Fig. 4A). We distinguished six types of plant surfaces in this study (Fig. 8): (1) smooth (*s*), (2) waxy (*w*), (3) mixed waxy and hairy (*wh*), (4) hairy (*h*), (5) glandular hairy (*g*), as well as (6) mixed hairy and glandular hairy (*hg*). *Smooth* plant surfaces (*s*) lack both wax crystals and trichomes. Their surface topography is determined only by cell irregularities and leaf veins. *Waxy* plant surfaces (*w*) bear epicuticular crystalline waxes, defined in our description according to the terminology of Barthlott et al. (1998). Some surfaces have both types of surface structures: *hairs and wax* crystals (*wh*). *Hairy* surfaces (*h*) are defined as those, covered with non-glandular trichomes, whereas *glandular hairy* surfaces (*g*) bear glandular trichomes. Some surfaces bear both these trichome types (*hg*). Most of the species studied belong to the mixed non-glandular hairy and glandular hairy type (*hg*). 13 species have this surface type on both sides of the leaf; 17 or 16 species have it on either the adaxial or abaxial surface, respectively. Eleven species were covered with epicuticular wax crystals (*w*) on the adaxial and 6 species—on the abaxial leaf surface. Several species bear waxes on both sides of the leaf (*B. oleracea*, *B. semperflorens*, *C. persicum*, *S. hoffmanii*, *L. sativa*, and *V. tricolor*). Three species' adaxial leaf sides and five species' abaxial surfaces were covered with both crystalline waxes and trichomes (*wh*). The hairy plant surface type (*h*) as well as the glandular hairy one (*hg*) are each represented in 4 species adaxially and 6 species abaxially. Only *S. media* has no outgrowths on either side of the leaf (*s*). At the family level, no characteristic surface type is distinguished. For example, representatives of Asteraceae examined bear epicuticular crystalline waxes (*L. sativa*), a combination of epicuticular crystalline waxes and either non-glandular (*G. jamesonii*) or glandular trichomes (*C. officinale*).

Although many attempts have been undertaken by previous authors to classify a great variety of trichomes (e.g. Hummel and Staesche 1962; Uphof 1962; Esau 1965; Napp and Zinn 1973; Johnson 1975; Payne 1978; Metcalfe and Chalk 1988), there is no generally

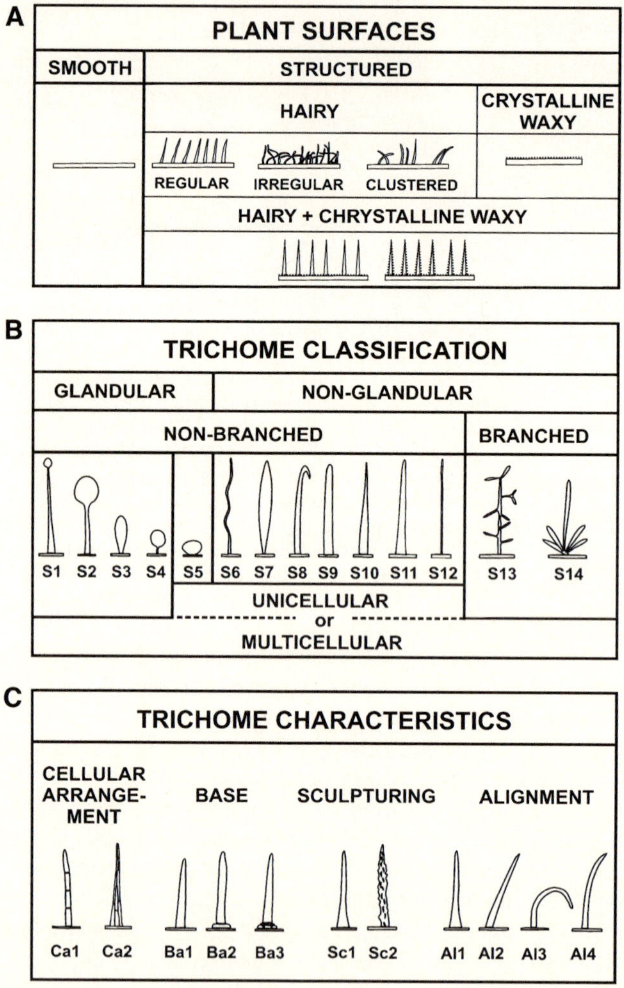

Fig. 4 (**A**, **B**) Diagrams of plant surfaces and trichome types found on the tested leaves. The diagrams are based on the trichome arrangement, cellular arrangement in the single trichome, its function, and shape: (*S1*) cone-shaped with glandular head; (*S2*) capitate; (*S3*) clavate, club-shaped; (*S4*) stalked glandular heads; (*S5*) peltate; (*S6*) spiral-shaped; (*S7*) spindle-shaped; (*S8*) hook-shaped; (*S9*) finger-shaped; (*S10*) dart-shaped; (*S11*) cone-shaped; (*S12*) thread-shaped; (*S13*) tree-shaped, dendroid; (*S14*) star-shaped, stellate. (**C**) Diagram showing the characteristics of trichomes found on the tested leaf surfaces. The diagram is based on the trichome alignment, surface texture, base structure and cellular arrangement. (*Al1*) perpendicular; (*Al2*) inclined; (*Al3*) curved; (*Al4*) bent; (*Sc1*) smooth; (*Sc2*) nodose; (*Ba1*) without base/socket; (*Ba2*) unicellular socket; (*Ba3*) multi-cellular socket; (*Ca1*) uniseriate; (*Ca2*) multiseriate

accepted terminology. To describe the trichomes found on the plants used in our study, their indumentum, cellular arrangement, function, branching, shape, surface sculpturing, base, and orientation were considered (Fig. 4).

Trichomes occur regularly or irregularly, sparsely, dispersed, numerously or densely. They may be uni- or multi-cellular. The latter may consist of either one (uniseriate, Fig. 4C, *Ca1*) or two and more rows of cells (multiseriate, Fig. 4C, *Ca2*). We also distinguished non-glandular and glandular trichomes. Glandular trichomes contain secretions, often accumulated in rounded heads. Capitate glandular trichomes (Fig. 4B, *S1* and *S2*) bear conspicuous uni- or multicellular stalks having glandular heads on their tips (diameter of the head is smaller than the stalk length). Peltate glandular trichomes (Fig. 4B, *S5*) are uni- or multicellular globular glandular heads. Stalked glandular heads (Fig. 4B, *S4*) have very short stalks, the length of which is smaller than the diameter of their heads. Clavate glandular trichomes (Fig. 4B, *S3*) are club-shaped and mainly multi-cellular. Non-glandular trichomes can be distinguished according to their various shapes. Besides the cone-shaped ones (Fig. 4B, *S11*), frequently occurring in plant surfaces studied, trichomes may have shapes similar to spirals (Fig. 4B, *S6*), threads (Fig. 4B, *S12*), hooks (Fig. 4B, *S8*), fingers (Fig. 4B, *S9*), darts (Fig. 4B, *S10*) or spindles (Fig. 4B, *S7*). Trichomes are usually aligned in one preferred direction. Some of them form a right angle with the leaf surface (perpendicular orientation) (Fig. 4C, *Al1*). Inclined trichomes are sloped (Fig. 4C, *Al2*), bent trichomes are slightly arcuated (Fig. 4C, *Al4*), whereas curved ones have a stronger deflection, up to 180° and more (Fig. 4C, *Al3*). The trichome surface may be smooth (Fig. 4C, *Sc1*) or rough at the microscale level with nodose knobby irregular outgrowths (protuberances) (Fig. 4C, *Sc2*). Trichomes may emerge only from the leaf epidermis (Fig. 4C, *Ba1*), or be based on a mostly convex uni- (Fig. 4C, *Ba2*) or multicellular socket (Fig. 4C, *Ba3*). A detailed description of studied plant surfaces is given in the Appendix and Figs. 5–7.

The quantitative analyses of trichomes on 14 abaxial leaf surfaces used in traction force tests, showed that trichomes differ strongly in their shape, anatomy, orientation, length, diameter, density, and aspect ratio (Table 1). Even on one single surface, trichomes may be very heterogeneous. The averaged trichome length ranged from 262.1 μm in *C. annuum* to 650.4 μm in *P. ambiguus*. The thinnest trichomes were found in *F.* × hybrida (15.4 μm), and the thickest ones in *C. hirsuticaulis* (87.7 μm). On *F.* × hybrida, the lowest density of trichomes was recorded (2.6 mm^{-2}), and in the domatia of *C. annuum* the highest (70.6 mm^{-2}). The trichomes of *C. annuum* and *C. hirsuticaulis* had the lowest aspect ratio (6.1 and 7.3, respectively), whereas those of *F.* × hybrida and *Gerbera jamesonii* × hybrida had the highest (22.6 and 16.1, respectively).

The values of measured adhesion force varied in different plant species (Table 1). The highest force was recorded in *N. sylvestris* (median: 207.0 μN), and the lowest one in *G. jamesonii* (median: 4.5 μN) (Kruskal–Wallis one way ANOVA on ranks, Tukey Test, $H_{13,279} = 143.989$, $P \leq 0.001$). Referring to plant surface types, the adhesion force on hairy surfaces (*h*, median: 55.2 μN), glandular ones (*g*, median: 43.5 μN), as well as non-glandular and glandular hairy ones (*hg*, median: 33.2 μN) were significantly higher than on surfaces covered with wax crystals (*w*, median: 5.1 μN) and both wax crystals and trichomes (*wh*, median: 4.5 μN) (Kruskal–Wallis one way ANOVA on ranks, $H_{5,279} = 82.399$, $P \leq 0.001$). Further significant differences were found between *h* and *hg* surfaces.

Inversion experiment

Bug attachment to plant surface varied, depending on the plant species, surface type, leaf side, and insect sex (Fig. 8, Table 2). The attachment ability did not depend on the plant

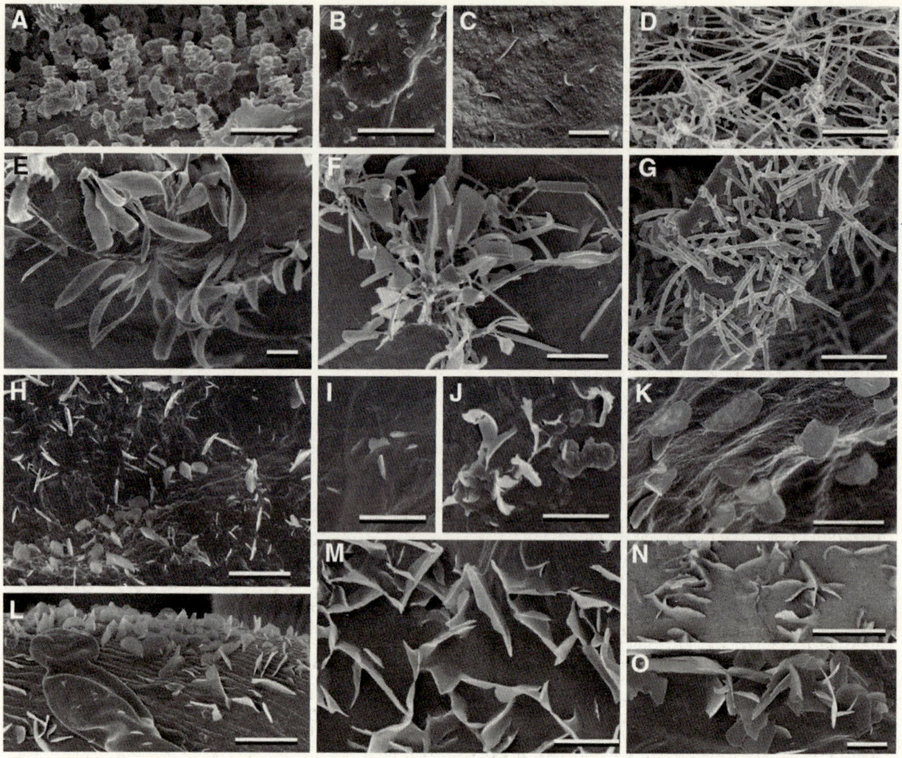

Fig. 5 Cryo-SEM micrographs of plant surfaces covered with epicuticular crystalline waxes used in the experiments with *Dicyphus errans*. (**A**) *Aristolochia elegans*, abaxial surface covered with transversely ridged rodlets. (**B**) *Begonia semperflorens*, abaxial surface, polygonal granules. (**C**) *Begonia semperflorens*, adaxial surface, membraneous entire platelets. (**D**) *Brassica oleracea*, adaxial surface, filamentous crystals. (**E**) *Lactuca sativa*, adaxial surface, rosettes of membraneous platelets. (**F**) *Hibiscus rosa-chinensis*, adaxial side, clusters of platelets. (**G**) *Tropaeolum majus*, abaxial side, tubules. (**H**) *Fuchsia* × hybrida, adaxial side, platelets. (**I**) *Capsicum annuum*, adaxial side, group of platelets. (**J**) *Rosa* × hybrida, adaxial side, platelets. (**K**) *Cyclamen persicum*, abaxial side, plates. (**L**) *Malus domestica*, abaxial side, cluster of platelets on the trichome. (**M**) *Malus domestica*, abaxial side, irregular membraneous platelets. (**N**) *Syngonium auritum*, abaxial surface irregular membraneous platelets. (**O**) *Viola tricolor*, adaxial side, grouped polygonal plates surrounding stomata. Scale bar = 2 μm

family. *D. errans* attached strongly (categories 2 and 3) to the adaxial leaf surfaces of 17 plant species, and to the abaxial leaf surfaces of 22 plant species. Generally, bugs attached better to pubescent plant substrates.

Significantly different attachment ability of bugs was shown for different plant surface types (Table 2). Surfaces covered with crystalline wax (*w*) strongly impeded attachment: in 91.7% of tests, the bug fell down immediately after inversion (category 0) or after inversion followed by light shaking (category 1). On surfaces with a combination of crystalline waxes and trichomes (*wh*), the attachment success was greater (e.g. *G. jamsonii* and the abaxial leaf side of *M. domestica*) than on waxy surfaces (*w*) without trichomes. As shown in Table 2, on the *wh* surface, categories 2 and 3 were significantly larger (category 2 + 3: 40%) than on the *w* surface (category 2 + 3: 8.3%) (χ^2 analysis of contingency tables, $\chi^2_{3,799} = 435.103$, $P \leq 0.001$). On the smooth surface (*s*) of *S. media*, bugs attached

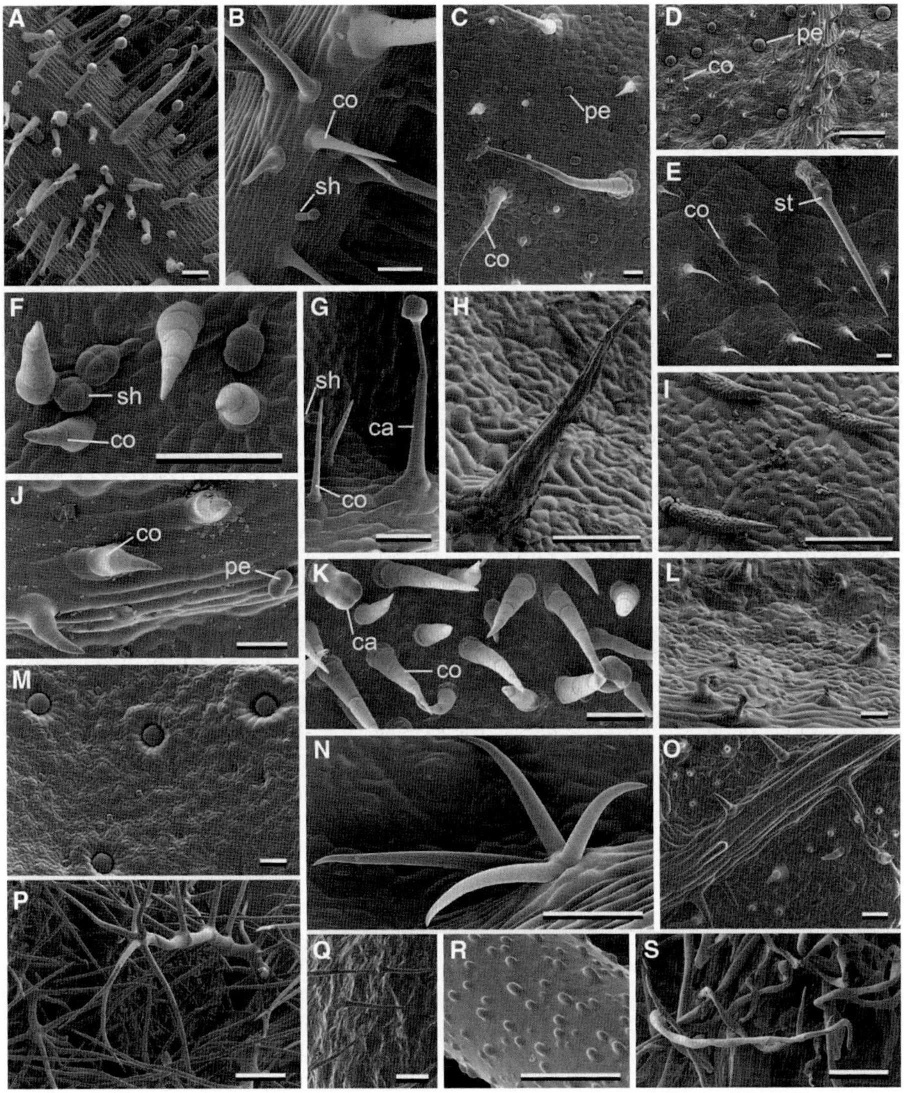

Fig. 6 Cryo-SEM micrographs of pubescent leaf surfaces used in the inversion test with *Dicyphus errans*. (**A**) *Datura innoxia*, abaxial side, capitate glandular trichomes. (**B**) *Borago officinalis*, abaxial side. (**C**) *Cissus njegerre*, adaxial side. (**D**) *Cissus njegerre*, abaxial side. (**E**) *Urtica dioica*, adaxial side. (**F**) *Lagenaria siceria*, adaxial side. (**G**) *Pelargonium zonale*, abaxial side. (**H**) *Cuphea lanceolata*, abaxial side, multiseriate trichome. (**I**) *Cuphea lanceolata*, adaxial side, spindle-shaped, procumbent, nodose trichomes. (**J**) *Ocimum basilicum*, adaxial side. (**K**) *Lagenaria siceria*, abaxial side. (**L**) *Calendula officinalis* 'Orangestrahlen', abaxial side, capitate trichomes. (**M**) *Ocimum basilicum*, abaxial side, peltate trichomes. (**N**) *Hibiscus rosa-chinensis*, abaxial side, star-shaped trichome. (**O**) *Tropaeolum majus*, abaxial side, cone-shaped trichomes and crystalline waxy coverage. (**P**) *Verbascum thapsus*, abaxial side, felt-like indumentum with tree-shaped trichomes. (**Q**) *Oenothera biennis*, adaxial side, cone-shaped trichomes. (**R**) *Oenothera biennis*, adaxial side, knobby surface, detail of a nodose trichome. (**S**) *Malus domestica*, abaxial side, thread-shaped and spiral-shaped, twisted trichomes. *co*, cone-shaped trichome; *sh*, stalked glandular head; *pe*, peltate trichome; *st*, stinging trichome; *ca*, capitate trichome. Scale bars: **A–Q**, **S** = 100 μm; **R** = 10 μm

Springer

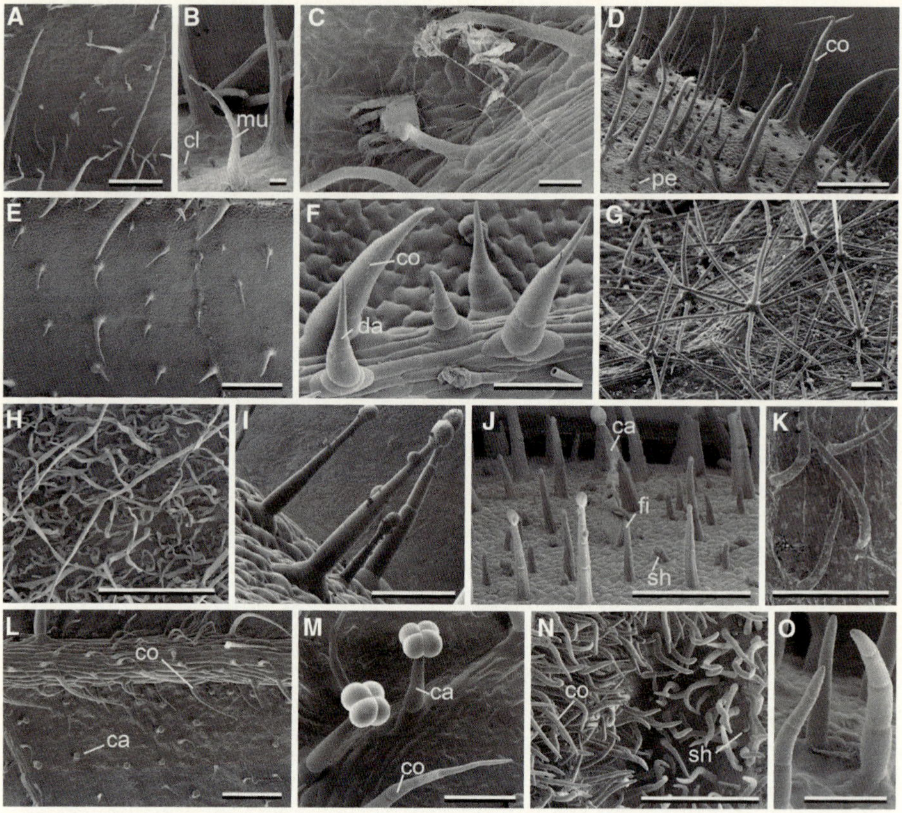

Fig. 7 Cryo-SEM micrographs of the abaxial leaf surfaces of pubescent plants used in the traction force test with *Dicyphus errans* females. (**A**) *Cleome hirsuticaulis.* (**B**) *Cleome hirsuticaulis*, multi-cellular and clavate trichomes in detail. (**C**) *Capsicum annuum* 'Yolo Wonder', finger-shaped trichomes. (**D**) *Plectranthus ambiguus.* (**E**) *Cucumis sativus* L. 'Rawa', cone-shaped trichomes marginally orientated. (**F**) *Cucumis sativus* 'Rawa'. (**G**) *Solanum melongena* L. 'Black Beauty', star-shaped trichomes. (**H**) *Gerbera jamesonii* × hybrida, felt-like indumentum. (**I**) *Petunia* × hybrida, capitate trichomes. (**J**) *Nicotiana sylvestris.* (**K**) *Fuchsia* × hybrida 'Winston Churchill', cone-shaped trichomes. (**L**) *Lycopersicon esculentum* 'Grit'. (**M**) *Lycopersicon esculentum*, capitate and cone-shaped trichomes. (**N**) *Digitalis purpurea* 'Excelsior'. (**O**) *Digitalis purpurea*, finger-shaped trichome in detail. *ca*, capitate; *cl*, clavate; *co*, cone-shaped; *da*, dart-shaped; *fi*, finger-shaped; *mu*, multi-cellular; *pe*, peltate; *sh*, stalked glandular head. Scale bars: **A**, **D**, **E**, **H**, **J**, **L**, **N** = 500 μm; **B**, **C**, **F**, **G**, **I**, **K**, **M**, **O** = 100 μm

in 70% of the trials in category 2 and 30% in category 3. Except for the abaxial side of *C. annum* and *F.* × hybrida, as well as the adaxial side of *B. officinalis* and *O. basilicum*, inversions of the bug on surface types *hg*, *h*, and *g* resulted in the highest frequency of attachment in categories 2 and 3. On the majority of *hg* surfaces, bugs attached most frequently in category 2 (31.5%) and 3 (45.8%).

Females (61.3%) attached significantly better than males (58.0%) (χ^2 analysis of contingency tables, $\chi^2_{3,799} = 9.041$, $P \leq 0.029$).

The frequency of the optimum attachment (category 3) was significantly higher on the abaxial (41.5%) than on the adaxial (27.5%) leaf surfaces (χ^2 analysis of contingency tables, $\chi^2_{3,799} = 20.891$, $P \leq 0.001$).

Table 1 Trichome length, diameter, density per unit area, aspect ratio and measured adhesion force of 14 plant surfaces. Different letters in the right column indicate statistically significant differences between values (Kruskal–Wallis one way ANOVA on ranks, $H_{13,259} = 143.989$, $P \leq 0.001$ and Tukey test, $P < 0.05$)

No.	Plant species, cultivar	Trichome characteristics				Adhesion force [μN] median, signific., min.-max.
		Length [μm] mean ± sd (n)[1]	Diameter [μm] mean ± sd (n)[1]	Density [mm^{-2}] mean ± sd (n)[2]	Aspect ratio	
1	*Brassica oleracea var. gongylodes* 'Gigant'	–	–	–	–	6.6 [ad] (3.1–61.1)
2	*Capsicum annuum* 'Yolo Wonder'	262.1 ± 311.3 (38)	43.2 ± 3.7 (12)	70.6 ± 0.6 (5)	6.1	36.7 [bcde] (7.0–131.0)
3	*Cleome hirsuticaulis*	643.7 ± 254.3 (57)	87.7 ± 130.2 (18)	4.2 ± 1.7 (5)	7.3	83.4 [bc] (4.9–303.0)
4	*Cucumis sativus* 'Rawa'	439.5 ± 196.4 (150)	38.2 ± 13.3 (42)	7.0 ± 0.7 (5)	11.5	85.8 [bc] (4.4–192.0)
5	*Cyclamen persicum* 'Sierra F1'	–	–	–	–	4.7 [a] (1.9–321.0)
6	*Digitalis purpurea* 'Excelsior'	416.0 ± 143.4 (470)	40.4 ± 38.0 (15)	58.0 ± 2.6 (5)	10.3	16.2 [ac] (4.3–44.8)
7	*Fuchsia* x hybrida 'Winston Churchill'	348.1 ± 373.0 (60)	15.4 ± 7.7 (7)	2.6 ± 0.6 (5)	22.6	10.0 [ac] (0.9–317.0)
8	*Gerbera jamesonii* × hybrida	415.6 ± 260.3 (16)	25.8 ± 24.5 (27)	12.0 ± 0.8 (5)	16.1	4.5 [a] (2.8–205.0)
9	*Ibicella lutea*	399.7 ± 233.7 (36)	37.9 ± 14.0 (24)	17.4 ± 3.9 (5)	10.5	41.0 [bcd] (6.4–401.0)
10	*Lycopersicon esculentum* 'Grit'	550.6 ± 880.3 (328)	74.6 ± 30.4 (22)	40.2 ± 8.2 (5)	7.4	36.4 [bcd] (6.3–183.0)
11	*Nicotiana sylvestris*	323.8 ± 100.3 (100)	36.9 ± 9.5 (50)	25.6 ± 4.0 (5)	8.8	207.0 [b] (107.0–510.0)
12	*Petunia* × hybrida	355.3 ± 155.7 (218)	44.3 ± 13.5 (10)	13.1 ± 6.4 (5)	8.0	137.0 [b] (28.3–289.0)
13	*Plectranthus ambiguus*	650.4 ± 311.7 (338)	81.3 ± 98.9 (105)	8.1 ± 1.4 (5)	8.0	5.6 [ae] (1.4–47.7)
14	*Solanum melongena* 'Black Beauty'	468.0 ± 164.5 (175)	39.5 ± 7.7 (24)	3.8 ± 1.4 (5)	11,8	48.1 [bc] (26.4–165.0)

1) n, sample size of trichomes measured

2) n, sample size of areas with counted trichomes

Traction force experiment

The median force values varied considerably from 0.07 mN to 1.3 mN, depending on the plant substrate (Fig. 9). These values corresponded to between 1.8 and 34.2 times of the bug's body weight (3.8 ± 0.6 mg). The lowest force was measured on *B. oleracea* var. *gongylodes*, densely covered with wax crystals, whereas the highest force was observed on *D. purpurea* having a mixed hairy and glandular hairy surface, and on the hairy surface of *S. melongena*.

Bugs performed better on pubescent surfaces than on waxy ones. Their force was significantly stronger on *L. esculentum*, *D. purpurea*, *P. ambiguus*, and *S. melongena* (Kruskal–Wallis one way ANOVA on ranks, $F_{14,149} = 105.095$, $P \leq 0.001$). The highest

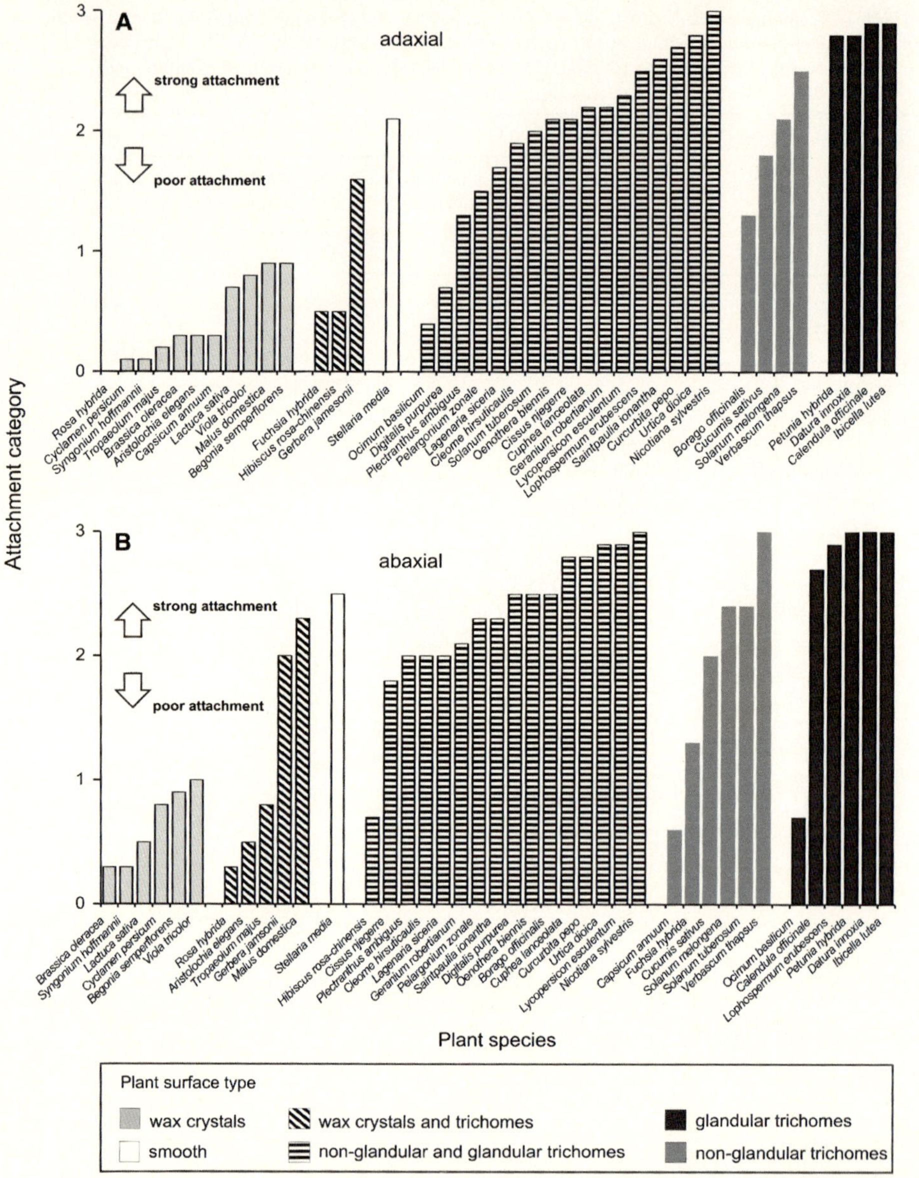

Fig. 8 Attachment ability of *Dicyphus errans* imagines in inversion tests with 40 plant species belonging to 25 plant families and 6 plant surface types; (**A**) Adaxial leaf side. (**B**) Abaxial side. (0) fell immediately after inverting, (1) fell after inverting and slight shaking (f = 2 Hz, A = 50 mm), (2) fell after inverting and strong shaking (f = 4 Hz, A = 100 mm), and (3) did not fall at all. (Mean scores, sample size: N = 10 adult *D. errans*, n = 5 repetitions per leaf surface)

traction force was generated on *hg* substrates, followed by *wh*, *g*, and *h*. There was no significant difference between plants covered with trichomes and no correlation was found between measured force and the plant family. Qualitative properties of trichomes (shape, orientation, alignment etc.) did not influence the traction force, however, leaves covered

Table 2 Relative [%] and absolute (in parenthesis) frequencies of attachment ability categories of *Dicyphus errans* obtained in the inversion tests for various plant surface types, sex, and leaf side, as well as statistical differences between compared groups (χ^2 Analysis of Contingency Tables)

Compared groups	Frequencies of categories of attachment ability				Statistics
	0	1	2	3	
Plant surface types					
smooth (*s*)	0.0 (0)	0.0 (0)	70.0 (14)	30.0 (6)	$\chi^2_{3,799} = 435.103$, $P \leq 0.001$
waxy (*w*)	60.0 (108)	31.7 (57)	8.3 (15)	0.0 (0)	
waxy + hairy (*wh*)	32.9 (23)	27.1 (19)	30.0 (35)	10.0 (13)	
hairy + glandular hairy (*hg*)	7.9 (26)	14.8 (49)	31.5 (104)	45.8 (151)	
hairy (h)	8.9 (8)	24.4 (22)	40.0 (36)	26.7 (24)	
glandular hairy (*hg*)	4.5 (5)	5.5 (6)	10.0 (11)	80.0 (88)	
Sex					
female	18.0 (72)	20.8 (83)	23.3 (93)	38.0 (152)	$\chi^2_{3,799} = 9.041$, $P = 0.029$
male	24.5 (98)	17.5 (70)	27.0 (108)	31.0 (124)	
Leaf surface					
adaxial	26.0 (104)	20.5 (82)	26.0 (104)	27.5 (110)	$\chi^2_{3,799} = 20.891$, $P \leq 0.001$
abaxial	16.5 (66)	17.8 (71)	24.3 (97)	41.5 (166)	

(0) fell immediately after inversion, (1) fell after inversion and light shaking (f = 2 Hz, A = 50 mm), (2) fell after inversion and strong shaking (f = 4 Hz, A = 100 mm), and (3) did not fall at all

with felt-like trichomes in *G. jamesonii* and *S. melongena* strongly enhanced the bugs' traction.

Considering the trichome morphometrical data, there was a significant correlation between the absolute traction force and (1) the length and (2) the diameter of trichomes (Fig. 10A, B) (linear regression, one way ANOVA). However, there was no correlation between the force and both trichome density and aspect ratio (Fig. 10C, D).

The comparision of traction experiment results and adhesive properties of pubescent (*h*, *hg*, and *g*) plants (see section "Plant surface characterisation") did not show any influence of plant adhesion on bug traction force (linear regression and one way ANOVA, Fig. 11A). However, if we consider only plant species from the *g* and *hg* surface types, a significantly negative correlation between the traction force and plant adhesion force became apparent (linear regression and one way ANOVA, Fig. 11B).

As shown in Table 3, the comparision of the traction force measured on the glass surface before and after testing on the plant substrate did not show pretarsus failure. In some cases, the force values on the glass were enhanced after the plant surface test (*C. hirsuticaulis*, *F.* × hybrida, *I. lutea*, *S. melongena*), although no clear trend could be recognized. A significant decrease of the traction force from 0.12 to 0.09 mN after walking on the leaf was observed only on the waxy *B. oleracea* (paired *t*-test, $T_{4,14} = 4.517$, $P = 0.011$). Bug median traction force on the glass surface before testing on the plant surface was 0.23 mN (6.1 times of body weight). SEM-micrographs demonstrate the degree of pretarsi contamination after walking on a plant surface (Fig. 12, Table 3). It ranged widely from 0 to 90% of the entire pretarsus area. The largest contaminated areas were found after walking on leaves of *B. oleracea* covered with wax crystals (60–90%). After walking on pubescent plant surfaces (*h*, *g*, *hg*) the contamination degree ranged between 0 and 80%.

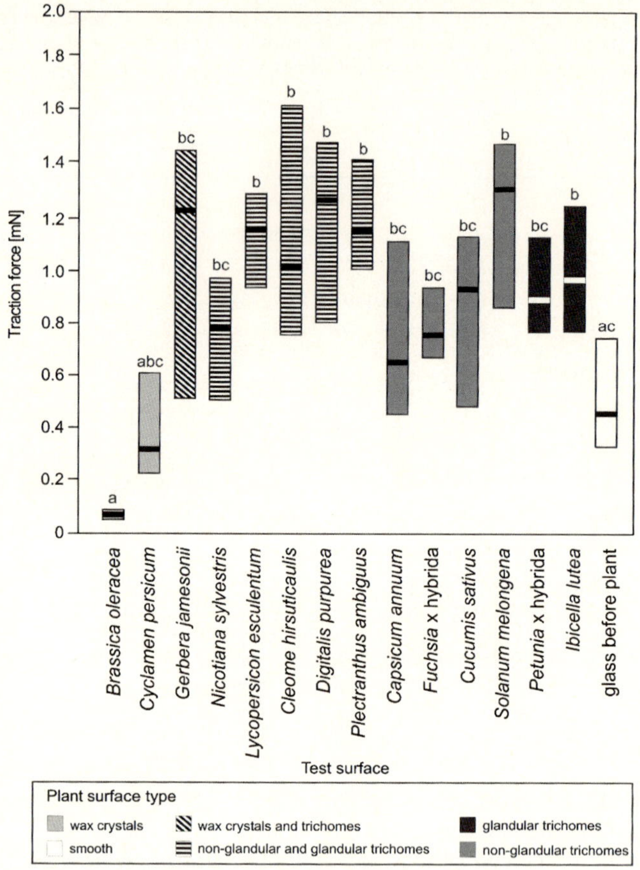

Fig. 9 Traction force generated by *Dicyphus errans* females on different plant species and surface types (median values are indicated by bold crossline). Different letters show significant differences between traction force values on different plant and glass surfaces (Kruskal–Wallis one way ANOVA on Ranks, $H_{14,149} = 105.095$, $P \leq 0.01$ and Dunn's method, $P < 0.05$)

Discussion

The attachment system of *Dicyphus errans*

Morphological and behavioural differences in gripping smooth and pubescent plant surfaces have been previously shown in mirid bugs (Wheeler 2001). There are two basic strategies for adhereing to surfaces. On the one hand, curved elongated claws may interlock with rough substrates and plant trichomes. On the other hand, flexible pseudopulvilli enable mirid bugs to also attach to smooth surfaces (Wigglesworth 1959; Schuh 1976; Jordan 1962). *D. errans* bears an attachment system of the smooth type according to Gorb and Beutel (2001) and Gorb (2001). Structures found on the dorsal side of pseudopulvilli in *D. errans* (Fig. 3C) were previously reported for true bugs to be used to transport an adhesive secretion to the smooth ventral side of pulvilli (Ghazi-Bayat 1979; Ghazi-Bayat and Hasenfuß 1980, 1981). The secretion is essential in adhesion enhancement (Schuh and Slater 1995). While attaching to smooth surfaces, the claws are held at a very small angle to

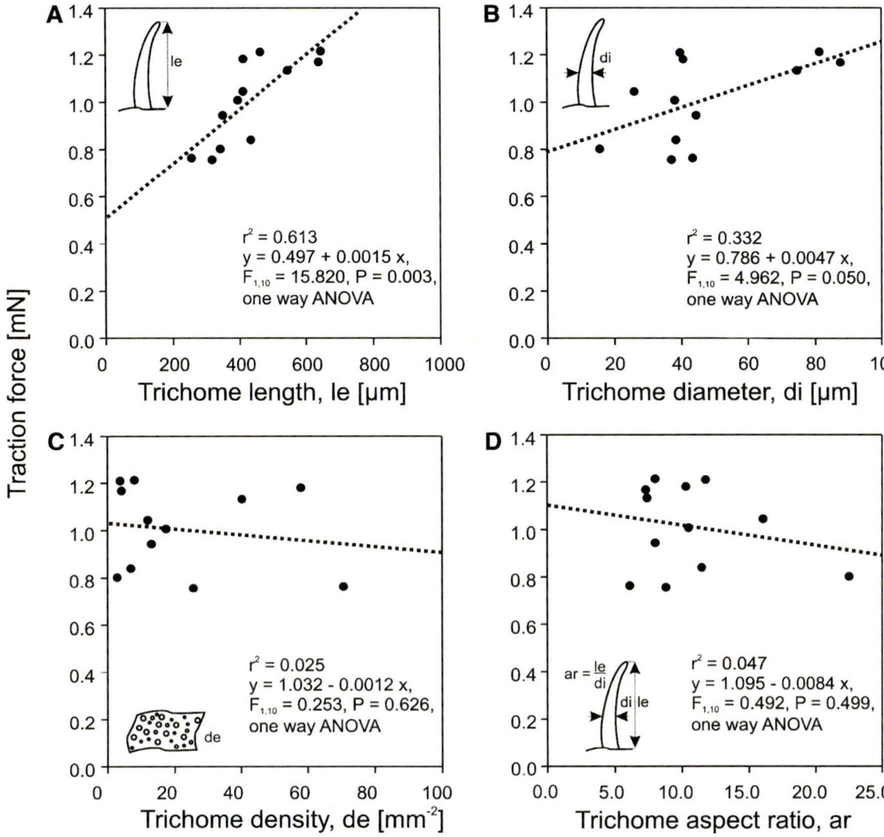

Fig. 10 Traction force versus morphometrical variables of trichomes. Mean force values versus the averaged trichome length (**A**), diameter (**B**), density (**C**), and aspect ratio (**D**). Dotted lines indicate linear regressions

the ventral side of the legs and the paired claws forced apart. Thus, the ventral side of pseudopulvilli can come into contact with a substrate and peel off using the claws as a kind of lever. The therefore required flexibility of these structures is indicated by the presence of resilin in pseudopulvilli and in the membranous cuticle surrounding the unguitractor plate (Fig. 3A, B). Resilin has been previously reported in the tarsi of cockroaches (Frazier et al. 1999), ants (Federle et al. 2001), flies (Niederegger and Gorb 2003), wasps (Frantsevich and Gorb 2004), and several other insect species (Gorb 1996). The elastic structural protein, resilin, is suggested to be responsible for the storage of elastic energy in joints, tendons, wings and other parts of the insect body (Anderson and Weis-Fogh 1964).

While adhering to a trichome, one part of the paired claws obviously interlocks with the middle region of the trichome, using the border of the basal claw plate and the ventro-distal claw surface. Four hypothesis of mirid bug interlocking mechanism on trichomes are proposed: (1) hooking the chiseled-sharpened ventro-distal claw surface and claw tooth to the trichome, (2) clinging to the trichome using the ventro-proximal surface and margins of the basal lobe-like thickening of the claw, (3) gripping the trichome using the ventral claw surface and pseudopulvillus (Southwood 1986), (4) surrounding the trichome with the claw concavity without using either the claw tooth or sharpened claw regions. An extremely

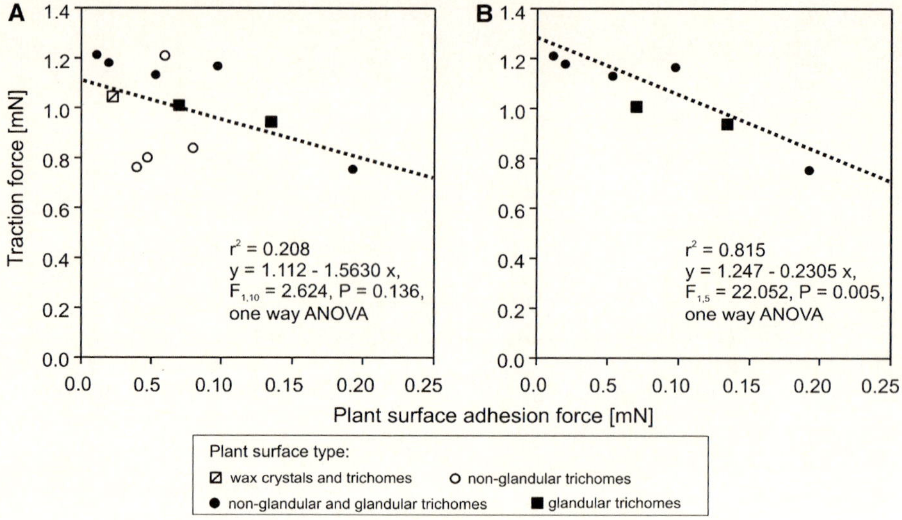

Fig. 11 Mean bug traction force versus mean plant surface adhesion force. (**A**) Data pooled together for all 12 pubescent plant species used. (**B**) Data for plant species with glandular trichomes or with a mixture of non-glandular and glandular trichomes only. Dotted lines indicate linear regressions

small pretarsus makes live observations very difficult. Our experiment with Cryo-SEM showed that the assumed original interlocking site can be seen in the micrographs as a small horizontal groove (claw print) on the side wall of the capitate trichome (Fig. 3F), because the trichome and/or claw were slightly dislocated during the freezing and vacuum procedures. The "clawprint" groove indicates that sharpened claws grasp the flexible trichome surface to get a grip on the smooth wall, leaving an imprint. In contrast to a previous report (Southwood 1986), the use of pseudopulvilli by *D. errans* as an interlocking mechanism with trichomes could not be observed.

It could be conjectured that there should be an optimum range of trichome diameter for interlocking with curved claws. But morphometrical variables of the pretarsus do not indicate obvious specialisation of claw curvature to particular trichome geometry. It seems that claws are able to attach to a wide spectrum of plant trichomes. The diameter, measured in the middle of trichomes (39.5 μm) in the plant with the highest traction force (*S. melongena*, see Fig. 9) does not correspond to the diameter of a circle fitting the concavity of the claw (16.6 μm). We suggest three mechanisms explaining this. (1) Probably, the diameter of only the middle of trichomes is not crucial for mirid bug claw interlocking, but rather the mean diameter of the entire trichome. (2) A bug's claw may move up and down along the trichome until it finds the optimum diameter. (3) The already mentioned flexibility of the claw enables the bug to grip a range of trichomes having different diameters.

Bug attachment abilities

In the inversion experiment, forces acting on the insect were applied in a normal direction to the surface. *D. errans* was attached to the underside of a glass surface and tried to detach from it. Attachment of *D. errans* on substrates covered with wax crystals was poor. This result confirms previously reported observations for some Anthocoridae (Lauenstein 1976), Chrysomelidae (*Phaedon cochleariae* Fab., Stork 1980b, *C. fastuosa*, Gorb and Gorb

Table 3 Pretarsus contamination (in % of the entire pretarsus area) and measured traction force on glass surface before and after the walking on plant surface in the traction force experiment with *Dicyphus errans*

No.	Plant species, cultivar	Pretarsus contamination [% of the pretarsus area]	Traction force [mN]		
			On glass before on plant surface mean ± sd	On glass after on plant surface mean ± sd	Statistical differences between forces on glass (before vs. after)
1	*Brassica oleracea* var. *gongylodes* 'Gigant'	60–90	0.12 ± 0.05	0.09 ± 0.04	$T = 4.517$, $P = 0.011$
2	*Capsicum annuum* 'Yolo Wonder'	0–50	0.46 ± 0.15	0.43 ± 0.15	$T = 0.536$, $P = 0.620$
3	*Cleome hirsuticaulis*	0–50	0.19 ± 0.09	0.20 ± 0.09	$T = 1.262$, $P = 0.276$
4	*Cucumis sativus* 'Rawa'	0–80	0.14 ± 0.06	0.13 ± 0.10	$W = 5.000$, $P = 0.625$
5	*Cyclamen persicum* 'Sierra F1'	0–50	0.25 ± 0.10	0.17 ± 0.03	$T = 2.026$, $P = 0.113$
6	*Digitalis purpurea* 'Excelsior'	0–50	0.26 ± 0.08	0.13 ± 0.04	$T = 0.098$, $P = 0.927$
7	*Fuchsia* × hybrida 'Winston Churchill'	0–50	0.28 ± 0.17	0.35 ± 0.11	$T = 1.934$, $P = 0.125$
8	*Gerbera jamesonii* × hybrida	0–80	0.29 ± 0.08	0.26 ± 0.09	$T = 1.385$, $P = 0.238$
9	*Ibicella lutea*	0–80	0.21 ± 0.14	0.31 ± 0.33	$W = 7.000$, $P = 0.438$
10	*Lycopersicon esculentum* 'Grit'	50–80	0.34 ± 0.13	0.31 ± 0.23	$T = 0.469$, $P = 0.664$
11	*Nicotiana sylvestris*	0–80	0.23 ± 0.12	0.22 ± 0.08	$T = 0.421$, $P = 0.695$
12	*Petunia* × hybrida	0–50	0.17 ± 0.08	0.13 ± 0.04	$T = 1.338$, $P = 0.252$
13	*Plectranthus ambiguus*	50–80	0.29 ± 0.07	0.22 ± 0.08	$T = 2.480$, $P = 0.068$
14	*Solanum melongena* 'Black Beauty'	0–50	0.35 ± 0.15	0.36 ± 0.15	$W = 5.000$, $P = 0.625$

Statistics in the right column; *T*, paired *t*-test; *W*, Wilcoxon signed rank test; *P* = probability value

2002), Formicidae (*Oecophylla smaragdina* Fab., Federle et al. 2000), as well as several predatory insect species (Eigenbrode 2004). Gorb and Gorb (2002) proposed four hypotheses to explain reduced insect attachment on plant surfaces covered with crystalline epicuticular wax. These facts might explain some differences in attachment ability of bugs on different waxy surfaces in our inversion experiment.

On a smooth leaf surface of *S. media*, *D. errans* could attach properly. This may be explained by (1) small leaves where claws could cling to the leaf margin and (2) the ability of the mirid bug to attach to smooth surfaces using pseudopulvilli.

However, the results of the inversion experiment confirm strong adaptation of *D. errans* to a life on pubescent plants. The best attachment was found on hairy and even glandular hairy surfaces, e.g. *Lycopersicon esculentum*, where many other insects are not able to walk (e.g. Kennedy 2003). For example, the leaf beetle *Chrysolina fastuosa* L. (Chrysomelidae) was not

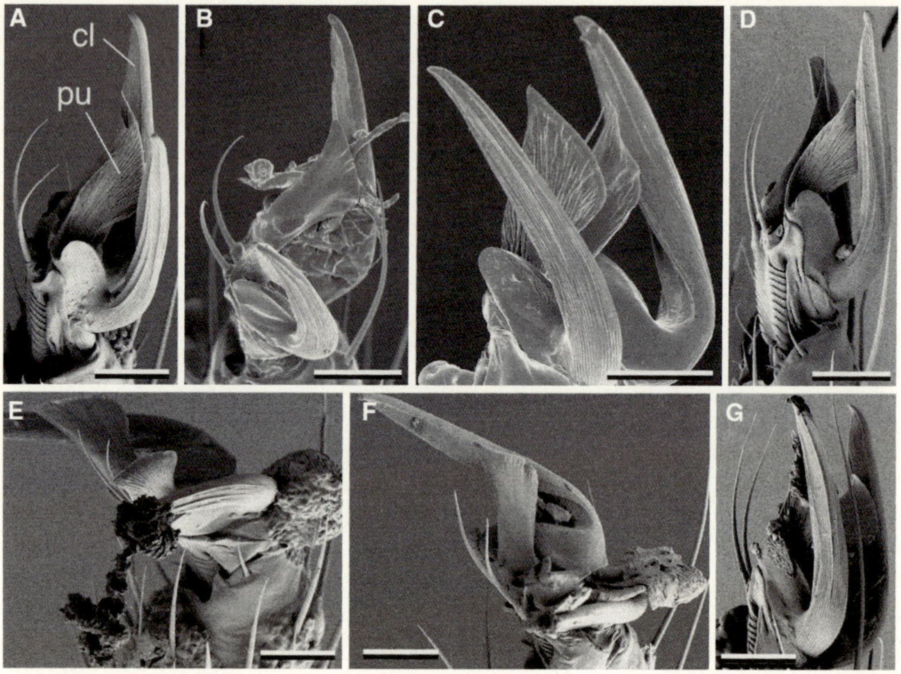

Fig. 12 SEM-micrographs of pretarsi of *Dicyphus errans* females after walking on different plant surfaces. (**A**) *Solanum melongena*, almost no claw contamination. (**B**, **C**) *Ibicella lutea*, heavy (**B**) and light (**C**) contamination. (**D**) *Cucumis sativus*, light contamination. (**E**) *Cleome hirsuticaulis*, heavy contamination. (**F**) *Lycopersicon esculentum*, heavy contamination. (**G**) *Plectranthus ambiguus*, heavy contamination. *cl*, claw; *pu*, pseudopulvilli. Scale bar = 25 μm

able to attach to the upper surface of the leaf of this plant in inversion experiments (Gorb and Gorb 2002). Beetle tarsi, covered with adhesive hairs, glued together and were effectively disabled by the secretion of glandular trichomes.

General significantly stronger attachment of *D. errans* was found on abaxial leaf sides, than has been observed in inversion experiments with *C. fastuosa* (Gorb and Gorb 2002). These surfaces are usually rougher than adaxial ones. For *Rumex obtusifolius* L. (Polygonaceae), Hosoda and Gorb (2004) quantitatively demonstrated lower surface roughness on the adaxial leaf side. According to our data, abaxial surfaces bear trichomes more frequently and denser than adaxial ones. Intense coverage with trichomes correlates with the stronger bug attachment.

Inversion experiments with *C. fastuosa* (Gorb and Gorb 2002) and with *D. errans* in this study do not show that size and density of trichomes have a significantly quantitative effect on attachment. However, this test is not sophisticated enough to reveal differences between diverse surfaces covered with trichomes. Subsequent traction force tests demonstrated a higher resolution for quantifying substrate influence on bug attachment and locomotion.

In the traction force experiment, forces acting on the insect were applied in a lateral direction to the surface. *D. errans* generated force walking and pulling horizontally. Traction force corresponds to the friction between pretarsi of the insect and the substrate. High traction force, measured on hairy plant surfaces, confirms the results of our inversion experiment as well as previous assumptions by Southwood (1986) and Voigt (2005) that

D. errans is adapted to, and has a close association with pubescent plant species. Considering trichome morphometry data, the trichome length and width seem to be the most essential variables, explaining measured forces (Fig. 10A, B). These variables effect claw interlocking. Longer and thicker trichomes are found to enhance bug traction force. We have not found any force dependence on trichome aspect ratio (Fig. 10D), although the aspect ratio should correlate with trichome bending stiffness and therefore influence the attachment strength of tarsi in contact. Trichome density has no significant influence on the attachment of *D. errans* (Fig. 10C) in contrast to some other entomophagous and phytophagous insect species, which are negatively affected by dense pubescence (Levin 1973; Fordyce and Agrawal 2001; Valverde et al. 2001). The ability of *D. errans* to attach on a wide range of pubescent plant surfaces with different trichome density may be a possible explanation of this effect.

Surfaces covered with glandular trichomes are the most adhesive ones. During locomotion, a bug must overcome the adhesion of the plant itself. We suppose that in contrast to less adhesive plant surfaces, on highly adhesive or rather sticky ones, bugs would move more carefully and therefore less powerfully. For this reason we measured plant surface adhesion and estimated its influence on the bug's traction force (Fig. 11). If we consider all 12 pubescent plant species studied, no relationship between plant adhesive properties and generated traction force was found (Fig. 11A). However, a significant negative correlation was found, if only *g* and *hg* surfaces are considered (Fig. 11B). Thus, we suggest that adhesive properties of the substrates, which are fully or partly covered with glandular trichomes, are remarkable influencing factors for bugs' traction force. In contrast to this finding previous observations show that under field conditions, *D. errans* prefers to live on adhesive glandular plants (Voigt 2005). These plants provide many small arthropods, trapped by glandular trichomes, as prey for the bug and protect the latter from natural enemies, which are unable to walk on such surfaces (Voigt 2005). There may be a trade-off between higher energy effort and risk of walking on glandular plant surfaces, and the access to freshly captured prey at minimal competition and danger of predators.

Contamination of feet (Fig. 12) by slipping on glandular trichomes during traction could also be a reason for less traction force generation on strongly adhesive surfaces. However, no considerable impact on the function of the attachment system walking on glass was found after walking on pubescent and glandular plant surfaces. *D. errans* possesses special cuticular protuberances and corresponding grooming behaviour for very extensive body grooming. These adaptations lead to the reduction of contamination and allow a life on pubescent and even glandular plant surfaces (Voigt 2005; Voigt et al. 2006). During the traction force experiment, insects were observed grooming themselves between individual tests. The generated traction force on glass (6.6 times of the bug body weight) indicates that attachment performance and ability to move in the bug is also rather strong on smooth surfaces, corresponding to the strong attachment on smooth plant substrate (*S. media*) in our inversion experiment.

The traction force on glass before and after walking on a plant surface significantly differed only in *Brassica oleracea*, which is densely covered with filamentous wax crystals (Table 3). SEM-micrographs showed contamination of 60–90% of the total tarsus area with wax crystals, which may support the contamination-hypothesis (Gorb and Gorb 2002). Foot contamination by wax crystals has previously been reported, in particular, for the beetles *P. cochleariae* (Stork 1980b), *Adalia bipunctata* L. (Gorb et al. 2005), and *C. fastuosa* (Gorb and Gorb 2006b).

Conclusions

Plant surfaces are important attachment substrates for phytophagous insects and their natural enemies. The omnivorous mirid bug *D. errans* inhabits pubescent, particularly glandular plant surfaces, which are avoided by many other entomophagous arthropods. The attachment ability of *D. errans* depends significantly on the plant surface type. Epicuticular crystalline waxes reduce bug attachment, whereas non-glandular and glandular trichomes, as well as the combination of both, promote attachment. Not only the presence of trichomes, but also their geometrical variables length and diameter influence attachment ability. The adhesion force of surfaces covered with glandular trichomes, as well as with a combination of non-glandular and glandular trichomes, has a negative correlation with the mirid bug's traction force. Direct mechanical contact between bug pretarsi and trichomes was visualized. We found that trichomes may provide suitable interlocking sites for attachment and locomotion of small specialised predatory insects, such as *D. errans*. Continued investigations on the anatomy and material properties of trichomes and bug pretarsi as well as ecomorphological studies will aid in further understanding of mechanical interactions between the bug and different types of plant surfaces.

Acknowledgements The first author thanks C. Neinhuis and the staff of the Botanical Garden as well as V. Pohris, M. Müller, and the staff of the Chair for Forest Protection at the Institute of Silviculture and Forest Protection, Faculty of Forestry, Geo and Hydro Sciences, Department of Forestry, Dresden University of Technology (Dresden, Germany) for providing space for the rearing of test plants and insects and for valuable discussions. The bug species were determined by K. Arnold (Geyer, Germany). The German companies Ernst Benary Samenzucht GmbH (Hann/Muenden), Quedlinburger Saatgut GmbH (Quedlinburg), Bruno Nebelung GmbH & Co. (Everswinkel), Cyclamen-Sprünken (Straelen), Eich Jungpflanzen Vertriebs GmbH (Grolsheim), Florensis Deutschland GmbH and Selecta Klemm GmbH & Co. KG (Stuttgart) provided seeds and seedlings. V. Kastner (Max-Planck Institute of Developmental Biology, Tübingen, Germany) helped with the experiments and linguistic corrections of the manuscript, and J. Schuppert (Max-Planck Institute for Metals Research, Stuttgart, Germany) with Cryo-SEM techniques. M. Varenberg (Max-Planck Institute for Metals Research, Stuttgart, Germany) and P. Perez Goodwyn (Graduated School of Agriculture, Kyoto University, Japan) are acknowledged for motivating discussion. The Federal Ministry of Education and Research, Germany (project BIOKON Phase II, 01RS0411) and the German National Academic Foundation (doctoral scholarship, E2002D0730) provided funding for the project.

Appendix

Below we provide a detailed description of the plant surfaces tested, according to the structural types shown in Fig. 4. Names of plant species are arranged alphabetically for quick access.

Aristolochia elegans Mart. (Aristolochiaceae) is covered with transversely ridged wax crystal rodlets, very densely distributed, almost without gaps, on the abaxial leaf side. There, sparsely distributed, inclined unicellular, hook-shaped trichomes are also found (Fig. 5A).

Begonia × *semperflorens* Link & Otto (Begoniaceae) has a smooth and shiny leaf surface. On the adaxial side, membraneous whole wax platelets are sparsely arranged (Fig. 5C). On the abaxial side, dispersed polygonal wax granules were found (Fig. 5B).

Borago officinalis L. (Boraginaceae) bears a densely hirsute indumentum, consisting of perpendicular and inclined, unicellular, tapered, cone-shaped, and strongly nodose trichomes on both leaf sides (Fig. 6B). Longer trichomes have a convex multicellular base, while shorter ones possess a convex unicellular base. Moreover, dispersed short capitate trichomes (stalked glandular heads) were found (Fig. 6B, see Fig. 4B, *S4*).

Brassica oleracea L. *var. gongylodes* L. 'Gigant' (Brassicaceae): the pruinouse surface of this plant is densely covered with filamentous wax crystals (polygonal rodlets and threads) (Fig. 5D).

Calendula officinale L. 'Orangestrahlen' (Compositae) has different trichomes on both leaf sides: (1) uniseriate, 2- to 10-celled and multiseriate, multicellular capitate trichomes with glandular heads (Fig. 6L), (2) rounded, cone-shaped, multicellular, glandular trichomes without a head, and (3) sparsely dispersed capitate trichomes (stalked glandular heads, see Fig. 4B, *S4*).

Capsicum annuum L. 'Yolo Wonder' (Solanaceae) has a smooth appearing adaxial leaf surface that bears very sparsely dispersed wax crystals, arranged in groups from 3 to 10 platelets (Fig. 5I). Groups of numerous curved, finger-shaped, unicellular trichomes occur in the leaf domatia on the abaxial leaf surface (Fig. 7C).

Cissus njegerre Gilg (Vitaceae), a tropical liana, has a pilose indumentum. On the adaxial leaf side, following trichomes occur: (1) regularly dispersed, bent, uniseriate, 5- to 11-celled ones, (2) perpendicular, 2- to 3-cellular, cone-shaped, nodose trichomes with multicellular convex base, and (3) several dispersed, 1- and 4-celled, peltate trichomes (Fig. 6C). The abaxial leaf surface is covered with regularly dispersed, strongly curved, cone-shaped, uniseriate, 2-celled trichomes with one convex basal cell, as well as with bent, 5- to 11-cellular, nodose trichomes with a multi-cellular convex base, mostly with distally oriented tips (Fig. 6D). The trichomes are more densely concentrated along leaf veins. Moreover, perpendicular 2- to 4-cellular capitate trichomes are scattered all over the leaf lamina. The abaxial leaf surface is also covered with 1- and 4-cellular peltate trichomes.

Cleome hirsuticaulis J. F. Macbr. (Capparidaceae): the leaves have a pilose indumentum, which is most dense along leaf veins and on the abaxial side (Fig. 7A). It consists of dispersed, both inclined and bent, multiseriate, multicellular, cone-shaped glandular trichomes with small glandular heads and a multicellular convex base. Very long and slightly shorter trichomes of this type were found. Multicellular clavate trichomes as well as uniseriate, 3- to 10-cellular trichomes (Fig. 7B) seldom occur.

Cucumis sativus L. 'Rawa' (Curcurbitaceae): the hirsute indumentum shows regularly dispersed, perpendicular (or inclined), uniseriate, uni- to 4-cellular, tapered cone-shaped (or dart-shaped) trichomes with a multi-cellular base (Fig. 7E, F). The trichome density is higher on the adaxial leaf side. Trichomes are orientated towards the leaf margin.

Cuphea lanceolata W. T. Aiton (Lythraceae): regularly dispersed, unicellular, spindle-shaped, nodose trichomes were found on the leaf. They are procumbent and located close to the leaf surface (Fig. 6I). Only the trichome tip and the distal half are bent away from the lamina at an angle of approximately 45°. On the abaxial side, trichomes occur more densely along leaf veins and sparsely in the intercostal areas. Besides, long inclined, multiseriate, multicellular trichomes with small glandular heads are scattered on both leaf sides (Fig. 6H).

Curcurbita pepo L. 'Hokkaido' (Curcurbitaceae) possesses a dense (adaxial side), and sparse (abaxial side) hirsute pubescence, consisting of regularly dispersed, perpendicular and inclined, uniseriate, 3- to rarely 5-cellular, tapered cone-shaped trichomes with thickened bases, as well as sparse 2-celluar capitate trichomes. Trichomes are most densely concentrated along leaf veins.

Cyclamen persicum Mill. 'Sierra F1' (Primulaceae) the leaves bear, on the adaxial side, sparsely distributed, single, entire wax platelets. On the abaxial lamina surface, these can be also clustered in rosettes (Fig. 5K).

Datura innoxia Mill. (Solanaceae) shows a pilose indumentum, composed of dispersed, uniseriate, 3- to 5-cellular, inclined, cone-shaped, nodose and capitate trichomes (Fig. 6A). The adaxial pubescence is denser than the abaxial one.

Digitalis purpurea L. 'Excelsior' (Scrophulariaceae) has dispersed, bent, uniseriate, 3- to 5-cellular, nodose, cone-shaped, partially finger-shaped trichomes, distributed very densely along leaf veins (Fig. 7N, O). Sparsely dispersed, inclined capitate trichomes (stalked 2-cellular glandular heads, Fig. 7N) and capitate ones with 4-cellular heads were also found.

Fuchsia × hybrida Hort. ex Vilm 'Winston Churchill' (Onagraceae): the adaxial leaf side possesses sparsely dispersed, inclined (on veins also bent), unicellular, thread-like and spiral-shaped, nodose trichomes with a rounded tip and a convex unicellular base. The surface is also covered with individual, entire, crystal wax platelets, often arranged in clusters (Fig. 5H). On the abaxial side, trichomes are inclined, bent and curved, unicellular, cone- and spiral-shaped, and nodose with a convex unicellular base, concentrated along the leaf vein and sparsely dispersed in the intercostal areas (Fig. 7K).

Geranium robertianum L. (Geraniaceae) bears bent, 2- to 4-cellular, capitate, nodose trichomes with a multicellular base as well as sparse, inclined, unicellular, cone-shaped, nodose trichomes on the abaxial leaf side. Capitate trichomes (stalked glandular heads, see Fig. 4B, *S4*) cover the entire surface. On the abaxial leaf side, dense, curved, unicellular, thread-shaped trichomes as well as inclined 4-cellular, capitate, nodose trichomes were found on the leaf veins. Additionally, stiff cone-shaped, nodose trichomes with a multicellular base and capitate trichomes (stalked glandular heads) are present.

Gerbera jamesonii × hybrida Adlam (Compositae): the adaxial leaf side is covered with sparse, inclined and bent uniseriate 4-cellular thread-shaped trichomes as well as dispersed clusters of membranous crystalline wax platelets. The felt-like indumentum of the abaxial surface along and around the leaf veins consists of inclined, bent, curved, uniseriate, multi-cellular, spiral-shaped, often twisted trichomes with convex unicellular bases (Fig. 7H). Inclined, bent, curved, uniseriate, multicellular, spiral-shaped trichomes as well as occasional, entire crystal wax platelets also appear sparsely in the intercostal areas.

Hibiscus rosa-chinensis L. (Malvaceae) shows the following surface structures on the adaxial leaf surface: (1) inclined and bent, unicellular trichomes on leaf veins, (2) sparsely distributed, 4-cellular peltate trichomes in the intercostal areas as well as wax crystals (clusters of membraneous and entire platelets) (Fig. 5F). On the abaxial side, there are inclined and bent, unicellular trichomes, and 2- to 8-cellular stellate trichomes on the leaf veins (Fig. 6N) as well as sparsely distributed 4-cellular peltate trichomes in the intercostal areas.

Ibicella lutea (Lindl.) Van Eselt. (Martynaceae) is densely covered with the following trichomes: (1) glandular-pubescent, perpendicular and inclined, uniseriate, 3- to 8-cellular, capitate with convex unicellular base, (2) perpendicular, uniseriate, 4-cellular, cone-shaped, (3) perpendicular, 3-cellular, finger-shaped, (4) multicellular clavate, and (5) peltate.

Lactuca sativa L. 'Attraktion' (Asteraceae): membraneous wax platelets, arranged as rosettes, are found on the adaxial leaf surface (Fig. 5E). On the abaxial surface there are clusters of irregular wax platelets.

Lagenaria siceria (Mol) Stadley (Curcurbitaceae) exhibits a pilose coverage, consisting of (1) capitate trichomes with an unicellular stem and 4-cellular head, (2) perpendicular and inclined, uniseriate 2- to 5-cellular, dart-shaped, nodose trichomes with thickened basal cell, (3) dispersed 4-cellular peltate trichomes as well as (4) stalked 4-cellular heads

(Fig. 6K). Trichomes are located sparsely on the adaxial side and densely on the abaxial surface, and concentrated especially along leaf veins.

Lophospermum erubescens D. Don (Scrophulariaceae): the leaf surface is covered with dispersed perpendicular, 3- to 4-celled, cone-shaped glandular trichomes of different length, with glandular head or with rounded tip, and always with one convex basal cell. Clavate trichomes are regularly dispersed.

Lycopersicon esculentum Mill. 'Grit' (Solanaceae) bears regularly dispersed (1) 2-cellular, bent and curved, tapered, cone-shaped, nodose trichomes with convex unicellular bases, (2) inclined and bent, uniseriate, 3- to 5-cellular, cone-shaped, nodose glandular trichomes, (3) sporadic uniseriate, 5- to 12-cellular, cone-shaped trichomes with a multi-cellular base, and (4) capitate trichomes with 4-cellular glandular heads; these are mainly concentrated along leaf veins (Fig. 7L, M).

Malus domestica Borkh. (Rosaceae): the leaf surface appears adaxially smooth with sparsely occurring irregular, membraneous wax platelets arranged in clusters (Fig. 5M), whereas on the abaxial side, a felt-like pubescence was found (Fig. 6S). The inclined and bent unicellular thread-shaped and spiral-shaped, partly twisted trichomes are concentrated along the leaf veins. Besides, sparsely occurring irregular wax crystal platelets arranged in clusters are present on both the trichomes and lamina surface (Fig. 5L).

Nicotiana sylvestris Spegazz. et Gomez (Solanaceae) is covered with (1) dispersed inclined, 2- to 6-cellular, capitate, (2) cone-shaped, glandular trichomes, (3) stalked glandular heads, (4) perpendicular, 2- to 5-cellular finger-shaped trichomes with a convex unicellular base. They all were concentrated along the leaf veins (Fig. 7J).

Ocimum basilicum L. (Lamiaceae): the surface is smooth with some dispersed 1-, 2- and 4-cellular peltate trichomes (Fig. 6M). Cone-shaped, nodose, tapered trichomes are present along the adaxial veins (Fig. 6J).

Oenothera biennis L. (Onagraceae): regularly dispersed, inclined and bent, unicellular, cone-shaped, nodose, tapered trichomes as well as clavate ones were found on the adaxial leaf side (Fig. 6Q, R). The abaxial, felt-like indumentum consists of (1) curved thread-shaped trichomes, (2) dispersed, inclined and bent unicellular cone-shaped tapered nodose trichomes, and (3) densely distributed clavate trichomes.

Pelargonium zonale (L.) L'Hérit. (Geraniaceae): the adaxial lamina side is predominantly covered with regularly dispersed, inclined and bent, unicellular, cone-shaped, tapered trichomes with a slightly thickened base. Inclined, 2- to 4-cellular capitate trichomes with uni- and multicellular bases as well as regularly dispersed, capitate trichomes (stalked glandular heads, Fig. 4B, *S4*) were also found. The abaxial leaf surface bears mainly (1) regularly dispersed, inclined 2- to 4-cellular capitate trichomes with a uni- and multicellular base, but also (2) regularly dispersed capitate trichomes (stalked glandular heads, see Fig. 4B, *S4*) as well as (3) irregularly dispersed, inclined and bent, unicellular, cone-shaped, tapered trichomes with slightly thickened bases (Fig. 6G).

Petunia × hybrida Vilm. (Solanaceae): the pilose, glandular indumentum consists of dispersed perpendicular and inclined uniseriate 1- to 3-, rarely 5-cellular capitate trichomes (Fig. 7I).

Plectranthus ambiguus (Bolus) Codd. (Lamiaceae) has a pilose indumentum on both leaf sides with (1) regularly dispersed, inclined and bent, on the veins especially curved, uniseriate, 4- to 15-cellular, cone-shaped, tapered, nodose trichomes with convex multicellular base, (2) dispersed 4-cellular peltate trichomes, (3) capitate trichomes (stalked glandular heads, see Fig. 4B, *S4*), and (4) sparsely occurring perpendicular 2- to 4-cellular capitate trichomes (Fig. 7D).

Rosa × hybrida L. 'Bernstein' (Rosaceae) bears adaxially, sparsely dispersed, irregular wax crystal platelets arranged in small groups (Fig. 5J). On the abaxial leaf side, very sparsely dispersed, irregular wax platelets, and unicellular peltate trichomes are present.

Saintpaulia ionantha Wendl. (Gesneriaceae) possesses a pilose indumentum, consisting of (1) densely dispersed, inclined, bent and curved, uniseriate, 2- to 12-cellular, cone-shaped, tapered trichomes, rarely with small rounded glandular heads, and (2) 1- to 4-cellular peltate trichomes.

Solanum melongena L. 'Black Beauty' (Solanaceae): the felt-like hairy coverage exhibits multicellular, stellate nodose trichomes, consisting of 5–9 branches, with the central branch perpendicularly oriented and other branches, arising perpendicularly from the central one, in a ray-like manner around it (Fig. 7G). Trichomes are very densely distributed along veins and on the abaxial side. Inclined, unicellular, cone-shaped trichomes were found only on the adaxial leaf surface.

Solanum tuberosum L. 'Acapella' (Solanaceae) is covered with regularly dispersed, inclined, uniseriate, 3- to 5-cellular, nodose trichomes as well as clavate ones, which are denser on the abaxial side than on the adaxial one.

Stellaria media (L.) Vill. (Caryophyllaceae): the leaf lamina is smooth without any crystalline waxes or trichomes. Cell irregularities determine the surface topography.

Syngonium auritum Schott (Araceae): leaves appear shiny with membraneous and irregular wax platelets, arranged in rosettes. Crystals on the abaxial side are much more densely distributed around the stomata (Fig. 5N).

Tropaeolum majus L. (Tropaeolaceae) is densely covered with tubular wax crystals on the glaucous adaxial leaf surface. The abaxial side bears scattered, perpendicular, and inclined, uniseriate, 2- to 6-cellular, cone-shaped trichomes with a convex unicellular base. The entire surface here, including the trichomes, is densely covered with tubular wax crystals (Fig. 5G, 6O).

Urtica dioica L. (Urticaceae) has (1) inclined, unicellular, cone-shaped, nodose trichomes, (2) stinging trichomes, and (3) bent, unicellular, cone-shaped, nodose trichomes with thickened bases. Trichomes are orientated distally.

Verbascum thapsus L. (Scrophulariaceae): the felt-like indumentum consists abaxially of bent multicellular branched, tree-shaped (dendroid) trichomes (Fig. 6P). Additionally, 4- to 6-cellular stellate trichomes are located on the adaxial side of the leaf.

Viola tricolor L. (Violaceae) has a smooth appearing surface with polygonal wax crystal plates arranged in groups, partially covering the stomata on the abaxial leaf side (Fig. 5O).

References

Andersen SO, Weis-Fogh T (1964) Resilin, a rubber-like protein in arthropod cuticle. Adv Insect Physiol 2:1–65

Andow DA, Prokrym DR (1990) Plant structural complexity and host-finding by a parasitoid. Oecologia 82:62–165

Andres MR, Connor EF (2003) The community-wide and guild-specific effects of pubescence on the folivorous insects of manzanitas *Arctostaphylos* spp. Ecol Entomol 28:383–396

Arzet H-R (1972) Suchverhalten und Nahrungsverbrauch der Larven von *Chrysopa carnea* Steph. Dissertation, Georg-August-Universität, Göttingen, Germany, pp. 93

Avé DA, Gregory P, Tingey WM (1987) Aphid repellent sesquiterpenes in glandular trichomes of *Solanum berthaultii* and *S. tuberosum*. Entomol Exp Appl 44:131–138

Barthlott W, Neinhuis C, Cutler D, Ditsch F, Meusel I, Theisen I, Wilhelmi H (1998) Classification and terminology of plant epicuticular waxes. Botan J Linn Soc126:237–260

Becerra JX (2003) Synchronous coadaption in an ancient case of herbivory. PNAS 100:12804–12807

Begon M, Townsend CR, Harper JL (2006) Ecology: from individuals to ecosystems, 4th edn. Blackwell Publishing Ltd., 738 pp

Belcher DW, Thurston R (1982) Inhibition of movement of larvae of the convergent lady beetle by leaf trichomes of tobacco. Environ Entomol 11:91–94

Benedict HJ, Leigh TF, Hyer AH (1983) *Lygus hesperus* (Heteroptera: Miridae) oviposition behaviour, growth, and survival in relationship to cotton trichome density. Environ Entomol 12:331–335

Bernays EA, Chapman RF (1994) Host-plant selection by phytophagous insects. Contemporary topics in entomology, 2nd edn. Chapman & Hall, pp. 312

Betz O (2002) Performance and adaptive value of tarsal morphology in rove beetles of the genus *Stenus* (Coleoptera, Staphylinidae). J Exp Biol 205:1097–1113

Brennan EB, Weinbaum SA (2001) Stylet penetration and survival of three psyllid species on adult leaves and 'waxy' and 'de-waxy' juvenile leaves of *Eucalyptus globulus*. Entomol Exp Appl 100:355–363

Carter MC, Sutherland D, Dixon AFG (1984) Plant structure and the searching efficiency of coccinellid larvae. Oecologia 63:394–397

Chapman RF (1977) The role of leaf surface in food selection by acridids and other insects. Coll Int CNRS 265:133–149

Coll M, Smith LA, Ridgway RL (1997) Effect of plants on the searching efficiency of a generalist predator: the importance of predator-prey spatial associations. Entomol Exp Appl 83:1–10

Coll M (1998) Living and feeding on plants in predatory heteroptera. In: Coll M, Ruberson JR (eds) Predatory Heteroptera: their ecology and use in biological control. Proc Entomol Soc Am, pp 89–130

Cortesero AM, Stapel JO, Lewis WJ (2000) Understanding manipulating plant attributes to enhance biological control. Biol Control 17:35–49

Dalin P, Björkman C, Eklund K (2004) Leaf beetle grazing does not induce willow trichome defence in the coppicing willow *Salix viminalis*. Agr Forest Entomol 6:105–109

Dixon AFG (1986) Habitat specifity and foraging behaviour of aphidophagous insects. In: Hodek I (ed) Ecology of aphidophaga. Proceedings of the 2nd symosium held at Zvíkovské Podhradí, September 2–8, 1984, Series Entomologica 35, pp 151–154

Dos Santos TM, Júnior ALB, Soares JJ (2003) Influência de tricomas do algodoeiro sobre os aspectos biológicos e capacidade predatória de *Chrysoperla externa* (Hagen) alimentada com *Aphis gossypii* Glover. Bragantia, Campinas 62:243–254

Duffey SS (1986) Plant glandular trichomes: their partial role in defense against insects. In: Juniper B, Southwood R (ed) Insects and the plant surface. Edward Arnold Publishers, London, pp 151–172

Ecole CC, Picanço MC, Guedes RNC, Brommonschenkel SH (2001) Effect of cropping season and possible compounds involved in the resistance of *Lycopersicon hirsutum* f. *typicum* to *Tuta absoluta* (Meyrick) (Lep., Gelechiidae). J Appl Entomol 125:193–200

Eigenbrode SD (1996) Plant surface waxes and insect behaviour. In: Kerstiens G (eds) Plant cuticles—an integral functional approach. BIOS Publ., Oxford, pp 201–222

Eigenbrode SD (2004) The effects of plant epicuticular waxy blooms on attachment and effectiveness of predatory insects. Arthr Struct Dev 33:91–102

Eigenbrode SD, Castagnola T, Roux M-B, Steljes L (1996) Mobility of three generalist predators is greater on cabbage with glossy leaf wax than on cabbage with a wax bloom. Entomol Exp Appl 81:335–343

Eisner T, Eisner M, Hoebeke ER (1998) When defense backfires: detrimental effect of a plant's protective trichomes on an insect beneficial to the plant. Proc Natl Acad Sci USA 95:4410–4414

Elsey KD (1974) Influence of plant host on searching speed of two predators. Entomophaga 19:3–6

Esau K (1965) Plant anatomy, 2nd edn. John Wiley & Sons, pp 767

Evans HF (1976) Mutual interference between predatory anthocorids. Ecol Entomol 1:283–286

Federle W, Brainerd WL, McMahon T, Hölldobler B (2001) Biomechanics of the movable pretarsal adhesive organ in ants and bees. PNAS 98:6215–6220

Federle W, Maschwitz U, Fiala B, Riederer M, Hölldobler B (1997) Slippery ant-plants and skilful climbers: selection and protection of specific ant partners by epicuticular wax blooms in *Macaranga* (Euphorbiaceae). Oecologia 112:217–224

Federle W, Rohrseitz K, Hölldobler B (2000) Attachment forces of ants measured with a centrifuge: better "wax-runners" have poorer attachment to a smooth surface. J Exp Biol 203:505–512

Fordyce JA, Agrawal AA (2001) The role of plant trichomes and caterpillar group size on growth and defence of the pipevine swallowtail *Battus philenor*. J Anim Ecol 70:997–1005

Frantsevich L, Gorb S (2004) Structure and mechanics of the tarsal chain in the hornet, *Vespa crabro* (Hymenoptera: Vespidae): implications on the attachment mechanism. Arthr Struct Dev 33:77–89

Frazier SF, Larsen GS, Neff D, Quimby L, Carney M, DiCaprio RA, Zill SN (1999) Elasticity and movements of the cockroach tarsus in walking. J Comp Physiol A 185:157–172

Gassmann AJ, Hare JD (2005) Indirect cost of a defensive trait: variation in trichomes type affects the natural enemies of herbivorous insects on *Datura wrightii*. Oecologia 144:62–71

Gaume L, Perret P, Gorb E, Gorb S, Labat J-J, Rowe N (2004) How do plant waxes cause flies to slide? Experimental tests of wax-based trapping mechanisms in three pitfall carnivorous plants. Arthr Struct Dev 33:103–111

Gepp J (1977) Bewegungsbehinderung von Arthropoden durch Trichome an Bohnenpflanzen (*Phaseolus vulgaris* L.). Anz Schädlingskde Pflanzenschutz Umweltschutz 50:8–12

Ghazi-Bayat A (1979) Zur Oberflächenstruktur der tarsalen Haftlappen von *Coreus marginatus* (L.) (Coreidae, Heteroptera). Zool Anz 203:345–347

Ghazi-Bayat A, Hasenfuss I (1980) Die Oberflächenstrukturen des Prätarsus von *Elasmucha ferrugata* (Fabricius) (Acanthosomatidae, Heteroptera). Zool Anz 205:76–80

Ghazi-Bayat A, Hasenfuss I (1981) Über den Transportweg der Haftflüssigkeit der Pulvilli bei *Coptosoma scutellatum* (Geoffr.) (Plataspididae, Heteroptera). Nachrichtenblatt der Bayerischen Entomologen 30:58

Gibson RW (1971) Glandular hairs providing resistance to aphids in certain wild potato species. Ann Appl Biol 68:113–119

Gibson RW, Pickett JA (1983) Wild potato repels aphids by release of aphid alarm pheromone. Nature 302:608–609

Gilbert LE (1971) Butterfly-plant coevolution: has *Passiflora adenopoda* won the selectional race with Heliconiine butterflies? Science 172:585–586

Gingras D, Boivin G (2002) Effect of plant structure, host density and foraging duration on host finding by *Trichogramma evanescens* (Hymenoptera: Trichogrammatidae). Environ Entomol 31:1153–1157

Goffreda JC, Mutschler MA, Avé DA, Tingey WM, Steffens JC (1989) Aphid deterrence by glucose esters in glandular trichome exudate of the wild tomato, *Lycopersicon pennelli*. J Chem Ecol 15:2135–2147

Gorb E, Gorb S (2006a) Combination of the surface profile and chemistry reduces the attachment of the beetle *Gastrophysa viridula* on the *Rumex obtusifolius* leaf surface. Proceedings of the 5th plant biomechanics conference, August 28th–September 1st, 2006, vol 2. Stockholm, Sweden, pp 537–542

Gorb E, Gorb S (2006b) Do plant waxes make insect attachment structures dirty? Experimental evidence for the contamination hypotheses. In: Herrel A, Speck T, Rowe NP (eds) Ecology and biomechanics. CRC Press, Taylor & Francis, Boca Raton, Florida, pp 147–162

Gorb EV, Gorb SN (2002) Attachment ability of the beetle *Chrysolina fastuosa* on various plant surfaces. Entomol Exp Appl 105:13–28

Gorb E, Haas K, Henrich A, Enders S, Barbakadze N, Gorb S (2005) Composite structure of the crystalline wax layer of the slippery zone in the pitchers of the carnivorous plant *Nepenthes alata* and its effect on insect attachment. J Exp Biol 208:4651–4662

Gorb E, Kastner V, Peressadko A, Arzt E, Gaume L, Rowe N, Gorb S (2004) Structure and properties of the glandular surface in the digestive zone of the pitcher in the carnivorous plant *Nepenthes ventrata* and its role in insect trapping and retention. J Exp Biol 207:2974–2963

Gorb SN (1996) Design of insect unguitractor apparatus. J Morphology 230:219–230

Gorb SN (1999) Serial elastic elements in the damselfly wing: mobile vein joints contain resilin. Naturwissenschaften 86:552–555

Gorb SN (2001) Attachment devices of insect cuticle. Kluwer Academic Publishers, Dordrecht, pp 305

Gorb SN, Beutel RG (2001) Evolution of locomotory attachment pads of hexapods. Naturwissenschaften 88:530–534

Goundoudaki S, Tsitsipis JA, Margaritopoulos JT, Zarpas KD, Divanidis S (2003) Performance of the tobacco aphid *Myzus persicae* (Hemiptera: Aphididae) on oriental and Virginia tobacco varieties. Agr Forest Entomol 5:285–291

Gregory P, Ave DA, Bouthyette PJ, Tingey WM (1986) Insect-defensive chemistry of potato glandular trichomes. In: Juniper B, Southwood R (eds) Insects and the plant surface. Edward Arnold Publishers, London, pp 173–184

Grevstad FS, Klepetka BW (1992) The influence of plant architecture on the foraging efficiencies of a suite of ladybird beetles feeding on aphids. Oecologia 92:399–404

Grimaldi D, Engel MS (2004) Evolution of the insects. Cambridge University Press, New York, pp 755

Haddad NM, Hicks WM (2000) Host pubescence and the behavior and performance of the butterfly *Papilio troilus* (Lepidoptera: Papilionidae). Environ Entomol 29:299–303

Hare JD (2005) Biological activity of acyl glucose esters from *Datura wrightii* glandular trichomes against three native insect herbivores. J Chem Ecol 31:1475–1491

Hare JD, Elle E (2002) Variable impact of diverse insect herbivores on dimorphic *Datura wrightii*. Ecology 83:2711–2720

Hare JD, Smith II JL (2005) Competition, herbivory, and reproduction of trichome phenotypes of *Datura wrightii*. Ecology 86:334–339
Hoffmann GD, McEvoy PB (1985) The mechanism of trichome resistance in *Anaphalis margaritacea* to the meadow spittlebug *Philaenus spumarius*. Entomol Exp Appl 39:123–129
Hosoda N, Gorb SN (2004) Critical roughness for insect attachment: experimental evidences for the beetle *Gastrophysa viridula*. Abstracts of 2 Bremer Bionik-Kongress, 29.-30.10.2004, Bremen
Hoxie RP, Wellso SG, Webster JA (1975) Cereal leaf beetle response to wheat trichome length and density. Environ Entomol 4:365–370
Hummel K, Staesche K (1962) Die Verbreitung der Haartypen in ihren natürlichen Verwandtschaftsgruppen. In: Zimmermann W, Ozenda PG (eds) Handbuch der Pflanzenanatomie, Gebrüder Bornträger, Berlin, 2. v. neub. Aufl., Band IV, Teil 5. Abteilung: Histologie, Gebrüder Borntraeger, Berlin-Nikolassee, pp 207–292
Jeffree CF (1986) The cuticle, epicuticular waxes and trichomes of plants, with reference to their structure, functions and evolution. In: Juniper B, Southwood R (eds) Insects and the plant surface. Edward Arnold Publishers, London, pp 23–64
Jervis M, Kidd N (1996) Insect natural enemies. Practical approaches to their study and evaluation. Chapman & Hall, London, 491 pp
Johnson B (1953) The injurious effects of the hooked epidermal hairs of french beans (*Phaseolus vulgaris* L.) on *Aphis craccivora* Koch. Bull Entomol Res 44:779–788
Johnson B (1956) The influence on aphids of the glandular hairs on tomato plants. Plant Pathology 5:131–132
Johnson HB (1975) Plant pubescence: an ecological perspective. Bot Rev 41:233–258
Jolivet P (1998) Interrelationship between insects and plants. CRC Press, Boca Raton, Florida, 309 pp
Jordan KHC (1962) Landwanzen. Die neue Brehm-Bücherei, Bd. 294, A.-Ziemsen-Verlag, Wittenberg-Lutherstadt, 116 pp
Juniper BE (1995) Waxes on plant surfaces and their interactions with insects. In: Hamilton RJ (ed) Waxes: chemistry, molecular biology and functions. Oily, West Ferry, Dundee, pp 157–174
Juniper BE, Jeffree CE (1983) Plant surfaces. Edward Arnold Limited, London, 93 pp
Kareiva P, Sahakian R (1990) Tritrophic effects of a simple architectural mutation in pea plants. Nature 345:433–434
Kashyap RK, Kennedy GG, Farrar RR Jr (1991) Behavioral response of *Trichogramma pretiosum* Riley and *Telenomus sphingis* (Ashmead) to trichome/methyl ketone mediated resistance in tomato. J Chem Ecol 17:543–556
Keller MA (1987) Influence of leaf surfaces on movements by the hymenopterous parasitoid *Trichogramma exiguum*. Entomol Exp Appl 43:55–59
Kennedy CEJ (1986) Attachment may be a basis for specialization in oak aphids. Ecol Entomol 11:291–300
Kennedy GG (2003) Tomato, pests, parasitoids, and predators: tritrophic interactions involving the genus *Lycopersicon*. Annu Rev Entomol 48:51–72
Kowalewski E, Robinson RW (1978) White fly species resistance in *Cucumis* species. Cucurbit Genetics Cooperative Report 1:38
Kullenberg B (1946) Studien über die Biologie der Capsiden. Zool Bidrag Uppsala 23:1–522
Lapointe SL, Tingey WM (1986) Glandular trichomes of *Solanum berthaultii* confer resistance to green peach aphid (Homoptera: Aphididae). J Econ Entomol 79:1264–1268
Lauenstein G (1976) Untersuchungen zu Biologie und Verhaltensweisen der Räuberischen Blumenwanze *Anthocoris nemorum* L. (Het.: Anthocoridae). Dissertation, G.-August-Universität Göttingen, Germany
Lee YI, Kogan M, Larsen JR Jr (1986) Attachment of the potato leafhopper to soybean plant surfaces as affected by morphology of the pretarsus. Entomol Exp Appl 42:101–107
Lees AD, Hardie J (1988) The organs of adhesion in the aphid *Megoura viciae*. J Exp Biol 136:209–228
Leite GLD, Picanço M, Della Lucia TMC, Moreira MD (1999) Role of canopy height in the resistance of *Lycopersicon hirsutum* f. *glabrum* to *Tuta absoluta* (Lep., Gelechiidae). J Appl Entomol 123:459–463
Levin DA (1973) The role of trichomes in plant defense. Quart Rev Biol 48:3–15
Malakar R, Tingey WM (2000) Glandular trichomes of *Solanum berthaultii* and its hybrids with potato deter oviposition and impair growth of potato tuber moth. Entomol Exp Appl 94:249–257
McKinney KB (1938) Physical characteristics on the foliage of beans and tomatoes that tend to control some small insect pests. J Econ Ent 31:630–631
Metcalfe CR, Chalk L (1988) Anatomy of the dicotyledons. Leaves, stem, and wood in relation to taxonomy with notes on economic uses, vol II, 2nd edn. Oxford University Press, 330 pp
Müller C (2006) Plant-insect interactions on cuticular surface. In: Riederer M, Müller C (eds) Biology of plant cuticle. Blackwell Publishing, Oxford, pp 398–422

Musetti L, Neal JJ (1997) Resistance to the pink potato aphid, *Macrosiphum euphorbiae*, in two accessions of *Lycopersicon hirsutum* f. *glabratum*. Entomol Exp Appl 84:137–146
Napp-Zinn K (1973) Anatomie des Blattes. II Blattanatomie der Angiospermen. A. Entwicklungsgeschichte und topographische Anatormie des Angiospermenblattes. In: Linsbauer K (ed) Handbuch der Pflanzenanatomie, Bd. VIII, Teil 2 A, Bornträger-Verlag, Berlin
Niederegger S, Gorb S (2003) Tarsal movements in flies during leg attachment and detachment on a smooth substrate. J Ins Physiol 49:611–620
Obrycki JJ (1986) The influence of foliar pubescence on entomophagous species. In: Boerthel DJ, Eikenbarry RD (eds) Interactions of plant resistance and parasitoids and predators of insects. J. Wiley & Sons, New York, 200 pp
Obrycki JJ, Tauber MJ (1984) Natural enemy activity on glandular pubescent potato plants in the greenhouse: An unreliable predictor of effects in the field. Environ Entomol 13:679–683
Payne WW (1978) A glossary of plant hair terminology. Brittonia 30:239–255
Peeters PJ (2002) Correlations between leaf structural traits and the densities of herbivorous insect guilds. Biol J Linnean Soc 77:43–65
Pillemer EA, Tingey WM (1978) Hooked trichomes and resistance of *Phaseolus vulgaris* to *Empoasca fabae* (Harris). Entomol Exp Appl 24:83–94
Putman WL (1955) Bionomics of *Stethorus punctillum* Weise (Coleoptera: Coccinellidae) in Ontario. Can Entomol 86:9–33
Quilici S, Iperti G (1986) The influence of host plant on the searching ability of first instar larvae of *Propylea quadrodecimpunctata*. In: Hodek I (ed) Ecology of aphidophaga. Series Entomologica 35:99–111
Rabb RL, Bradley JR (1968) The influence of host plants on parasitism of eggs of the tobacco hornworm. J Econ Entomol 61:1249–1252
Ranger CM, Hower AA (2002) Glandular trichomes on perennial alfalfa affect host-selection behavior of *Empoasca fabae*. Entomol Exp Appl 105:71–81
Ringlund K, Everson EH (1968) Leaf pubescence in common wheat, *Triticum aestivum* L., and resistance to the cereal leaf beetle, *Oulema melanopus* (L.). Crop Science 8:705–710
Roberts JJ, Foster JE (1983) Effect of leaf pubescence in wheat on the bird cherry oak aphid (Homoptera: Aphidae). J Econ Entomol 76:1320–1322
Rudgers JA, Strauss SY, Wendel JF (2004) Trade-offs among anti-herbivore resistance traits: insights from Gossypieae (Malvaceae). Am J Bot 91:871–880
Schoonhoven LM, van Loon JA, Dicke M (2005) Insect-plant biology, 2nd edn. Oxford University press, New York, 421 pp
Schuh RT (1976) Pretarsal structure in the Miridae (Hemiptera) with a cladistic analysis of relationships within the family. Am Mus Novit 2601:1–39
Schuh RT, Slater JA (1995) True bugs of the world (Hemiptera: Heteroptera): Classification and natural history. Cornell University Press, London, 336 pp
Seidenstücker G (1967) Eine Phyline mit *Dicyphus*-Kralle (Heteroptera, Miridae). Reichenbachia Bd. 8. 27:215–220
Sengonca C, Gerlach S (1984) Einfluss der Blattoberfläche auf die Wirksamkeit des räuberischen Thrips, *Scolothrips longicornis* (Thysan.: Thripidae). Entomophaga 29:55–61
Shanower TG, Romeis J, Peter AJ (1996) *Pigeonpea* plant trichomes: Multiple trophic level interactions. In: Ananthakrishnan TN (ed) Biotechnological perspectives in chemical ecology of insects. Oxford & IBH, New Delhi, pp 76–88
Shah MA (1980) Beutesuchverhalten von Coccinelliden. Dissertation, Universität Hohenheim, Germany
Shah MA (1982) The influence of plant surfaces on the searching behaviour of coccinellid larvae. Entomol Exp Appl 31:377–380
Simmons AT, Gurr GM (2004) Trichomes-based host plant resistance of *Lycopersicon* species and the biocontrol agent *Mallada signata*: are they compatible? Entomol Exp Appl 113:95–101
Simmons AT, Gurr GM (2005) Trichomes of *Lycopersicon* species and their hybrids: effects on pests and natural enemies. Agr Forest Entomol 7:265–276
Simmons AT, Gurr GM, McGrath D, Martin PM, Nicol HI (2004) Entrapment of *Helicoverpa armigera* (Hübner) (Lepidoptera: Noctuidae) on glandular trichomes of *Lycopersicon* species. Australian J Entomol 43:196–200
Simmons AT, Gurr GM, McGrath D, Nicol HI, Martin PM (2003) Trichomes of *Lycopersicon* spp. and their effect on *Myzus persicae* (Sulzer) (Hemiptera: Aphidiae). Australian J Entomol 42:373–378
Smith CM (2005) Plant resistance to arthropods. Springer, Dordrecht, 423 pp
Southwood TRE (1973) The insect/plant relationship—an evolutionary perspective. Symp Roy Entomol Soc London 6:3–30

Southwood R (1986) Plant surfaces and insects-an overview. In: Juniper B, Southwood R (ed) Insects and the plant surface. Edward Arnold Publishers, London, pp 1–22

Stork NE (1980b) Role of wax blooms in preventing attachment to brassicas by the mustard beetle, *Phaedon cochleariae*. Entomol Exp Appl 28:100–107

Stork NE (1986) The form of plant waxes: a means of preventing insect attachment? In: Juniper B, Southwood R (eds) Insects and the plant surface. Edward Arnold Publishers, London, pp 346–347

Thurston R (1970) Toxicity of trichome exudates of *Nicotiana* and *Petunia* species to tobacco hornworm larvae. J Econ Entomol 63:272–274

Thurston R, Webster JA (1962) Toxicity of *Nicotiana gossei* domin to *Myzus persicae* (Sulzer). Entomol Exp Appl 5:233–238

Thurston R, Smith WT, Cooper BP (1966) Alkaloid secretion by trichomes of *Nicotiana* species and resistance to aphids. Entomol Exp Appl 9:428–423

Uphof JCT (1962) Plant hairs. In: Zimmermann W, Ozenda PG (eds) Handbuch der Pflanzenanatomie, Gebrüder Bornträger, Berlin, 2. v. neub. Aufl., Band IV, Teil 5, Abteilung: Histologie, Gebrüder Borntraeger, Berlin-Nikolassee, pp 1–206

Valverde PL, Fornoni J, Núnez-Farfán J (2001) Defensive role of leaf trichomes in resistance to herbivorous insects in *Datura stramonium*. J Evol Biol 14:424–432

Van Dam NM, Hare JD (1998) Differences in distribution and performance of to sap-sucking herbivores on glandular and non-glandular *Datura wrightii*. Ecol Entomol 23:22–32

Voigt D (2004) Relevanz der Pflanzen im Räuber-Beute-(Wirts-)Pflanzen-Komplex der Weichwanze *Dicyphus errans* Wolff (Heteroptera, Miridae, Bryocorinae). Mitteilungen der BBA, Berlin-Dahlem 396:480

Voigt D (2005) Untersuchungen zur Morphologie, Biologie und Ökologie der räuberischen Weichwanze *Dicyphus errans* Wolff (Heteroptera, Miridae, Bryocorinae). Dissertation, TU Dresden, Germany, http://nbn-resolving.de/urn:nbn:de:swb:14-1138036391273-82564

Voigt D, Pohris V, Wyss U (2006) Zur Nahrungsaufnahme von *Dicyphus errans* Wolff (Heteroptera, Miridae, Bryocorinae): Nahrungsspektrum, Potenzial und Verhalten. Mitt Dtsch Ges Allg Angew Ent 15:305–308

Wagner E (1955) Bemerkungen zum System der Miridae (Hemiptera, Heteroptera). Dtsch Entomol Z 2:230–242

Walters PJ (1974) A method for culturing *Stethorus spp*. (Coleoptera: Coccinellidae) on *Tetranychus urticae* (Koch) (Acarina: Tetranychidae). J Aust Ent Soc 13:245–246

Wheeler AG (2001) Biology of the plant bugs (Hemiptera: Miridae): Pests, predators, opportunists. Cornell University Press, London, pp 507

Wigglesworth VB (1959) Physiologie der Insekten. Reihe Exp. Biologie, Bd. 14, Birkhäuser Verlag, Basel & Stuttgart, 2. Aufl

Zvereva EL, Kozlov MV, Niemelä P (1998) Effects of leaf pubescence in *Salix borealis* on host-plant choice and feeding behaviour of the leaf beetle, *Melasoma lapponica*. Entomol Exp Appl 89:297–303

Zwölfer H (2003) Insekten und Pflanzen. In: Dettner K, Peters W (eds) Lehrbuch der Entomologie. Spektrum Akademischer Verlag, Heidelberg, 2. neub. Aufl, pp 499–520

Arthropod-Plant Interactions
DOI 10.1007/s11829-010-9088-1

ORIGINAL PAPER

Locomotion in a sticky terrain

Dagmar Voigt · Stanislav Gorb

Received: 6 October 2009 / Accepted: 2 February 2010

Abstract The mirid bug *Pameridea roridulae* lives mutalistically on the protocarnivorous plant *Roridula gorgonias*. The latter resembles an effective, three-dimensional flypaper trap which captures numerous flying insects. We have recently shown that *P. roridulae* bugs are not trapped by the plant, because they are covered with a layer of epicuticular grease, which is considerably thicker than in other insects. The present study demonstrates that the bugs' morphology and locomotory characteristics also contribute to their specialisation for life on the adhesive plant surface. A structural analysis of the mirid bug's attachment system, and an experimental study on its attachment ability were carried out. In traction force tests, maximum forces of 8.8 mN were measured on adaxial *R. gorgonias* leaves, corresponding to 126 times the bug's body weight. On smooth surfaces, generated forces were only 47 times the bug's body weight. Compared to closely related mirid bug species avoiding contact with plant adhesive secretion, *P. roridulae* is distinctly stronger and heavier, and holds its body close to the plant substrate. Two locomotion strategies on the glandular hairy plant surfaces are suggested for mirid bug species from the tribus Dicyphini: (1) avoidance strategy, characterised by the slim body held at a large distance from the plant surface by using long, slender legs, and (2) defense strategy, where trapping of the heavy bugs, situated close to the plant surface, is overcome by generating strong forces during locomotion and by having a thick anti-adhesive epicuticular greasy layer on the bugs' cuticle.

Handling editor: Heikki Hokkanen.

D. Voigt (✉) · S. Gorb
Evolutionary Biomaterials Group, Department of Thin-Films and Biological Systems,
Max-Planck Institute for Metals Research, Heisenbergstraße 03, 70569 Stuttgart, Germany
e-mail: voigt@mf.mpg.de

S. Gorb
e-mail: sgorb@zoologie.uni-kiel.de

D. Voigt · S. Gorb
Department of Functional Morphology and Biomechanics, Zoological Institute,
Christian Albrechts University of Kiel, Am Botanischen Garten 1–9, 24098 Kiel, Germany

Keywords Attachment · Biomechanics · Bryocorinae · Heteroptera · Insect-plant interactions · Miridae · Traction force · Trichomes

Introduction

South African bugs *Pameridea roridulae* Reuter (Heteroptera, Miridae, Bryocorinae, Dicyphini) are obligately associated with protocarnivorous plants *Roridula gorgonias* Planch. (Roridulaceae; Reuter 1907; Dolling and Palmer 1991; Ellis and Midgley 1996; Anderson and Midgley 2002; Anderson 2006). The host plant is densely covered with glandular trichomes of three different types (short, middle-sized, and long tentacle-shaped ones; Fenner 1904; Bruce 1907; Voigt et al. 2009). It is a three-dimensional, meshed, sticky, flypaper trap that has been observed to capture numerous insects of considerable body mass (Marloth 1903, 1910). Nevertheless, *P. roridulae* live and walk confidently on the sticky plant surface. The surface of these mirid bugs is covered with an anti-adhesive, greasy, epicuticular layer that prevents the bugs' adherence to the plant secretion (Voigt and Gorb 2008) and enables them to live in a digestive mutualism with *R. gorgonias* (Ellis and Midgley 1996). Mirid bugs feed on the insects captured by plant trichomes and defecate on their leaves (Marloth 1903; Loyd 1934). The nitrogen in the bugs' faeces is absorbed, for nutrition, through the thin leaf cuticle of the plant, since the plants themselves produce no enzymes to digest captured insects (Marloth 1910; Lloyd 1934; Ellis and Midgley 1996; Anderson and Midgley 2002, 2003). Furthermore, juveniles of *Pameridea* are primary pollinators of *Roridula*'s flowers (Anderson et al. 2003).

Recent coevolutionary studies on mutualistic interactions resulted in phylogenetic and geographical associations, suggesting cospeciation, coadaptation, and long-term coexistence in the *Roridula-Pameridea* complex (Anderson et al. 2004; Anderson 2006). In host choice experiments with *Roridula dentata* L. (Roridulaceae) and *Pameridea marlothi* Poppius (Heteroptera, Miridae), bugs preferred host plants over unrelated host plant species and closely related sister species (Anderson et al. 2004). The presence of an intermediate number of *P. marlothi* caused positive growth rates in *R. dentata* (Anderson and Midgley 2007).

Only a few data about mutualistic effects on bug's fitness and specific adaptations are currently available. Since relationships between insects and plants are highly diverse associations, they need broad studies of various factors and features at different conditions for a better understanding and evaluation (Thompson 1988). Besides genetic, fossil, taxonomic, geographical, ecological, and biochemical evidence, insights into the functional morphology and ethology may indicate associations and coadaptations (Futuyma and Slatkin 1983).

Similar to *R. gorgonias* and *P. roridulae*, relationships between mirid bugs and carnivorous plants have been reported from representatives of the plant genera *Byblis* and *Drosera* in Australia (e.g. Schuh 1995). These bug species are representatives of the mirid bug subfamilies Orthotylinae and Bryocorinae, which are known to be specialised in living on glandular hairy plants (Reuter 1913; Kullenberg 1946; Falkingham 1995; Dolling and Palmer 1991; Wheeler 2001; Sugiura and Yamazaki 2006; Voigt et al. 2007). Such mirid bugs have been reported to bear elongated, curved, sharp claws and free, rather large pseudopulvilli (Reuter 1913; Wagner 1955). Besides specialised claws, they have slim bodies as well as long, slender legs (Roberts 1930; Southwood 1986; Voigt et al. 2007). In contrast to *P. roridulae*, other related species have been observed avoiding contact with sticky glandular plant secretions by means of morphological adaptations.

Additionally, mirid bugs may show a particular behaviour. For example, *Dicyphus errans* Wolff (Heteroptera, Miridae, Bryocorinae, Dicyphini) stalks along the plant surface touching only trichome tips and stops frequently to groom itself (Southwood 1986; Voigt et al. 2004). For *D. errans*, non-glandular and glandular hairy plant substrates have been experimentally demonstrated to be important attachment substrates (Voigt et al. 2007), whereas geometrical variables of trichomes (length and diameter) significantly influenced the bug's attachment ability. Trichomes have been found to provide suitable interlocking sites for attachment and locomotion of this tiny mirid bug. It clings to single trichomes using one part of the paired claw.

Preliminary observations on *P. roridulae* let us suggest a different behavior strategy. This insect keeps its body close to the plant substrate and frequently contacts adhesive trichome secretion that covers almost the entire surface of *R. gorgonias* and adheres occasionally to bristles on the mirid bug′s body surface (Voigt and Gorb 2008). It runs quickly between glandular trichomes, moves sometimes with jump-like movements or short flights. However, it grooms itself infrequently. Compared to *D. errans*, morphological differences are also ascertainable in *P. roridulae*; the latter appears burlier. Although the length of the body and legs is similar to that in *D. errans*, the width differs between the species and is distinctly larger in *P. roridulae* (Table 1).

Possibly, not only the anti-adhesive, greasy, epicuticular layer enables these bugs to live on the sticky plant surface, but also other morphological and behavioral characteristics contribute to it′s ability. The stronger build of *P. roridulae* may result in a larger muscle mass, due to which mirid bugs could generate stronger forces during locomotion on the sticky substrate, where they have possibly to free themselves from trichomes' viscous secretion. As a specialist species, it is strongly associated with its single host plant (Anderson et al. 2004; Anderson 2006). Similar to related species of Miridae, they have to attach properly to the surface, not only during locomotion and feeding, but also during copulation, molting and oviposition of eggs into the plant tissue (Wheeler 2001).

Thus, the question arises, how do *P. roridulae* bugs attach and walk on their host plant? Do their attachment devices and attachment ability differ from those of related mirid bugs? Are they specialised for the surface of *R. gorgonias*? Previously, pretarsal structures in the genus *Pameridea* have been described to be similar to those of the genus *Dicyphus*, bearing short, curved claws, triangular pseudopulvilli and a straight, bristle-like parempodia (Dolling and Palmer 1991). However, up to now, no illustrations of these structures are available in the literature.

The present study should contribute to the better understanding of the complex mutualistic relationship between the protocarnivorous plant and the associated mirid bugs, in

Table 1 Comparison of morphometric variables previously referred for *Dicyphus errans* (Wagner 1970/71) and *Pameridea roridulae* (Dolling and Palmer 1991)

Variable (mm)	*Dicyphus errans*	*Pameridea roridulae*
Body length	4.7	4.9
Body width	1.2	2.5
Length of anterior leg	5.3	6.2
Width of anterior tibia	0.1	0.3
Width of anterior femur	0.3	0.8

particular at the interface between the plant surface and insect attachment system. Using light and cryo-scanning electron microscopy (cryo-SEM), we analyzed the pretarsal structures of *P. roridulae*. Its forces, generated in a traction experiment, were measured on various substrates: glandular hairy adaxial and abaxial leaf surfaces of the bug's host plant, smooth adaxial and sparsely hairy abaxial leaves of *Rumex obtusifolius* L. (Polygonaceae) and glass.

Materials and methods

Insects and plants

Shrubs of *R. gorgonias* (seeded, potted, 1–3 year old) with *P. roridulae*, were obtained from a private glasshouse culture (Klaus Keller, Augsburg, Germany). They were kept under laboratory conditions during experiments (23.7 ± 1.7°C, 47.3 ± 10.0% RH, 16 h photoperiod), and fed with wingless, adult *Drosophila melongaster* Meigen (Diptera, Drosophilidae; Zoo-Schöniger, Stuttgart, Germany).

Leaves from plants of *Rumex obtusifolius* L. (Polygonaceae) at eight-leaf stage were collected from the wayside in a humid mixed forest in Stuttgart-Büsnau (Baden-Württemberg, Germany). Its adaxial leaf surface is uneven because of the prominent venation and irregularly shaped convex epidermal cells and has a maximum height of about 9 μm (Holloway 1967; Gorb and Gorb 2009). The cells are relatively large (50.0 ± 9.04 μm long, 30.7 ± 7.45 μm wide), covered with a cuticle and a smooth, amorphous, epicuticular wax layer (0.1 ± 0.05 μm thick). Sporadically, very sparse small wax crystals occur in some places on the surface. The crystals have irregular scale-like or granular shapes and vary in size (0.5 ± 0.16 μm long, 0.3 ± 0.07 μm wide, and 0.1 ± 0.0 μm thick) (Gorb and Gorb 2009). The abaxial leaf side is roughly similar to the adaxial, however, convex cells are striated and tiny (Holloway 1967).

Observations and structural studies

The attachment positions of five male and five female freely walking *P. roridulae* to the plant surface were studied (1) visually without optical instruments, (2) visually with a stereomicroscope Olympus SZX 12 with a DF PLAPO 1xPF objective (Olympus Corp., Tokyo, Japan), and (3) using digital recordings. Images were taken using (1) a digital single lens reflex camera Canon EOS 20D combined with the Canon macro lens EF 100 mm, 1:2.8, USM and the Canon macro twin lite MT-24EX (Canon Inc., Japan), and (2) a Nikon Coolpix E995 digital camera adapted to the stereomicroscope with a C-Mount adapter and MDC 2 relay lens MXA 29005 (Nikon Corp., Tokyo, Japan).

To analyze the morphology of the mirid bug's attachment system and the interaction with the plant surface, a cryo-SEM Hitachi S-4800 (Hitachi High-Technologies Corp., Tokyo, Japan) equipped with a Gatan ALTO 2500 cryo-preparation system (Gatan Inc., Abingdon, UK) was used. Fresh samples of mirid bug legs and plant leaves were cut out using a razor blade, mounted on metal holders by Tissue-Tek® O.C.T.™ Compound (Sakura Finetek Europe B. V., Zoeterwoude, Netherlands), frozen in the preparation chamber at −140°C, sputter-coated with gold–palladium (6 nm thickness) and examined in a frozen state in the cryo-SEM at 3 kV and −120°C.

To visualize in detail how mirid bugs' claws interlock with trichomes on the host plant surface, leaf samples of *R. gorgonias* with attached *P. roridulae* were prepared, clamped perpendicularly on holders, and examined in the cryo-SEM as described above. Additionally, from SEM micrographs of 10 randomly selected pretarsi of adult specimens (males and females pooled together), the inner length of the claw, diameter of a circle fitting the inner curvature of the claw, and length of the setiform parempodia were estimated, using Sigma Scan Pro 5 (SPSS, Inc., Chicago, IL, USA) software.

Traction force experiment

Traction force measurements were carried out according to Gorb et al. (2004) and Voigt et al. (2007). Insects were anaesthetised with CO_2 and attached to a 10 cm long human hair with a molten wax droplet. Their forewings were glued together. Bugs were weighed using an analytical balance AG 204 Delta Range (Mettler Toledo GmbH, Greifensee, Switzerland). Following anaesthesia, insects were allowed to recover for 1 h. Then, the free end of the hair was connected to the force sensor.

Fresh leaves were removed from the middle part of *R. gorgonias* plants and attached to a horizontal glass plate using double-sided tape. Pieces (5×5 cm^2) were cut out with a razor blade from the intercostal leaf areas of the non-host plant *R. obtusifolius*. Using a FORT-10 force transducer (10 g capacity, Biopac Systems Ltd., Santa Barbara, CA, USA), the traction force was measured in males and females of *P. roridulae*. Each specimen was tested (1) walking distally on the adaxial and abaxial leaf surface of *R. gorgonias* (contact angle of water could not be determined), (2) walking proximally on the adaxial and abaxial leaf surface of *R. gorgonias*, (3) walking distally on the non-host plant *R. obtusifolius* (contact angle of water 40–70°, surface energy 35.43 mN*m^{-1}; Gorb and Gorb 2009), and (4) walking on a glass surface (contact angle of water 55°, surface energy 42.9 mN*m^{-1}). The succession of different substrates (1–4) was randomly organised during the experiment. Before the exchange of substrates, bugs were allowed to groom and to walk over KIMTECHScience™ precision wipes (Kimberly-Clark Europe Ltd, Surrez, UK).

Using AcqKnowledge 3.7.0 software (Biopac Systems Ltd, Goleta, CA, USA), force–time curves were recorded to estimate the maximal traction force produced by a single bug during five consecutive runs on a test substrate. On each plant surface, five males and five females were tested individually. In total, 70 individual tests with 10 different specimens were carried out ($N = 7$ substrates, $n = 10$ individual tests). Kruskal–Wallis one-way ANOVA on ranks followed by all pair-wise multiple comparision procedures (Dunn's test) was applied to determine differences in forces between test surfaces (software SigmaStat 3.1.1®, Systat Software, Inc., Richmond, California, USA). Force measurements in various walking directions on the same leaf surface of *R. gorgonias* were included because trichomes are oriented distally and thus some anisotropical effects could occur. In contrast to *R. gorgonias*, the dock plant *R. obtusifolius* represents a non-host plant having a totally different surface. Since no structural anisotropy was found on dock plant leaf surfaces, traction was tested only for mirid bugs walking distally on the leaf.

Results

The attachment system of *Pameridea roridulae*

Mirid bugs were able to move quickly and confidently, and to attach properly to the whole plant surface (stems, both leaf sides and inflorescence; Fig. 1a–h). Frequently they were observed resting in axillae (Fig. 1b, d, e). Predominant interlocking sites were trichome

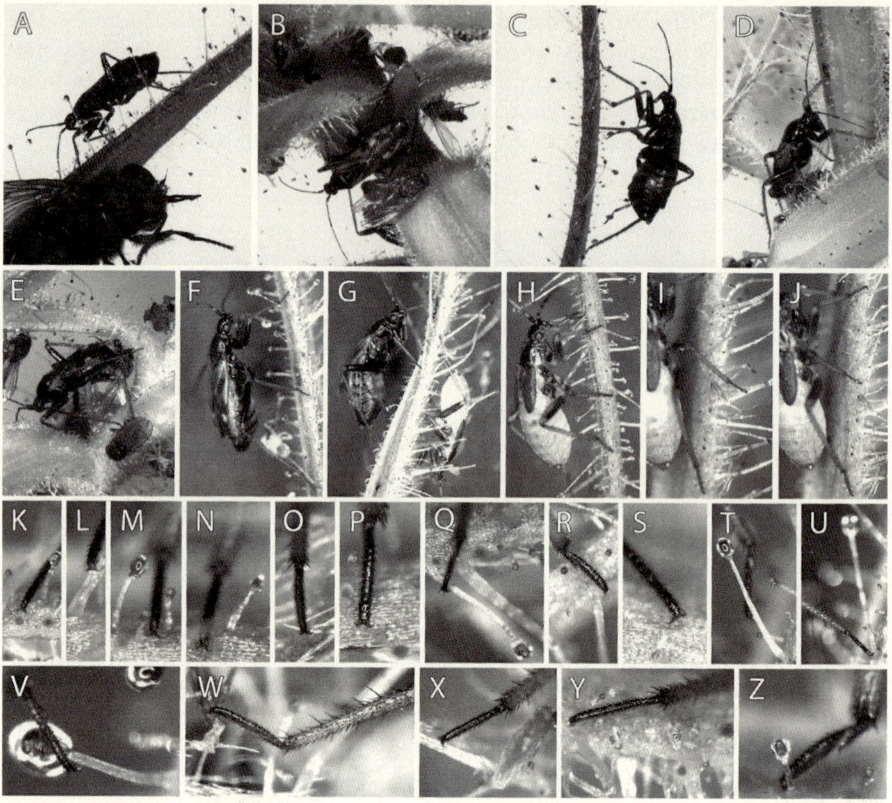

Fig. 1 Images of various positions of *Pameridea roridulae* bugs and their tarsi on the surface of the host plant *Roridula gorgonias*. **a–c** A female bug, adhering to trichomes, close to a captured blow fly (**a**); at the plant stem in an axilla (**b**); and close to the leaf tip (**c**). **d**, **e** The male bug in diverse similar attachment positions. **f** A female, adhering to the smooth plant surface between trichomes; pretarsi almost orientated parallel to the plant surface. **g** A female and 5th instar nymph clinging to the plant surface at opposite leaf sides. **h–j** 5th instar nymph demonstrating similar behavior as in **f** (**h**), and pretarsi interlocked with trichomes (**i**, **j**). **k–z** Details of pretarsi adhering to the plant surface at various positions. *Note*: claw interlocking close to the base of a single trichome (**k**, **l**, **q**, **p**, **r**, **u**, **x**, **y**, **z**), at higher sites of the trichome stem (**t**, **v**), and attachment to the smooth plant surface between trichomes (**m**, **n**, **s**). The body length of the bugs is about 5 mm

stems, namely their bases, up to the lower one-third of the trichome (Fig. 1f–n, p, q–s, u, w, y, z). Infrequently, claws clung more distally to the trichomes (Fig. 1v). The leg span (distance between left and right leg pretarsi) fits approximately to the width of the leaf lamina. Thus, bugs grip trichomes at both leaf margins at once, and hold their body above the leaf lamina. However, bugs also attached to the smooth plant epidermis between trichomes, positioning their tarsus almost parallel and very close to the surface (Fig. 1f). Claw interlocking, where claw curvature fits the trichome stem, occurred on filiform thread-shaped (Fig. 1o), small (Fig. 1v) and middle-sized trichomes (Fig. 1t). On thicker tentacle-shaped ones, the position of claws was similar to that on smooth surfaces (Fig. 1x). Claws moved apart under load, applied on the flat surface, allowing pseudo-pulvilli to get a grip on the plane surface.

Cryo-SEM studies of the pretarsus clarified its position on the plant surface (Fig. 2a) and the structure of attachment devices (Fig. 2b–f). Paired curved claws are 23.1 ± 6.77 μm (mean ± SD) long and connected to the unguitractor plate at their bases (Fig. 2c). They have slightly tapered tips (claw teeth, 2.6 ± 0.51 μm long) and widened basal lobe-like structures. The latter ones bear paired, trapezium-shaped pseudopulvilli and between them, two setiform parempodia (13.8 ± 1.0 μm long). Dorsally, a stronger (guard) seta is visible (Fig. 2f). The claw surface bears longitudinal striae. The inner side of claws appears sharp-edged. Adhering to a single short trichome, its stem is clamped beween the inner side of one claw part and a single pulvillus (Fig. 2d, e). The average diameter of an ideal circle fitting the concavity of claws is 16.1 ± 4.14 μm.

Traction force of *Pameridea roridulae*

Plant substrates used in the traction force experiment varied in the structure of their surfaces (Fig. 3). The adaxial leaf side of *R. gorgonias* is densely covered with filiform, thread-shaped, distally pointing trichomes and interdispersed short glandular trichomes (Fig. 3a). The abaxial leaf side bears short and medium-sized glandular trichomes, and the

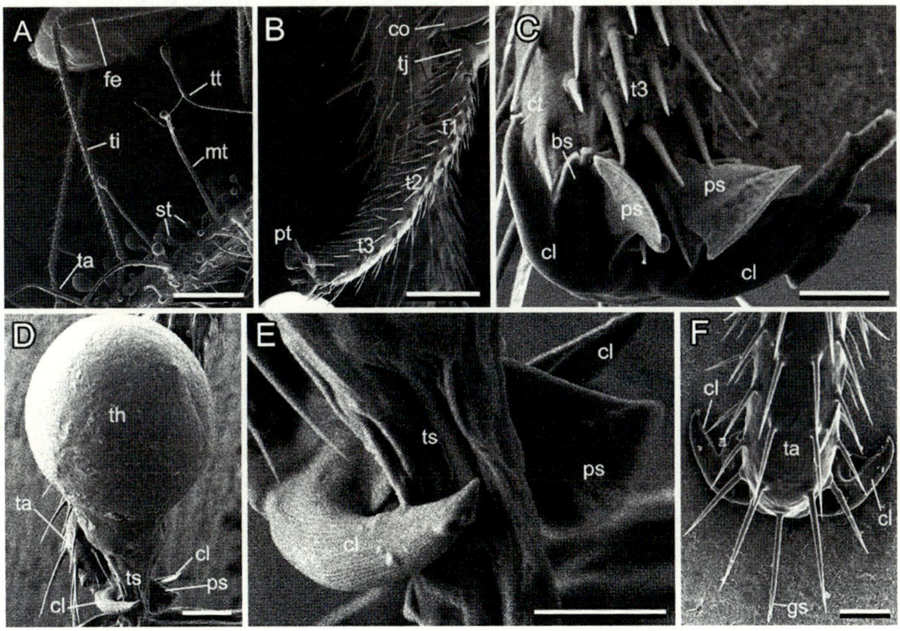

Fig. 2 Cryo-SEM micrographs of the attachment devices in *Pameridea roridulae* bugs. **a** Female's abdominal tip and hind legs, adhering to the glandular hairy abaxial plant surface of *Roridula gorgonias*. **b** Female tarsus. **c** Female pretarsus (unguitractor plate hidden by tarsomere 3 and other pretarsal structures). **d, e** Male pretarsus interlocked with a short glandular trichome. The middle of the trichome stem is clamped between a single claw, its basal lobe-like structure and the pseudopulvillus. **f** A male pretarsus attached to the smooth metallic surface of a SEM sample holder. Pseudopulvilli are pressed against the substrate beneath the outspread claws. The guard seta is clearly visible. (*bs*) basal lobe-like structure; (*cl*) claw; (*co*) grooming comb; (*ct*) claw tooth; (*fe*) femur; (*gs*) guard seta; (*mt*) middle-sized trichome; (*ps*) pseudopulvillus; (*pt*) pretarsus; (*st*) short trichome; (*ta*) tarsus; (*t1*) 1st tarsomere; (*t2*) 2nd tarsomere; (*t3*) 3rd tarsomere; (*th*) trichome glandular head; (*ts*) trichome stem; (*tt*) tentacle-shaped trichome; (*ti*) tibia; (*tj*) tibia-tarsus joint. *Scale bars*: **a** = 1,000 μm; **b, d** = 50 μm; **c, e, f** = 20 μm

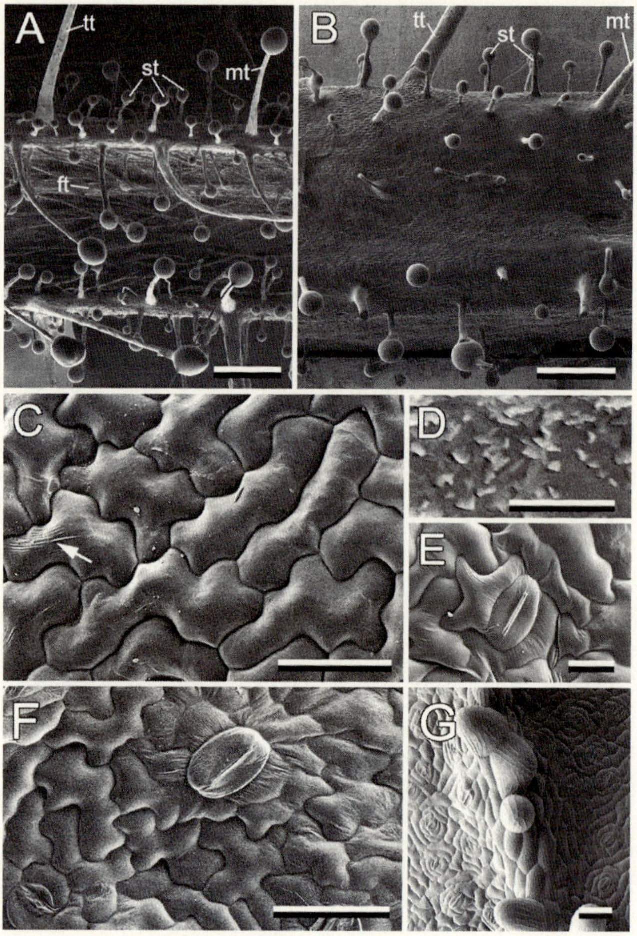

Fig. 3 Cryo-SEM micrographs of plant substrates used in traction force experiment with *Pameridea roridulae* bugs. **a**, **b** The leaf surface of *Roridula gorgonias* bears various types of trichomes on the adaxial (**a**), and on the abaxial leaf side (**b**). **c**–**e** The uneven adaxial leaf surface of *Rumex obtusifolius* shows irregular convex cells and stomata (**c**, **e**), sparse cuticular foldings (*arrow*) and wax granules (**d**). **f**, **g** The abaxial leaf surface is similar to the adaxial, however, prominent veins are covered with small papilla-shaped trichomes (**b**). (*ft*) filiform, thread-shaped trichomes; (*mt*) middle-sized trichomes; (*st*) short trichomes; (*tt*), tentacle-shaped trichomes. *Scale bars*: **a**, **b**, **c**, **f**, **g** = 50 μm; **d** = 1 μm; **e** = 20 μm

spaces between them are rather smooth, having slight cell irregularities (Fig. 3b). Leaf margins are densely covered with long, medium-sized and short trichomes.

Leaves of *R. obtusifolius* appear adaxially predominantly smooth, slightly textured by convex cells having sparse cuticular foldings (Fig. 3c), and a few discrete patches of epicuticular crystal wax granules and stomata (Fig. 3d–e). On the abaxial leaf side, convex cells and venation are more prominent. Larger veins are covered with papilla-shaped trichomes (Fig. 3f). Numerous stomata are found (Fig. 3g).

The body mass of *P. roridulae* ranged from 6.7 ± 0.35 mg (mean ± SD) in males to 7.0 ± 0.63 mg in females without a statistical difference between the sexes (*t*-test, $t = -0.864$, $P = 0.413$, $n = 10$). Traction forces differed significantly depending on the surface (Fig. 4). The highest forces, up to 8.8 mN, were generated on the adaxial leaf surface of *R. gorgonias*, corresponding to 126 times the bug's body mass. The lowest forces were measured on the adaxial leaf side of *R. obtusifolius* and on a glass surface (47 times the bug's body weight). Leaf side and pulling direction did not significantly influence the bug's performance on *R. gorgonias*. On *R. obtusifolius*, higher forces were generated on the rougher abaxial surface. Exept for the distal direction on the abaxial leaf side of *R. gorgonias*, the adaxial leaf side of *R. obtusifolius* and the glass surface, no statistical differences were found between sexes on the same surface (Table 2).

Discussion

Mirid bug behaviour

The locomotion mode of *P. roridulae* and the manner in which it attaches to plants affirm previous reports that this bug does not avoid contact with the sticky sites of the host plant surface (Voigt and Gorb 2008). Rather it frequently touches adhesive plant secretion droplets since it adheres, predominantly, to the trichome base at the lower third of the trichome stem. Thus, the body is held very close to the plant surface, unlike related species. For example, *Dicyphus errans* adheres to the glandular surface of *Ononis* sp., positioning the long tibiae and the especially long hind femuri almost vertically to the plant surface (Southwood 1986). Only the tarsal apexes come in contact with the plant surface and the body is held far away from it.

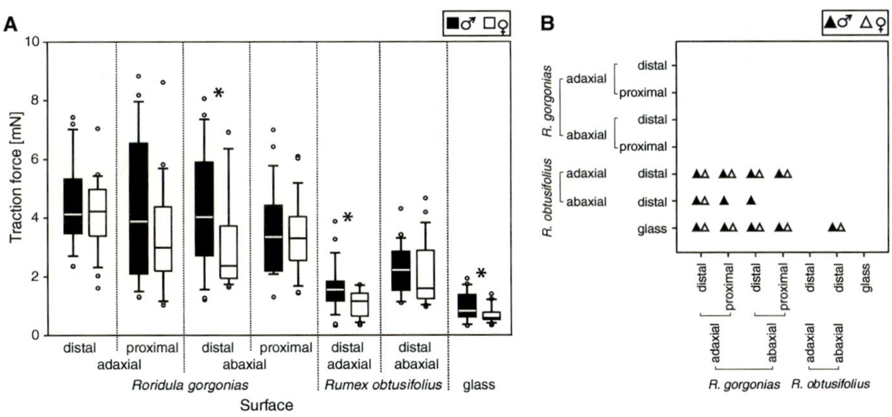

Fig. 4 Forces generated by *Pameridea roridulae* in traction experiments on leaf surfaces of *Roridula gorgonias* (traction in distal and proximal direction) and *Rumex obtusifolius* (traction in distal direction), and on cleaned glass (control). **a** *Box-and-whisker* diagram showing the traction force of single mirid bugs measured on various surfaces. The ends of the boxes define the 25 and 75th percentiles, with a line at the median and error bars defining the 10 and 19th percentiles. *Asterisks* show statistical differences between males and females (for statistical differences see Table 2). **b** Statistical differences between surfaces (Kruskal–Wallis one-way ANOVA on ranks, for males: $H_{6,173} = 101.5$, $P \leq 0.001$; for females: $H_{6,173} = 111.7$, $P \leq 0.001$; all pairwise multiple comparison procedures, Tukey test, $P < 0.05$ for both, separately males and females)

Table 2 Statistical differences of forces between males and females, generated in the traction experiment on various surfaces ($N_{♂♂} = 5$, $N_{♀♀} = 5$, $n = 5$). See also Fig. 2

Surface and traction direction	Statistical values
Roridula gorgonias, adaxial leaf side, distal	$t = 1.020$, $P = 0.313$
Roridula gorgonias, adaxial leaf side, proximal	$t = 1.797$, $P = 0.079$
Roridula gorgonias, abaxial leaf side, distal	$T = 761.000$, $P = 0.017$
Roridula gorgonias, abaxial leaf side, proximal	$t = 0.477$, $P = 0.636$
Rumex obtusifolius, adaxial leaf side, distal	$t = 2.893$, $P = 0.006$
Rumex obtusifolius, abaxial leaf side, proximal	$t = 0.561$, $P = 0.578$
Glass	$T = 756.000$, $P = 0.022$

t computed *t*-statistic in the *t*-test for normally distributed data, *T* computed *T*-statistic in the Mann–Whitney rank sum test for not normally distributed data, *P* computed p-values for *t* and *T*, respectively

Pretarsal adaptations to the trichome-covered terrain

Although the pretarsus of *P. roridulae* bears similar structures found in related bryocorine bug species (Schuh 1976; Cobben 1978; Wheeler 2001), claws in *P. roridulae* are distinctly shorter (mean length 23.1 μm) and thicker, e.g., compared to those of the mirid bug *D. errans* (mean length 74.0 μm, Voigt et al. 2007). The body length is similar in both mirid bug species (Table 1), but the body of *P. roridulae* appears stronger and thicker. The mean female body mass in *P. roridulae* is 7.0 mg corresponding to 1.8 times that of the female *D. errans*. The body mass, in combination with trichome stem flexibility, may explain why the bugs *P. roridulae* predominantly cling to the basal part of trichomes. Knowing trichome spring constants and bending behaviour (tentacle-shaped: 0.02 Nm^{-1}, bending at the tip; medium-sized: 0.08 Nm^{-1}, bending at the tip; short: 0.95 Nm^{-1}, bending at the base Voigt et al. 2009), the trichome's deflection, caused by the load of a mirid bug, can be estimated. Since the geometry of trichomes is similar, the deflection of short trichomes may also be considered as for the base of tentacle-shaped ones. Assuming a resting female (7 mg) having all six legs clinging to trichomes, a tentacle-shaped trichome would bend 1 mm at the tip but only about 20 μm at the base. For comparison, a female *D. errans* (3.8 mg) would cause a trichome deflection of 250 μm at the trichome's tip and about 7 μm at the base. Less trichome bending allows more effective bug attachment, as shown in previous studies on *D. errans*, where long and thin trichomes with a high aspect ratio resulted in lower traction forces (Voigt et al. 2007).

The heavier body of *P. roridulae* led us to assume a generally larger portion of muscles enabling this bug species to generate stronger forces compared to lighter related species. The assumption is confirmed by traction forces measured with *P. roridulae* on various surfaces. Traction force corresponds to the friction between insect pretarsi and the substrate. The proper adherence to trichomes and applying a pulling force to them while remaining in a certain position, is important in behavioural situations where mirid bugs' bodies move back and forth or up and down the substrate (Kullenberg 1946; Voigt et al. 2004; Voigt 2005). (1) During sucking the head and thorax are repeatedly moved up and down, supporting the rostrum bending inside the prey. (2) During the final phase of copulation the male and female couple their abdominal tips, heads facing in opposite directions, and move their bodies back and forth, as in a "tug of war". (3) The oviposition requires strong up and down movements of the female abdomen and therefore strong attachment to the plant substrate in order to insert the saw-shaped ovipositor into the plant tissue. (4) Molting nymphs and emerging imagines free themselves from old cuticle by pulling their body distally. For *P. roridulae*, the challenge is to perform these actions even upon its adhesive host plant terrain. Possibly, for this reason *P. roridulae* may generate higher traction forces (126 times higher than the body mass) than the related *D. errans* (34.2 times higher than the body mass) on both a glandular hairy plant surface as well as on glass (*P. roridulae*: 47, and *D. errans*: 6.1 times higher than the body mass). The latter values indicate that mirid bugs may also adhere sufficiently to microstructured and smooth surfaces such as those of *R. obtusifolius* and glass. Thus, smooth sites on *R. gorgonias* (trichome-free spaces on the leaf and wide, non claw-fitting trichome stems) provide suitable attachment substrates for *P. roridulae* bugs. Considering the lowest generated force on a smooth glass surface (females: 0.7 mN, median), a bug may still free itself from 10 tentacle-shaped trichomes adhering coincidentally to the bug's epicuticle (median adhesion force of a single trichome: 0.07 mN, Voigt and Gorb 2008).

Pameridea roridulae bugs were observed predominantly interlocking to trichomes using claws. Although their claws are much shorter (average length 23 μm) than those of *D. errans* (74 μm), the average diameter of an ideal circle fitting the concavity of claws in both species is about 16 μm. Since mean stem diameters of trichomes on the surface of *R. gorgonias* are from 75 μm in median-sized to 120 μm in tentacle-shaped ones (Voigt et al. 2009), claws of *P. roridulae* may only interlock with some sites of these trichome stems. Nevertheless, on rough surfaces, higher traction forces were measured compared to smooth ones. For example, on the abaxial leaf side of *R. obtusifolius*, small papilla-shaped trichomes provided interlocking sites for claws and therefore, a stronger traction force of mirid bugs was recorded than on the smooth adaxial leaf side. *P. roridulae* bugs performed best on their host plant. Since forces did not differ significantly between leaf sides and pull direction, leaves of *R. gorgonias*, in general, can be assumed to be a suitable attachment substrate for male and female bugs.

Sexual dimorphism in traction force generation

Compared to females, males generated significantly higher forces on the abaxial leaf side of *R. gorgonias*, on the adaxial leaf side of *R. obtusifolius*, and on glass. Differences between sexes have been previously found in traction force experiments with leaf beetles *Chrysolina polita* L. (Coleoptera, Chrysomelidae; Stork 1980) and ladybird beetles *Coccinella septempunctata* L. (Coleoptera, Coccinellidae; Gorb et al. 2008). Males of both species adhere stronger to smooth surfaces, whereas females performed better on rough ones. However, attachment systems of beetles (hairy) and mirid bugs (smooth) differ completely and cannot be compared outright in the context of our results (Beutel and Gorb 2001). Since significances between male and female force values occur on predominantly smooth substrates, one may conjecture that differences in material properties between male's and female's pseudopulvilli exist. These pad-like attachment structures in mirid bugs, adhering to smooth surfaces, need further micromechanical analyses for better understanding of their functionality depending on sex and species.

Two locomotion strategies of mirid bugs

Generally, in species of the order Heteroptera, a notable diversity of pretarsal structures has been reported and discussed in the evolutionary context (Schuh 1976; Cobben 1978). Some peculiar host plant relations have been previously assumed, but the adaptive value of these structures is difficult to interpret in direct functional connection with the substrate due to the absence of detailed analyses. Considering published data on both *D. errans* (Southwood 1986; Voigt et al. 2007) and *P. roridulae* (Voigt and Gorb 2008) in combination with results obtained in the present study, we may hypothesize two different strategies of locomotion and attachment on glandular hairy plant surfaces in species within one and the same mirid tribe Dicyphini of the subfamily Bryocorinae (Fig. 5). (1) The "avoidance strategy", as used by *D. errans*, consists of keeping the body at a large distance from the plant surface and glandular secretions of trichomes (Fig. 5a). This strategy relies on a light and slim body, long and slender, carefully-moving legs, whose tibiae are oriented almost vertically to the substrate. Pretarsi are equipped with long, narrow claws. Moreover, frequent resting and grooming are observed in these bugs, indicating that they do not have a protective layer on their epicuticle maintaining the body clean and free from adhesive plant secretion. (2) *P. roridulae* has a "defense strategy" allowing close contact with the plant surface and the trichomes' adhesive secretion (Fig. 5b). The body is heavy, with relatively thick legs that are predominantly used for attachment to the trichome bases by short claws. Mirid bugs with "defense strategy" generate stronger forces during locomotion, enabling them to walk, unhinderd, on a plant surface covered with sticky trichomes and to free themselves from adhesive secretions. Additionally, the greasy secretion, found in the *P. roridulae* epicuticle, contributes to reducing the adhesion of the plant adhesive to the bug. One may conjecture that the defense strategy is a specific adaptation of bugs *P. roridulae* for a life on the glandular plant surface of *R. gorgonias*. This would support recently published coevolutionary considerations on cospeciation, coadaptation, and long-term coexistence in the *Roridula-Pameridea* complex (Anderson et al. 2004; Anderson 2006). On the other hand, the avoidance strategy could be a typical feature of generalists, like *D. errans*, having a broad range of host plants.

To what extent the hypothesized strategies may be the case for other mirid bug species, is difficult to appraise, because neither detailed comparative nor experimental studies have been previously carried out. Representatives of the mirid bug subfamilies Orthotylinae and Bryocorinae are generally known to be specialised in living on glandular hairy plants

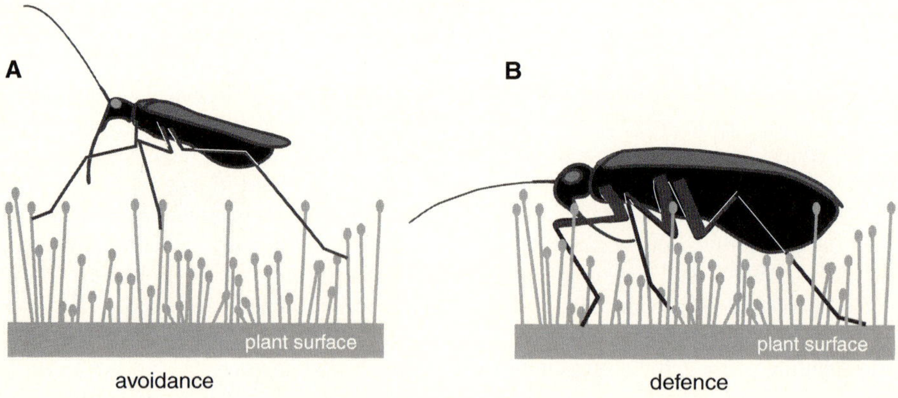

Fig. 5 Two assumed strategies of mirid bugs of the tribus Dicyphini (Heteroptera, Miridae, Bryocorinae) walking on plant surfaces, which are densely covered with glandular trichomes. **a** "Avoidance strategy": the light, slim body is held at a large distance from the plant surface and adhesive secretion of trichomes by using long, slender legs. Additionally, bugs frequently rest and groom to keep their body clean. **b** "Defense strategy": the heavy, tough body is situated close to the plant surface, frequently contacting trichomes and their viscous secretion. Bugs overcome the adhesion of the plant surface by generating strong forces during locomotion and having a thick anti-adhesive epicuticular greasy layer on their cuticle

(Reuter 1913; Kullenberg 1946; Cassis 1984; Falkingham 1995; Dolling and Palmer 1991; Wheeler 2001; Sugiura and Yamazaki 2006; Voigt et al. 2007). Several of them even live on insectivorous plants of the genera *Drosera* and *Byblis* (e.g. China and Carvalho 1951; China 1953; Southwood 1986; Falkingham 1995). These bugs also feed on trapped insects, but unlike *P. roridulae*, they avoid the contact with the sticky surface of their host plants (Russell 1953).

Only one previous report on different modes of locomotion on glandular plant surfaces is found in literature, however, for other true bug families than Miridae (Schwoerbel 1956). Stilt bugs *Gampsocoris punctipes* Germ. (Heteroptera, Berytidae) avoid the contact with glandular plant secretion of the spiny restharrow *Ononis spinosa* L. (Fabaceae) by means of their slim, small body and long, slender legs, whereas juveniles *Corizus hyoscami* L. (Heteroptera, Rhopalidae) appear stronger and walk unhamperedly between trichomes like a "snowplow" on the same plant surface (Schwoerbel 1956). Thus, for non-mirid species, the avoidance strategy is not only characteristic of generalists, because the specialist *G. punctipes* is obligatory associated with *O. spinosa*.

Concluding remarks

Pameridea roridulae lives in a complex, three-dimensional, sticky terrain where it moves without hindrance. In contrast to related species within the tribe Dicyphini of the family Bryocorinae, *P. roridulae* do not avoid adhesive secretions but rather walk between glandular trichomes and touch them very frequently. Previously, an anti-adhesive epicuticular layer protecting mirid bugs from being captured by the adhesive plant secretion has been described. The present results show that the mode of attachment and locomotion on the plant surface also contributes to the specialisation of *P. roridulae* to a life on the protocarnivorous flypaper trap *R. gorgonias*. There are direct, biomechanical interactions at the interface between the plant surface and insect attachment system. Dicyphine bugs use claws to cling to trichomes and use pad-like pseudopulvilli for adhesion to smooth hairless patches or to thick trichome stems. This strategy allows them to generate distinctly higher traction forces if compared to other mirid bug species. Thus, *P. roridulae* may even overcome contact with the adhesive secretion of numerous trichomes. Based on available data, we have hypothesized, here, two strategies of locomotion and attachment on glandular hairy plant surfaces within the mirid bug tribus Dicyphini: (1) avoidance strategy, characterised by the slim body held at a large distance from the plant surface by using long, slender legs, and (2) defense strategy, where trapping of the heavy bugs, situated close to the plant surface, is overcome by generating strong forces during locomotion and by having a thick anti-adhesive epicuticular greasy layer on the bugs' cuticle. The proper attachment and locomotion on the host plant surface may result in the better fitness of insects. Additionally, living on an adhesive plant trap has several long-term adventages for *P. roridulae* (e.g., protection, less competition, permanent food supply). The defense strategy could be specific to *Pameridea* species indicating specialized adaptations of bugs to the plant surface, which would support previously suggested coevolutionary processes in the *Roridula-Pameridea* complex. Further comparative broad screenings of pretarsal structures combined with attachment experiments on different substrates and analyses of epicuticular grease, including various species of mirid bugs and glandular hairy plants, will shed light on the evolutionary tendencies in this insect-plant mutualism.

Springer

Acknowledgments Klaus Keller (Augsburg, Germany) is kindly acknowledged for provision of plants and bugs. The bug species was determined by Kurt Arnold (Geyer, Erzgebirge, Germany). Victoria Kastner (Max-Planck Institute for Metals Research, Stuttgart, Germany) provided linguistic corrections of the manuscript. This study was supported by the Federal Ministry of Education and Research, Germany (BMBF project Inspirat 01RI0633D).

References

Anderson B (2006) Inferring evolutionary patterns from the biogeographical distributions of mutualists and exploiters. Biol J Linn Soc 89:541–549

Anderson B, Midgley JJ (2002) It takes two to tango but three is a tangle: mutualists and cheaters on the carnivorous plant *Roridula*. Oecologia 132:369–373

Anderson B, Midgley JJ (2003) Digestive mutualism, an alternate pathway in plant carnivory. Oikos 102:221–224

Anderson B, Midgley JJ (2007) Density-dependent outcomes in a digestive mutualism between carnivorous *Roridula* plants and their associated hemipterans. Oecologia 152:115–120

Anderson B, Midgley JJ, Stewart BA (2003) Facilitated selfing offers reproductive assurance: a constrained mutualism between a hemipteran and carnivorous plant. Am J Bot 90:1009–1015

Anderson B, Olivieri I, Lourmas M, Stewart BA (2004) Comparative population structures and local adaptation of two mutualists. Evolution 58:1730–1747

Beutel RG, Gorb SN (2001) Ultrastructure of attachment specializations of hexapods (Arthropoda): evolutionary patterns inferred from a revised ordinal phylogeny. J Zool Syst Evol Research 39:177–207

Bruce AN (1907) On the distribution, structure, and function of the tentacles of *Roridula*. Notes RBG Edin 17:83–98

Cassis G (1984) A systematic study of the subfamily Dicyphinae (Heteroptera: Miridae). Ph.D. dissertation, Oregon State University, Corvallis, USA. 389 pp

China WE (1953) Two new species of the genus *Cyrtopeltis* (Hemiptera) associated with sundews in Western Australia. West Aust Nat 4:1–8

China WE, Carvalho JCM (1951) A new ant-like mirid from Western Australia (Hemiptera, Miridae). Ann Mag Nat Hist 4:221–225

Cobben RH (1978) Evolutionary trends in Heteroptera. Part II. Mouthpart-structures and feeding strategies. Meded. Landbouwhogeschool Wageningen, 78-5, H. Veenman & Zonen, B. V., Wageningen. 407 pp

Dolling WR, Palmer JM (1991) *Pameridea* (Hemiptera: Miridae): predaceous bugs specific to the highly viscid plant genus *Roridula*. Syst Ent 16:319–328

Ellis AG, Midgley JJ (1996) A new plant-animal mutualism involving a plant with sticky leaves and a resisdent hemipteran insect. Oecologia 106:478–481

Falkingham C (1995) Carnivorous plants-carnivorous bugs. Is there a symbiotic relationship? Victiorian Naturalist 112:222–223

Fenner CA (1904) Beiträge zur Kenntnis der Anatomie, Entwicklungsgeschichte und Biologie der Laubblätter und Drüsen einiger Insektivoren. Flora Allg Bot Z 93:335–433

Futuyma DJ, Slatkin M (1983) Epilogue: the study of coevolution. In: Futuyma DJ, Slatkin M (eds) Coevolution. Sinauer Associates, Inc, Sunderland, pp 459–464

Gorb E, Gorb S (2009) Effects of surface topography and chemistry of *Rumex obtusifolius* leaves on the attachment of the beetle *Gastrophysa viridula*. Entomol Exp Appl 130:220–222

Gorb E, Kaster V, Peressadko A, Arzt E, Gaume L, Rowe N, Gorb S (2004) Structure and properties of the glandular surface in the digestive zone of the pitcher in the carnivorous plant *Nepenthes ventrata* and its role in insect trapping and retention. J Exp Biol 207:2947–2963

Gorb E, Hosoda N, Gorb S (2008) Nano-porous substrates reduce beetle attachment force. Proccedings of the 9th biennial conference on engineering systems design and analysis ESDA08, Haifa, Israel. pp 1–6

Holloway (1967) Wettability of plant surfaces. Ph.D. dissertation, University of London, UK

Kullenberg B (1946) Studien über die Biologie der Capsiden. Zool Bidrag Uppsala 23:1–522

Lloyd FE (1934) Is *Roridula* a carnivorous plant? Can J Res 10:780–786

Marloth R (1903) Some recent observations on the biology of *Roridula*. Ann Bot 17:151–158

Marloth R (1910) Further observations on the biology of *Roridula*. Trans Roy Soc South Afr 2:59–62

Reuter OM (1907) Ad cognitionem Capsidarum aethiopcarum scripsit. Öfv F Vet Soc Förh 49, 27 pp
Reuter OM (1913) Lebensgewohnheiten und Instinkte der Insekten bis zum Erwachen der sozialen Instinkte. Friedländer & Sohn, Berlin, 448 pp
Roberts JI (1930) The tobacco capsid (*Engytatus volucer* Kirk.) in Rhodesia. Bull Entomol Res 21:169–183
Russell MC (1953) Notes on insects associated with sundews (*Drosera*) at Lesmurdie. West Aust Nat 4:9–12
Schuh RT (1976) Pretarsal structures in the Miridae (Hemiptera) with cladistic analysis of relationships within the family. Am Mus Novit 2601, 39 pp
Schuh RT (1995) Plant bugs of the world (Insecta: Heteroptera: Miridae): Systematic catalog, distributions, host list, and bibliography. The New York Entomological Society, 1329 pp
Schwoerbel W (1956) Beobachtungen und Untersuchungen zur Biologie einiger einheimischer Wanzen. Zool Jb Syst 84:329–354
Southwood R (1986) Plant surfaces and insects-an overview. In: Juniper B, Southwood R (eds) Insects and the plant surface. Edward Arnold Publishers, London, pp 1–22
Stork NE (1980) Experimental analysis of adhesion of *Chrysolina polita* (Chrysomelidae: Coleoptera) on a variety of surfaces. J Exp Biol 88:91–107
Sugiura S, Yamazaki K (2006) Consequences of scavenging behaviour in a plant bug associated with a glandular plant. Biol J Linn Soc 88:593–602
Thompson JN (1988) Coevolution and alternative hypotheses on insect/plant interactions. Ecology 69:893–895
Voigt D (2005) Untersuchungen zur Morphologie, Biologie und Ökologie der räuberischen Weichwanze *Dicyphus errans* Wolff (Heteroptera, Miridae, Bryocorinae). Dissertation, TU Dresden, Germany, http://nbn-resolving.de/urn:nbn:de:swb:14-1138036391273-82564
Voigt D, Gorb S (2008) Insect trap as habitat: cohesion failure mechanism prevents adhesion of bugs *Pameridea roridulae* (Heteroptera, Miridae, Bryocorinae) to the sticky surface of the plant *Roridula gorgonias* (Roridulaceae). J Exp Biol 211:2647–2657
Voigt D, Wyss U, Mölck G (2004) Videodokumentation über die Biologie und das Verhalten der räuberisch lebenden Weichwanze *Dicyphus errans* Wolff (Heteroptera, Miridae, Dicyphinae). Mitt DPG 34:44–45
Voigt D, Gorb E, Gorb S (2007) Plant surface–bug interactions: *Dicyphus errans* stalking along trichomes. Arthropod-Plant Interactions 1:221–243
Voigt D, Gorb E, Gorb S (2009) Hierarchical organisation of the trap in the protocarnivorous plant *Roridula gorgonias* (Roridulaceae). J Exp Biol 212:3184–3191
Wagner E (1955) Bemerkungen zum System der Miridae (Hemiptera, Heteroptera). Dtsch Entomol Z 2:230–242
Wagner (1970/71) Die Miridae Hahn, 1831, des Mittelmeerraumes und der Makronesischen Inseln (Hemiptera, Heteroptera), Teil 1. Entomol Abh Mus Tierk Dresden 37, 484 pp
Wheeler AG (2001) Biology of the plant bugs (Hemiptera: Miridae): pests, predators, opportunists. Cornell University Press, London, 507 pp

Keyword Index